Advances in Watershed Hydrology

In Memoriam
Dr. Florisa Melone

Advances in Watershed Hydrology

Edited by

Tommaso Moramarco
Silvia Barbetta
Luca Brocca

Water Resources Publications, LLC — Since 1971 — WRP — Highlands Ranch

For Information and Correspondence:

Water Resources Publications, LLC
P. O. Box 630026, Highlands Ranch, Colorado 80163-0026, USA

Advances in Watershed Hydrology

Edited by
Tommaso Moramarco
Silvia Barbetta
Luca Brocca

Cover Designed by WRP

ISBN-13: 978-1-887-20185-8
ISBN-10: 1-887201-85-8

Library of Congress Control Number: 2015934502

This publication is printed and bound in Chelsea, Michigan, United States of America.

ACKNOWLEDGMENTS

Organizing an International Conference takes the efforts of multiple people. The Conference as well as this book could not have been possible without the efforts of Dr. Stefania Camici, Dr. Cristiano Corradini, Dr. Pamela Maccioni, Dr. Christian Massari, Dr. Angelica Tarpanelli and Dr. Graziano Zucco. The technical support of Mr Massimo Guadagno is also acknowledged.

The Florisa Melone International Memorial Conference was sponsored and co-sponsored by a number of organizations. The sponsors offered financial support for the conference and this is gratefully acknowledged. The co-sponsors offered support by promoting the conference through their newsletters. We wish to acknowledge the following sponsors and co-sponsors.

SPONSORS

Regione Umbria, Servizio Risorse Idriche e Rischio Idraulico,
Progetto POR-FESR, Perugia, Italy.
http://www.rischi.regione.umbria.it/Mediacenter/FE/

Ecosearch, Environmental Monitoring and Sampling, Città di Castello, Italy. http://www.ecosearch.info/home

Instrument Service Srl, Envoronmetal Monitoring Systems, Bolzano, Italy. http://www.instrumentservice.eu/it/

CO-SPONSORS

Gruppo Italiano di Idraulica, Italy. http://gii-idraulica.net/

International Association of Hydrological Sciences. http://iahs.info/

Società Idrologica Italiana, Italy. http://www.sii-ihs.it/

PREFACE

Dr. Florisa Melone (1954-2012) left us far too soon after three decades of outstanding scholar career spent at Research Institute for Geo-Hydrological Protection. She worked in environmental and water related areas, and made seminal contributions in the areas of hydro-meteorological monitoring based on ground and satellite observations, synthesis of the effect of spatial variability in hydraulic parameters and scaling, mathematical modeling of soil infiltration, rainfall spatial distribution analysis, modeling of surface water over watersheds, flood forecasting and hydraulic risk. In these areas she has published a multitude of highly cited original contributions.

The Research Institute for Geo-Hydrological Protection celebrated the life and work of Florisa by organizing the International Memorial Conference in Assisi, October 10-11 2013. It's not a coincidence that the Conference was organized in the holy place of the *Basilica of San Francis*: there everything talk about to deeper values of the life like the brotherhood, the peace, the friendship, the love for the neighbor, all values that Florisa fully shared in her life.

The Conference brought together some of the most renewed scholars came from different part of world like Africa, Asia, US, Europe, Italy of course, with the only aim to make memory of Florisa and her work and to present the new frontiers in the hydrology field. More than one hundred participants attended to the Conference wherein ten keynote lectures were delivered and a series of presentations were made. This result fostered the decision that a book illustrating aspects of past present and future in hydrology would have made honor to Florisa definitely.

The book takes the cue from the awareness that enhancements in hydrology have been achieved over last decades, however there is still much to do. In this direction, experimental analyses at the catchment scale are needed and they cannot disregard by intensive measurements campaigns of soil moisture in plots and high flow velocity in river sites, as well as the set-up of networks of sensors for soil moisture and rainfall monitoring. Indeed, the knowledge of soil moisture spatial variability antecedent the rainfall event is of paramount importance to investigate the flood formation process at catchment scale. Likewise, the identification of rainfall patterns has a high priority, considering that they can be driven by convective cells reaching maturity and producing high gradients of rain rate in few hours within limited areas and with heavy consequence in terms of flooding. Even the streamflow measurements during high flood represent an important matter considering the uncertainty for the high discharge estimation at river sites. Under this umbrella, the emerging technologies for the monitoring and the enhancing of modeling of the surface hydrological processes at catchment scale may be the step forward to understand the rainfall-to-runoff conversion process in headwater catchments and to develop a scalable system of operational tools leading up to suitable flood mitigation measures and reliable real-time warning systems. In addition to that, the efficacy of different best management practices is sought in relation to flood prevention and low impact development in a context of global changing as well. For this latter, the fundamental role of climate models and their capability in predicting future scenarios has to be thoroughly investigated. Therefore, the challenge might be to find answers to some questions such as: what is expected from the emerging technologies for the hydrological monitoring at catchment scale? How do we bridge the gap among monitoring, spatial variability and modeling? How to cope with the large uncertainties in precipitation forecasts due to limited understanding on complex atmospheric processes? How to enhance the prediction in ungauged basin and under climate change? What is the uncertainty in future climate scenarios and how the water resource management is affected by them? How to identify the best management practice, structural and non-structural, for flood prevention and mitigation?

A good opportunity may be offered by this book on some issues raised above and in particular on: a) the new perspective of hydro-meteorological monitoring by ground based and satellite, considering that a correct understanding of runoff dynamics cannot be separated from experimental observations; b) the discharge assessment in natural channels, essential for water resources management and flood forecasting; c) hydraulic risk analysis, of paramount importance for forward planning and civil protection activities; d) hydrological cycle, water systems and climate change.

The subject matter of the book is divided into four sections. The first section is dedicated to the Dr. Florisa Melone's life, including family, professional career and technical contributions.

Section II deals with hydrological processes and monitoring and is comprised of seven chapters. Chapter 2 presents a review of rainfall infiltration modeling at different spatial scales. Chapter 3 identifies new tracers and high-frequency monitoring strategies for hydrological processes monitoring. A review of satellite monitoring techniques for clouds and rainfall is given in Chapter 4. In Chapter 5 results achieved by using a multi-sensor system for flood detection and monitoring are illustrated. Chapter 6 along with Chapter 7 addresses the matter of streamflow monitoring during high floods by exploiting the entropy theory. In Chapter 8, performance of a Soil Water Balance Model is checked in an olive orchard catchment considering the effects of tree canopy as well.

Hydrological prediction and forecasting in ungauged basin is the subject matter of section III covering six chapters. Chapter 9 addresses how to select hydrologic model for gauged and ungauged small catchments. Chapter 10 is a review of the physical interpretations of the Muskingum parameters. Chapter 11 deals with the optimization methods in hydrology.

Chapter 12 addresses the prediction of flow-duration curves by geostatistical procedures. Results on high-resolution simulations with the WRF model for the 2010 Pakistan floods are presented in Chapter 13. Chapter 14 deals with uncertainty in hydraulic models using distributed information

The last section IV is on extreme events and global warning and encompasses 9 chapters. Chapter 15 presents a modeling system for a quick estimate of flood inundation and the economic resultant damages. Flood hazard detection by remote sensed information is described in Chapter 16. Chapter 17 proposes an analysis of geo-hazards in karst areas. Chapter 18 addresses the influence of urban water systems on water resources in a context of global warm. The uncertainty in the processing chain of water resources climate change impacts is analyzed in Chapter 19. Chapter 20 presents the impact of the ongoing climate change on springs in Apennines. Chapter 21 describes actions undertaken by Authority in charge of the climate change impact mitigation on water resources. In Chapter 22 flood risk to the population using historical record of flood events with human consequences is analyzed. Chapter 23 presents a methodological framework for assessing drought hazard using operational indicators.

Hopefully this volume gives different useful insights on advancements in hydrology and shed light on some of the topical problems like global warming and extreme events.

We would like to take this opportunity to express our deep gratitude to Dr. Florisa Melone. Florisa was a close friend, with whom we lived one of the most wonderful experiences that a researcher could have. We wouldn't dwell about the scientific capabilities of Florisa, they are already testified by the large number of her works. We would like instead to remark two important qualities of Florisa, rare to find in a human being in general and in a researcher, in particular, namely the humility and generosity. During her career, Florisa developed outstanding works in the areas of applied hydrology and for that she had a high visibility worldwide. Nevertheless, she remained humble and ready to cooperate with others. Besides, Florisa was a generous person always willing to help everyone and to put the troubles of others before her own; a heartfelt thanks to Florisa's family for their assistance in organizing the Memorial Conference and in particular to Claudio, her husband, and the two children, Giacomo e Michele; a special thanks to all young researchers of the Hydrology Group at IRPI; our gratitude is to all the book chapter contributors. This book reflects the affection and esteem that all of us have had and continue to have for Florisa.

Tommaso Moramarco
Silvia Barbetta
Luca Brocca
Perugia, Italy, August 2015

TABLE OF CONTENTS

CHAPTER 1

FLORISA MELONE: A PERSONAL REFLECTION

Vijay P Singh

Distinguished Professor &
Caroline & William N. Lehrer Distinguished Chair in Water Engineering
Department of Biological and Agricultural Engineering &
Department of Civil & Environmental Engineering
Texas A&M University, College Station, TX 77843, USA. Email: vsingh@tamu.edu

ABSTRACT

Dr. Florisa Melone symbolized what is good about humanity and in human beings, Fig. 1.1. In a life cut short by an unfortunate disease, Florisa left a lasting legacy and touched and enriched many lives. She was an outstanding scholar, a devoted wife, a caring mother, a lovely sister, a loyal friend, and, above all, an exemplary human being. This note is a personal reflection on some aspects of her life, including childhood, early and college education, marriage and family life, core beliefs and philosophy, sons' tribute to their mother, husband's tribute to his wife, professional career, technical contributions, and colleagues' tributes to collaborations with Florisa. The discussion is concluded with some personal memories.

Fig. 1.1 A portrait of Florisa.

1.1 CHILDHOOD

Dr. Florisa Melone was born on November 12, 1954, in Terni, Italy, where she spent her childhood and adolescence. Until she was about 2 years old, her family lived in the outskirts and later moved in the house in the downtown where her mother still lives. Her parents, Guido (father) and Flavia (mother), had two other children-both sons. Gianfranco was the oldest son, Giampiero the second oldest son, and Florisa was the youngest daughter. Figure 1.2 shows Florisa with her brothers when they were children. Both of her brothers are currently living in Terni; they are married and they each have a daughter, Rita and Silvia who also are married. Silvia has three children, making the mother of Florisa great-grandmother and Florisa great aunt. Unfortunately, Florisa's father Guido died in a car accident in 1990, leaving Flavia alone.

Fig. 1.2. Florisa two years old with her brothers, Gianfranco and Gianpiero.

1.2 EARLY EDUCATION

Florisa's academic career followed the usual path, involving five years of elementary school from 1959 to 1964, three years of middle school from 1965 to 1968 at the Institute Leonardo da Vinci, and then four years of high school from 1969 to 1973 at the Scientific Lyceum Galileo Galilei. Her elementary school was in Campitello, a small village close to Terni, while her other schools were in Terni, for her family had moved there. Figure 1.3 shows Florisa in a recital at her elementary school. In school, she exhibited an aptitude for excellence in academic and extracurricular activities. Figure 1.4 shows her with her class in elementary school and Figure 1.5 with her classmates in high school. Her two older brothers elected to interrupt their studies and started to work for lending a helping hand to the family, especially financially, while Florisa continued her studies which meant a huge sacrifice on the part of her whole family. She finished her high school in 1973 with honors, and went on to enroll in the faculty of mathematics at the University of Perugia, Perugia. She graduated on 3rd March 1978 with 110/110 cum laude.

Fig. 1.3. Recital (elementary school).

Fig. 1.4. Elementary school.

Fig. 1.5. Florisa with her classmates in high school.

1.3 MARRIAGE

During the first year at the University of Perugia, Florisa met Claudio with whom she got married on June 12, 1982, after an engagement period that lasted nearly 8 years. Since Florisa and Claudio had such a long companionship, they married with a lot of forethought, deliberation, planning-clearly marriage was not done in hurry by any means but it was a conclusion of the initial phase of their companionship and the beginning of her lives together.

Figure 1.6 portrays Florisa just before wedding and Fig. 1.7 is a portrait of Florisa and Claudio at the time of wedding. For them, the official wedding meant a new course in their lives with the convergence of different objectives, especially having a family and children. Enlightened as they were, Florisa and Claudio always gave

Fig. 1.6. Florisa just before wedding.

Fig. 1.7. A wedding portrait.

precedence to substance over form. Nevertheless they also believed that sometimes form and substance must be integrated. Formalizing the union between two people does not always equal the making of a relationship stronger, i.e., if the union is already stable a formal act is not needed, but it certainly is an act signifying the beginning in front of others and society where one lives. The wedding was a joy for them, for their parents, and for their friends and families.

1.4 FAMILY LIFE

About six months after their marriage, an unfortunate tragedy struck Florisa and Claudio. In January 1983 Claudio lost his mother who passed away due to a sudden illness. It was a blow for Florisa who had developed a filial relationship with her and a very sad period followed. She had to deal with the new situation and solve emotional and practical problems. Fortunately, in 1984, she was blessed with a son who she named Michele. The birth of Michele brought a new wind in her and was uplifting, as often happens when a new life comes to light. Figure 1.8 shows Florisa with Michele. She felt that the birth of her son was some kind of a compensation for the loss of her mother-in-law, even if obviously this was not true. With the birth of Michele, her life greatly changed. For one, she had to juggle between home, work and child. Because she would not settle for anything other than excellence, a very busy period followed. Luckily, she had her mother who was a great help on a multitude of occasions. Florisa and Claudio always thought of having another child and they were fortunately blessed with another son in 1989. They named second son Giacomo. Figure 1.9 show Florisa with Giacomo. During her pregnancy with Giacomo, the last three months of the gestation were difficult and even problematic. As a result, Florisa had to stay in bed for long hours each day in order to alleviate the possibility of losing the baby. In later years she confessed that this was a very depressing period and painful experience of her life, for all the feelings and fears that she felt about her child. Fortunately, everything went well and Michele had a brother. Figure 1.10 is a portrait of Florisa and Claudio with their sons.

Fig. 1.8. Florisa with Michele.

Fig. 1.9. Florisa with Giacomo.

Fig. 1.10. a) Florisa and family on vacation; b) Family on a skiing vacation.

Both of her sons are now grown up. As expected, they are well groomed, mature and handsome. They have made their choices and are living their lives. Michele is a graduate in mechanical engineering and now works in a private company, and Giacomo is dedicated to the field of computer graphics, and after his studies, is now waiting for a suitable employment.

Michele and Giacomo are two wonderful men. It is unfortunate that at such an early age they had to face and witness the illness and death of their mother who experienced a lot of suffering, but they did with courage and strength as their mother wanted. They are trying to cope with the tragedy of their mother's demise, overcome the irreparable loss, heal their emotional wounds, bear the pain and move forward. They have always been close to their mother and their father. Figure 1.10a is a portrait of family on vacation, and Figure 1.11 shows Florisa on a beach.

Fig. 1.11. Florisa on a beach.

1.5 CORE BELIEFS

As a human being, Florisa was one of the nicest persons one would encounter. She was exceptionally kind, caring and always willing to help others. She always listened to everyone and tried to figure out how she could give a helping hand. She was not religious in a traditional sense, but in her day-to-day life she always applied the Christian principles or principles of humanity, while not professing them. Those who knew her harbored the same impressions about her.

1.6 SONS' TRIBUTE TO THEIR MOTHER

Both Michele and Giacomo were near and dear to their mother. They harbor fond memories of their lives they shared with her. Elder son Michele writes about his mother as follows:

"To the person who taught me what is right and what is wrong, what is good and what is bad, what is important in life, to respect other people, to study, to learn, to draw, to be better and for everything that she gave me. I remember all the reproaches, all the advices, all the times that you helped me in the study, all the times that I was in a quandary and you helped me, but also the first time that you taught me to draw and paint. I remember all the holidays and the moments when we were together. I would like to remember you for everything that you realized in your life: raising two sons, taking care of a family extending to relatives, looking after a home, and working in a team that was for you like a second family. I would like to remember you for the strength, courage, wisdom, and lucidity to face the illness, and also when there was no hope. You were able to connect, to comfort, to understand, to love people and the last two days are the evidence. That everything you did would not be lost but that each one can take on as an example and inspiration from you and what you did. In particular, I would like to remember you for the niceness, lovability, gracefulness, harmony that you were able to give people."

Younger son Giacomo writes about his mother:

"I might say many things about my mother, but it is sufficient to look around and see how many people will attend the Conference who, in addition to presenting many technical works, would like to add something about her as a person. My mother always believed in humanity and human beings, who give their neighbor without requiring anything in return. She believed in what she did, not for money or power, but for the emotion that can give a generous gesture. I remember the joy that she felt when the CNR guys came to visit her during the illness and by a simple "thank you" made her smile. I can say that she taught us the love so much, by putting always the neighbor before herself, even during her illness. I think that only a few persons are able to understand and apply one of the best feelings of a human being: the love for the family and the neighbor. She taught us to never give up and I can say that she is successful with my father in what all parents hope and dream: to teach their children what life is."

1.7 HUSBAND'S TRIBUTE TO HIS WIFE

Florisa's husband Claudio writes: "A few days are left before the one-year anniversary of Florisa's demise and I did not realize yet that nearly twelve months have since passed. Mentally, it feels like time stopped that October 28 of the last year (2012). Inevitably, the loss of one's lifetime companion leaves permanent scars and signs and one of them is the different perception of the flow of time.

I met Florisa during my first year of university in 1974, when we followed together the same physics class that was common for the degree courses in Mathematics and Engineering. About a year later, our life story began and lasted for 38 years. Unfortunately in the journey we embarked upon together, she was taken away and our story broke on the way that you all know.

The period of 38 years is a very long time in a person's life and is full of memories, events and happenings, some really memorable, some leaving indelible impressions. Were I to choose among those I might have difficulty, so I will not dwell on specific events but say something generic. I would specifically like to mention two peculiarities of Florisa.

First, she was a special person who held in high esteem the human side of people with whom she had a relationship, both in the work environment and in the friendship or family. She always acted with great care for the needs of others and very often even putting them before her own. She was always ready to provide help to those who needed help, in small things as well as in the most important ones. I think this was a very beautiful side of her character, important and obvious to all who knew her not superficially. If someone needed help, she was always willing and in cases where she could not or was not able to be helpful, she felt very bad.

The other quality that I would like to highlight about Florisa's character is her determination combined with a strong will. She always showed great responsibility in the activities she did, from study to work to the management of the family and the house. In the last period of her life, she faced the disease in an exemplary way, overcoming the moments of despair with strength and courage that I was totally impressed. I think only a few people have faced the same physical and psychological hardships that she did with dignity and fortitude.

Regarding our lives together, what I would like to remember with a sense of fulfillment and gratitude is the fact that we grew up together as people. Our union began when we were quite young, about 20 years old and since then we really grew up together, helping each other and together we faced the beautiful and not so beautiful aspects that life showed us from time to time. It is really beautiful and gratifying to go back the memory lane and to envision myself now as if I was twenty, and understand how her closeness changed me, helping me to mature by calmly dealing with everyday living. I think that, in my turn, I had contributed so that she reached her serenity and I am particularly happy for that. I can say that we exchanged experiences and points of view, which we both enjoyed. What she liked most was to be able to continue our journey together to old age and to complete it. Unfortunately, our fate meant otherwise and we did not succeed.

Finally, I would like to end these few sentences with a consideration. It is common saying about a husband, a wife, a partner or a companion that "he or she is my better half." Here, Florisa was truly my better half and since she passed away, it is like my half has gone as well. I would like to affirm that this is not rhetoric and it is tough to go on without her, but I am helped by her masterful memory and the example that he gave to all of us and to me first.

A heartfelt thanks to all of you, who have wanted to honor and remember her, by organizing and participating in this wonderful event."

1.8 TRIBUTE FROM DR. TOMMASO MORAMARCO, A CLOSE CO-WORKER AND COLLEAGUE

Needless to say, it is not easy to talk about a person who passed away and with whom, in addition, one has shared the professional work load for more than 16 years. I would not dwell upon Florisa's scientific capabilities because they are well known to all and it suffices to have a look at her curriculum vita. I do not want to even talk about the scientific acknowledgments from national and international institutions of the Hydrology IRPI team that Florisa and I established together more than 12 years ago. In this short note, I would like instead to remark Florisa's qualities both as a colleague and friend.

As a colleague, in these many years of working together, I discovered how Florisa was endowed with a rare quality, difficult to find in a human being, in general, and in a researcher, in particular: the humility. During her career, Florisa developed outstanding works in the areas of applied hydrology and for that she had a high visibility towards the external world and despite that, she remained always humble and willing to listen and cooperate with others.

Florisa was also a close friend with deep generosity and loyalty. I have a particular memory of Florisa that I will keep forever in my heart that I like to share through this note. At the beginning of my stay in Perugia, on 6th January 1996, the feast day of Epiphany, my wife Ileana, around noon, felt very sick for a serious infection in her kidneys. At that time, I did not know any doctors in Perugia as well as the location of the hospital. Considering that it was a holiday and, in addition, the lunchtime, I phoned Florisa asking her direction for getting to the hospital. Although she had some guests at her home for the lunch of Epiphany, she left everything and came immediately to visit Ileana and to my great surprise she came with a doctor as well. She stayed with us as long as Ileana felt better. I guess that Florisa's guests had the Epiphany lunch without her. This is how Florisa was: a generous person always ready to help everyone if any. Florisa was agnostic, but I never met a person who has lived the Christian principles of humility and generosity, i.e., the love for the neighbor, like her. The proof is that even during her disease, she always put the problems of others before her own.

I miss Florisa, I miss her smiles, suggestions and encouragements, and above all her gentleness. I think my this feeling is shared by all those who knew her, and in particular the young researchers of the IRPI

Hydrology Team for whom Florisa was considered as a "second mother," always willing to accomplish any of their requests, solve doubts or difficulty. I would like to conclude with a sentence from Sirach's book of Bible (3, 17-19) that shows about the human beings what is well appreciated by God: "My child, conduct your affairs with gentleness and you will be loved more than a person of good repute. However great you become, humble yourself more, and you will find favor before the Lord." I deem that Florisa certainly has found favor before the Lord. Thanks Florisa!"

1.9 PROFESSIONAL CAREER

Dr. Florisa Melone received an M.S. degree in Mathematics, with excellence, from the University of Perugia, Italy, in 1978. She got a two-year scholarship of the National Research Council (CNR) in hydrological time series analysis, at the Institute for Hydrogeological Protection Research (IRPI) of CNR, from 1979 to 1980. She was a Researcher at CNR-IRPI from 1981 to 1996 and 1996-2011 Senior Researcher and 2001-2012 Head Researcher. Since 1981 she conducted research in the field of catchment hydrology addressing flood forecasting and hydraulic risk mitigation. In particular, the research topics she pursued were: hydro-meteorological monitoring, infiltration modelling, rainfall spatial variability, soil moisture spatial and temporal variability, rainfall runoff modelling, flood routing in natural channels, real-time flood forecasting, remote sensing for hydrological applications, water resources management, hydraulic risk analysis and climate change. She was a co-author of more 200 papers of which 57 appeared in refereed journals. She directed M. Sc. theses and she was tutor of research grants and scholarships in the hydrology research areas. She served as a referee for leading hydrology journals. She was the leader of the Hydrologic Section of IRPI for several years. She advised authorities for establishing measures for flood protection and also addressed forward planning. She was a coordinator and scientific person responsible for national projects on hydro-meteorological monitoring and flood forecasting and she actively participated in several research projects in the framework of Italian and European Programs. In the beginning Florisa's main collaborators were Professors Corrado Corradini and Lucio Ubertini at IRPI and from outside Professor Vijay P. Singh from the United States.

Several years later, Dr. Tommaso Moramarco became his main collaborator and then Dr. Luca Brocca and others joined her. A list of her collaborators is given in Table 1.1.

Table 1.1. Florisa's collaborators.

Name	Name	Name	Name
Barbetta, S.	Faruolo, M.	Martin, C.	Ponziani, F.
Bartalis, Z.	Ferraris, S.	Martinez-Fernández, J.	Ratto, S.
Berni, N.	Flammini, A.	Matgen, P.	Sahoo, B.
Bittelli, M.	Franchini, M.	Melzer, T.	Saltalippi, C.
Borga, M.,	Gumuzzio, A.	Morbidelli, R.	Singh, V.P.
Brocca, L.	Hahn, S.	Moramarco, T.	Smith, R.E.
Calamita, G.,	Hasenauer, S.	Naeimi, V.	Tarpanelli, A.
Camici, S.	Heitz, S.	Penna, D.	Tramutoli, V.
Caudaro, M.	Lacava, T.	Perumal, M.	Tullo, T.
Corato, G.	Lapenna, V.	Pergola, N.	Ubertini, L.
Corradini, C.	Latron, J.	Perrone, A.	Volk, M.
Didon-Lescot, J.-F.	Liersch, S.	Piscitelli, S.	Wagner, W.
Dorigo, W.	Llorens, P.		

1.10 TECHNICAL CONTRIBUTIONS

Being a mathematician by training Dr. Melone's research encompassed a broad range of topics dealing with surface water hydrology and unsaturated zone hydrology. Before discussing her technical contributions, it will be interesting to give a short background to her professional pursuits. Florisa started her professional pursuits in 1979, cooperating with Prof. Lucio Ubertini at the Institute for Hydrogeological Protection Research in Central Italy of National Research Council. The topic she developed at the beginning was the use of transfer plus noise models for simulation and prediction of discharge at different time scales. Because of her expertise in dealing with mathematical modeling she was able to incorporate important innovative elements in the classical approaches widely used in the literature at that time. Later, with the transfer of Dr. Corradini to Perugia she made a great contribution to the formation of a research group in both the experimental field (with the support by qualified technicians: Bani, Fastelli, Rosi) as well as the theoretical one. Professor Corradini writes: "Florisa was available to all the possible critical evaluation of the group activity and ready to address the activities along the lines considered more promising. For example, she preferred to break off from stochastic modeling, choosing the development of semi-distributed conceptual models. In this context Florisa brought forward mathematical elements and very well integrated them with the physical reality. She provided considerable contributions in a variety of studies as: modeling of rainfall spatial distribution and effects of orography, design of hydrometeorological networks for flood forecasting, definition of the characteristics of a meteorological radar for rainfall measurements, realization of adaptive models for flood forecasting, and applications of the geomorphologic instantaneous unit hydrograph and scale problems. The cooperation of Prof. Vijay Singh was a crucial element in developing a few of these subjects, particularly through the scientific discussions with Florisa. The confidential relationships established by Florisa with different visiting researchers were a basic element for improving the group level. Particularly useful was the cooperation with Dr. Roger Smith in the period from 1992 to 1999, when many valuable papers were realized on rainfall infiltration."

Her research accomplishments can be grouped into 10 categories. A short discussion in what follows highlights her contributions.

1.10.1 Rainfall Distribution and Generation

In collaboration with Corradini, Moramarco, Brocca, and others, Florisa made important contributions in the rainfall area. Her research focus was describing the spatial structure and distribution of rainfall in the Mediterranean area, development of a simple approach for stochastic generation of spatial rainfall patterns, and development of a new method for rainfall estimation using soil moisture observations. Her rainfall research is described in refereed journal articles given below:

1. Corradini, C., Melone, F. (1988). Spatial distribution of pre-warm front rainfall in the Mediterranean area. *Nordic Hydrology*, 19(1), 53-64.

2. Corradini, C., Melone, F. (1989). Spatial structure of rainfall in mid-latitude cold front-systems. *Journal of Hydrology*, 105(3-4), 297-316.

3. Tarpanelli, A., Franchini, M., Brocca, L., Camici, S., Melone, F., Moramarco, T. (2012). A simple approach for stochastic generation of spatial rainfall patterns. *Journal of Hydrology*, 472-473, 63-76.

4. Brocca, L., Moramarco, T., Melone, F., Wagner, W. (2013). A new method for rainfall estimation through soil moisture observations. *Geophysical Research Letters,* 40(5), 853-858.

1.10.2 Rainfall-Runoff Modeling

Florisa contributed significantly in the rainfall-runoff modeling area. Beginning with the development of instantaneous unit hydrographs for both gaged and ungaged basins, she investigated into the interaction of infiltration and surface runoff, assessment of urban flooding, improvement of runoff prediction by better accounting of soil moisture, and flood risk management. In collaboration with her colleagues at IRPI and University of Perugia and the United States, her research is widely published in refereed journals listed below:

1. Singh, V.P., Corradini, C., Melone, F. (1985). A comparison of some methods of deriving the instantaneous unit hydrograph. *Nordic Hydrology*, 16(1), 1-10.

2. Corradini, C., Melone, F., Singh, V.P. (1989). A simple approximation of the hydrograph downstream of a flooded area. *Nordic Hydrology*, 20(3), 179-190.

3. Corradini, C., Melone, F., Singh, V.P. (1995). Some remarks on the use of GIUH in the hydrological practice. *Nordic Hydrology*, 26(4-5), 297-312.

4. Corradini, C., Morbidelli, R., Melone, F. (1998). On the interaction between infiltration and Hortonian runoff. Journal of Hydrology, 204(1-4), 52-67.

5. Melone, F., Corradini, C., Singh, V.P. Simulation of the direct runoff hydrograph at basin outlet (1998). *Hydrological Processes,* 12(5), 769-779.

6. Melone, F., Corradini, C., Singh, V.P. (2002). Lag prediction in ungauged basins: An investigation through actual data of the upper Tiber river valley. *Hydrological Processes*, 16(5), 1085-1095.

7. Moramarco, T., Melone, F., Singh, V.P. (2005). Assessment of flooding in urbanized ungauged basins: A case study in the Upper Tiber area, Italy. *Hydrological Processes*, 19(10), 1909-1924.

8. Brocca, L., Barbetta, S., Melone, F., Moramarco, T. (2010). A continuous rainfall-runoff model derived from investigations in a small experimental basin. *IAHS-AISH Publication*, 336, 179-185.

9. Brocca, L., Melone, F., Moramarco, T., Wagner, W., Naeimi, V., Bartalis, Z., Hasenauer, S. (2010). Improving runoff prediction through the assimilation of the ASCAT soil moisture product. *Hydrology and Earth System Sciences*, 14(10), 1881-1893.

10. Brocca, L., Melone, F., Moramarco, T., Penna, D., Borga, M., Matgen, P., Heitz, S. (2011). Investigation of the hydrologic response of three experimental basins across Europe. *Bodenkultur*, 62(1-4), 31-37.

11. Brocca, L., Melone, F., Moramarco, T., Penna, D., Borga, M., Matgen, P., Gumuzzio, A., Martinez-Fernández, J., Wagner, W. (2013). Detecting threshold hydrological response through satellite soil moisture data. *Bodenkultur*, in press.

12. Brocca, L., Liersch, S., Melone, F., Moramarco, T., Volk, M. (2013). Application of a model-based rainfall-runoff database as efficient tool for flood risk management. *Hydrology and Earth System Sciences*, in press.

1.10.3 Flood Forecasting

In the area of flood forecasting she dealt with both real-time forecasting and prediction. She developed adaptive semi-distributed models for forecasting, incorporation of infiltration in forecasting, stage forecasting, variable parameter Muskingum stage forecasting and confidence intervals for real-time stage forecasts. Her collaborative research work is published in refereed journal articles given below:

1. Corradini, C., Melone, F. (1986). An adaptive model for on-line flood predictions using a piecewise uniformity framework. *Journal of Hydrology*, 88(3-4), 365-382.

2. Corradini, C., Melone, F., Ubertini, L. (1986). A semi-distributed adaptive model for real-time flood forecasting. *Water Resources Bulletin,* 22(6), 1031-1038.

3. Corradini, C., Melone, F., Singh, V.P. (1987). On the structure of a semi-distributed adaptive model for flood forecasting. *Hydrological Sciences Journal*, 32(2), 227-242.

4. Corradini, C., Morbidelli, R., Saltalippi, C., Melone, F. (2004). Flood forecasting and infiltration modeling. *Hydrological Sciences Journal*, 49(2), 227-236.

5. Moramarco, T., Barbetta, S., Melone, F., Singh, V.P. (2006). A real-time stage Muskingum forecasting model for a site without rating curve. *Hydrological Sciences Journal*, 51(1), 66-82.

6. Perumal, M., Moramarco, T., Barbetta, S., Melone, F., Sahoo, B. (2011). Real-time flood stage forecasting by variable parameter Muskingum stage hydrograph routing method. *Hydrology Research*, 42(2-3), 150-161.

7. Barbetta, S., Moramarco, T., Franchini, M., Melone, F., Brocca, L., Singh, V.P. (2011). Case study: Improving real-time stage forecasting Muskingum model by incorporating the rating curve model. *Journal of Hydrologic Engineering*, 16(6), 540-557.

8. Barbetta, S., Moramarco, T., Brocca, L., Franchini, M., Melone, F. (2013). Confidence interval of real-time forecast stages provided by the STAFOM-RCM model: The case study of the Tiber River (Italy). *Hydrological Processes*, doi:10.1002/hyp.9613, in press.

9. Barbetta, S., Moramarco, T., Brocca, L., Franchini, M., Melone, F. (2013). Confidence interval of real-time forecast stages provided by the STAFOM-RCM model: The case study of the Tiber River (Italy). *Hydrological Processes*, doi:10.1002/hyp.9613, in press.

1.10.4 Infiltration Modeling

In collaboration with Corradini, Smith, and others, Florisa investigated different aspects of infiltration, including modeling infiltration and redistribution in crusted soils, modeling infiltration for multistorm runoff and under complex rainfall sequences, modeling infiltration in layered soils under complex rainfall patterns, and representation of infiltration in adaptive rainfall-runoff models. Her collaborative research is published in refereed journal articles given below:

1. Corradini, C., Melone, F. (1992). Representation of infiltration in adaptive rainfall-runoff models. *Nordic Hydrology*, 23(5), 291-304.

2. Smith, R.E., Corradini, C., Melone, F. (1993). Modeling infiltration for multistorm runoff events. *Water Resources Research*, 29 (1), 133-144.

3. Corradini, C., Melone, F., Smith, R.E. (1994). Modeling infiltration during complex rainfall sequences. *Water Resources Research*, 30(10), 2777-2784.

4. Corradini, C., Melone, F., Smith, R.E. (1997). A unified model for infiltration and redistribution during complex rainfall patterns. *Journal of Hydrology*, 192(1-4), 104-124.

5. Smith, R.E., Corradini, C., Melone, F. (1999). A conceptual model for infiltration and redistribution in crusted soils. *Water Resources Research*, 35(5), 1385-1393.

6. Corradini, C., Melone, F., Smith, R.E. (2000). Modeling local infiltration for a two-layered soil under complex rainfall patterns. *Journal of Hydrology*, 237(1-2), 58-73.

7. Melone, F., Corradini, C., Morbidelli, R., Saltalippi, C. (2006). Laboratory experimental check of a conceptual model for infiltration under complex rainfall patterns. *Hydrological Processes*, 20(3), 439-452.

1.10.5 Flood Routing

In collaboration with Perumal and Moramarco, Florisa also delved into flow routing as published in a journal article shown below.

1. Perumal, M., Moramarco, T., Melone, F. (2007). A caution about the multilinear discrete lag-cascade model for flood routing. *Journal of Hydrology*, 338(3-4), 308-314.

1.10.6 Stage-Discharge Relation

Florisa made a significant contribution to the literature on rating curves. Her collaborative work is reported in two refereed journal articles given as follows:

1. Moramarco, T., Barbetta, S., Melone, F., Singh, V.P. (2005). Relating local stage and remote discharge with significant lateral inflow. *Journal of Hydrologic* Engineering, 10(1), 58-69.

2. Barbetta, S., Franchini, M., Melone, F., Moramarco, T. (2012). Enhancement and comprehensive evaluation of the rating curve model for different river sites. *Journal of Hydrology*, 464-465, 376-387.

1.10.7 Soil Moisture Variability

During the last five years of her professional life, Florisa, in collaboration with her colleagues at IRPI and University of Perugia and the United States, dealt with a wide range of aspects related to soil moisture. Beginning with an investigation into the spatial variability of soil moisture, she went on to investigating space-time soil moisture variability across scales, soil moisture estimation, estimation of antecedent wetness, soil moisture profiles under complex rainfall patterns, integration of soil moisture into rainfall-runoff modeling, and soil moisture measurement. Her collaborative work is published in refereed journal articles which are given below:

1. Brocca, L., Morbidelli, R., Melone, F., Moramarco, T. (2007). Soil moisture spatial variability in experimental areas of central Italy. *Journal of Hydrology*, 333(2-4), 356-373.

2. Brocca, L., Melone, F., Moramarco, T. (2008). On the estimation of antecedent wetness conditions in rainfall-runoff modeling. *Hydrological Processes*, 22(5), 629-642.

3. Melone, F., Corradini, C., Morbidelli, R., Saltalippi, C., Flammini, A. (2008). Comparison of theoretical and experimental soil moisture profiles under complex rainfall patterns. *Journal of Hydrologic Engineering*, 13(12), 1170-1176.

4. Brocca, L., Melone, F., Moramarco, T., Morbidelli, R. (2009). Antecedent wetness conditions based on ERS scatterometer data. *Journal of Hydrology*, 364(1-2), 73-87.

5. Brocca, L., Melone, F., Moramarco, T., Morbidelli, R. (2009). Soil moisture temporal stability over experimental areas in Central Italy. *Geoderma*, 148(3-4), 364-374.

6. Brocca, L., Melone, F., Moramarco, T., Singh, V.P. (2009). Assimilation of observed soil moisture data in storm rainfall-runoff modeling. *Journal of Hydrologic Engineering*, 14(2), 153-165.

7. Brocca, L., Melone, F., Moramarco, T., Morbidelli, R. (2010). Spatial-temporal variability of soil moisture and its estimation across scales. *Water Resources Research*, 46(2), W02516.

8. Brocca, L., Melone, F., Moramarco, T., Wagner, W., Hasenauer, S. (2010). ASCAT soil wetness index validation through in situ and modeled soil moisture data in central Italy. *Remote Sensing of Environment*, 114(11), 2745-2755.

9. Lacava, T., Brocca, L., Calice, G., Melone, F., Moramarco, T., Pergola, N., Tramutoli, V. (2010). Soil moisture variations monitoring by AMSU-based soil wetness indices: A long-term inter-comparison with ground measurements. *Remote Sensing of Environment*, 114(10), 2317-2325.

10. Camici, S., Tarpanelli, A., Brocca, L., Melone, F., Moramarco, T. (2011). Design soil moisture estimation by comparing continuous and storm-based rainfall-runoff modeling. *Water Resources Research*, 47(5), W05527.

11. Brocca, L., Hasenauer, S., Lacava, T., Melone, F., Moramarco, T., Wagner, W., Dorigo, W., Matgen, P., Martínez-Fernández, J., Llorens, P., Latron, J., Martin, C., Bittelli, M. (2011). Soil moisture estimation through ASCAT and AMSR-E sensors: An intercomparison and validation study across Europe. *Remote Sensing of Environment*, 115(12), 3390-3408.

12. Brocca, L., Moramarco, T., Melone, F., Wagner, W., Hasenauer, S., Hahn, S. (2012). Assimilation of surface- and root-zone ASCAT soil moisture products into rainfall-runoff modeling. *IEEE Transactions on Geoscience and Remote Sensing*, 50(7), 2542-2555.

13. Brocca, L., Tullo, T., Melone, F., Moramarco, T., Morbidelli, R. (2012). Catchment scale soil moisture spatial-temporal variability. *Journal of Hydrology*, 422-423, 63-75.

14. Lacava, T., Brocca, L., Coviello, I., Faruolo, M., Melone, F., Moramarco, T., Pergola, N., Tramutoli, V. (2012). Soil moisture variability estimation through AMSU radiometer. *European Journal of Remote Sensing*, 45, 89-97.

15. Calamita, G., Brocca, L., Perrone, A., Piscitelli, S., Lapenna, V., Melone, F., Moramarco, T. (2012). Electrical resistivity and TDR methods for soil moisture estimation in central Italy test-sites. *Journal of Hydrology*, 454-455, 101-112.

16. Brocca, L., Tarpanelli, A., Melone, F., Moramarco, T., Caudaro, M., Ratto, S., Ferraris, S., Berni, N., Ponziani, F., Wagner, W., Melzer, T. (2013). Soil moisture estimation in alpine catchments through modelling and satellite observations. *Vadose Zone Journal*, doi:10.2136/vzj2012.0102, in press.

17. Brocca, L., Camici, S., Melone, F., Moramarco, T., Martínez-Fernández, J., Didon-Lescot, J.-F., Morbidelli, R. (2013). Improving the representation of soil moisture by using a semi-analytical infiltration model. *Hydrological Processes*, doi:10.1002/hyp.9766, in press.

1.10.8 Flood Frequency Distributions

Florisa also delved into the area of flood frequency distributions as shown in a journal article given below:

1. Brocca, L., Melone, F., Moramarco, T. (2011). Distributed rainfall-runoff modelling for flood frequency estimation and flood forecasting. *Hydrological Processes*, 25(18), 2801-2813.

1.10.9 Hydraulic Modeling

In the hydraulics area, Florisa and her collaborators investigated into the variation of river discharge in ungaged basins, depth distribution in natural channels, uncertainty analysis for flow velocity estimation, and hydraulic modeling calibration. Her collaborative research is published in refereed journal articles given below:

1. Tarpanelli, A., Brocca, L., Melone, F., Moramarco, T. (2013). Hydraulic modelling calibration in small rivers by using coarse resolution synthetic aperture radar imagery. *Hydrological Processes*, 27(9), 1321-1330.

2. Tarpanelli, A., Brocca, L., Lacava, T., Melone, F., Moramarco, T., Faruolo, M., Pergola, N., Tramutoli, V. (2013). Toward the estimation of river discharge variations using MODIS data in ungauged basins. *Remote Sensing of Environment*, 136, 47-55.

3. Moramarco, T., Corato, G., Melone, F., Singh, V.P. (2013). An entropy-based method for determining the flow depth distribution in natural channels. *Journal of Hydrology*, 497, 176-188.

4. Corato, G., Melone, F., Moramarco, T., Singh, V.P. Uncertainty analysis of flow velocity estimation by a simplified entropy model (2013). *Hydrological Processes*, doi:10.1002/hyp.9590, in press.

1.10.10 Landslides

Florisa also dealt with landslide forecasting using soil moisture work that she and her co-workers did, as shown below:

1. Brocca, L., Ponziani, F., Moramarco, T., Melone, F., Berni, N., Wagner, W. (2012). Improving landslide forecasting using ASCAT-derived soil moisture data: A case study of the Torgiovannetto landslide in central Italy. R*emote Sensing*, 4(5), 1232-1244.

1.11 SUMMATION

Florisa is physically no more, but her spirits lives on. The ideals she lived for and practiced will be as inspiring as when she was alive. In many ways, she became a part of our lives and will therefore be with us as long we live. She is sorely missed, but we are grateful that we were blessed to have her with us as long as we did. She enriched our lives and the best tribute that we can pay her is that we live Florisa's way as much as we can.

1.12 ACKNOWLEDGMENTS

The author is most grateful to Florisa's family-her husband Claudio and her sons Michele and Giacomo-for providing the information on Florisa's personal life and writing their tributes to her. Tommaso Moramarco and Professor Corrado Corradini wrote their tributes and I am grateful to them. This article is a small token of his appreciation of and a small contribution to the celebration of the life of Dr. Florisa Melone - she was like a sister to him.

CHAPTER 2

A REVIEW OF LOCAL INFILTRATION MODELING: FROM REGULAR TO COMPLEX RAINFALL PATTERNS

C. Corradini, R. Morbidelli, C. Saltalippi, A. Flammini

Department of Civil and Environmental Engineering,
University of Perugia (Italy)
via G. Duranti, 93 06125 Perugia, Italy
corrado.corradini@unipg.it

ABSTRACT

An overview of the local infiltration modeling for soils under regular and erratic rainfalls is presented. For homogeneous soils, some remarks concerning the main classical approaches, widely used for isolated storms under the condition of soil infiltration capacity not exceeding the rainfall rate, are first performed. Then, an exhaustive analysis of the modeling formulated by Corradini et al. (1997) [Journal of Hydrology, 192:104-124] is carried out. The model represents appropriately the infiltration under any rainfall pattern, including also periods with redistribution of soil water and reinfiltration, and is therefore more general than that proposed earlier by Dagan and Bresler (1983) [Water Resources Research 19(2):413-420] using a similar starting point represented by a combination of depth-integrated forms of the continuity equation and Darcy's law. Considering that many natural soils can be schematized by two-layered soil profiles, after a short discussion about the characteristics of a soil profile with a sealing layer on the parent soil and the linked classical models, the more general formulation proposed by Corradini et al. (2000) [Journal of Hydrology, 237:58-73], that is applicable during arbitrary rainfall patterns over any two-layered soil combination where either layer may be more permeable, is examined.

2.1 INTRODUCTION

Rainfall infiltration represents the main loss in the process of surface runoff formation and the source of water that supplies the vadose zone and determines the recharge of aquifers. The major interest of infiltration modeling has been focused on the local (point) infiltration and, more recently, on its upscaling to the areal scale from the plot to the watershed. In this context a rigorous representation of point infiltration by accurate numerical solutions of the Richards equation, subject to real initial and boundary conditions with adequate explicit functional forms for the hydraulic soil properties (Smith et al. 1993), requires up to now a considerable computational effort and does not enable us to implement an upscaling procedure of comparable accuracy. On the other hand, the substitution of a real soil characterized by a widely recognized horizontal variability of soil hydraulic properties (Nielsen et al. 1973; Russo and Bresler, 1981) with a fictitious homogeneous soil appears to be unsuitable (Dagan and Bresler, 1983). On the basis of these elements, the flow modeling for surface and subsurface water, particularly in the light of their use in the hydrological practice, involves generally a representation of point infiltration by conceptual/semi-analytical models, if necessary, adapted for applications at the field scale.

Several approximate physically based formulations have been proposed for the estimate of point infiltration into homogeneous soils. Single-events characterized by continuous rainfalls of intensity, r, not lower than soil infiltration capacity, f_c, and initial moisture independent of depth have been considered. Among them, the following three approaches have been primarily used. The Philip infiltration equation for time to ponding t_p=0 (Philip, 1957; 1969), obtained as a series solution reduced to two terms of the Richards equation, then extended (see Chow et al. 1988) to conditions with an initial period characterized by unsaturated soil surface (pre-ponding) and a successive period involving saturated surface (post-ponding). Earlier for t_p=0 Green and Ampt (1911) proposed an equation derived from the Darcy law under the assumption of a rectangular vertical profile of soil water content, θ, while Mein and Larson (1973) extended it for applications to real rainfall events with t_p>0. Afterwards, under the assumption that soil water diffusivity varies rapidly near natural saturation, Smith and Parlange (1978) derived a single equation providing t_p and infiltration rate through the cumulative dynamic infiltration depth used as a surrogate for time, t. This model was then reformulated by Parlange et al. (1982) to represent a wide spectrum of soil hydraulic characteristics through the inclusion of an additional parameter. This parameter, variable between 0 and 1, reduces the new formulation to the limiting forms of Green and Ampt and Smith and Parlange.

Under natural conditions most rainfall patterns in a given location are complex and do not enable us to apply the aforementioned single event models because within the study period there are time intervals with $f_c>r$, while it is not possible to consider many separated storms as independent study cases because the condition of uniform initial soil moisture with depth is not satisfied. Thus, as specified by Smith et al. (1993), starting from the "Time Compression Approximation" (TCA) proposed by Reeves and Miller (1975) for post-hiatus rainfall generating immediate ponding, a simple procedure was applied to generalize this result. This is not supported by physical bases and consists in the assumption that a relation between f_c and the cumulative infiltration depth, I, well-recognized starting from ponding until $r \geq f_c$ (Smith, 1982; Smith and Parlange, 1989) is applicable also for $r<f_c$ under the continuity condition leading to estimate the successive ponding and post-ponding (Mils, 1980; Péschke and Kutílek, 1982; Verma, 1982). However, as it was shown by Smith et al. (1993) through the comparison with numerical solutions of the Richards equation, the procedure simplicity is not associated with accurate estimates of infiltration. This is due to the fact that the extended TCA neglects the soil moisture redistribution process that occurs whenever r falls below the f_c value and becomes very significant when extended periods with small values of r or with a rainfall hiatus are involved. Thus, infiltration under complex rainfall hyetographs should be estimated by models which provide also a simulation of the redistribution process. Dagan and Bresler (1983) developed a simplified model for homogeneous soils based on solutions of integrated forms of Darcy's law and continuity equation, obtained for different boundary conditions at the soil surface and assuming a rectangular shape for soil moisture vertical profile. Later, Corradini et al. (1997), starting from the same basis of Dagan and Bresler (1983), reformulated a previous model developed by Smith et al. (1993) and Corradini et al. (1994) obtaining a more advanced infiltration-redistribution-reinfiltration model applicable to any rainfall pattern over homogeneous soils. The framework of this model was then

adopted to set up a conceptual/semi-analytical approach representing the complex infiltration problem in two-layered soils: crusted soils (Smith et al. 1999) and upper layer more permeable than the underlying soil (Corradini et al. 2000). In this paper, the above overview is integrated through an in-depth analysis of the models by Corradini et al. (1997, 2000), with the objective to highlight the links between them and with the classical Green-Ampt equation. The topic of point infiltration due to the superimposition of the processes of rainfall, run-on and evapotranspiration is also addressed. Finally, the validation earlier performed of the two models by comparison with numerical simulations through the Richard equation and with experimental results is discussed.

2.2 THE MODELING FOR COMPLEX INFILTRATION STAGES IN HOMOGENEOUS SOILS

An appropriate model has to represent the process infiltration-redistribution during erratic rainfall sequences. Corradini et al. (1997) reformulated and improved an approach earlier proposed in two successive papers by the same authors (Smith et al. 1993; Corradini et al. 1994). The model can be easily described starting from the condition of a stepwise rainfall pattern characterized by successive cycles with constant rainfall rate larger than the saturated hydraulic conductivity and rainfall hiatus (Fig. 2.1). It was realized by combining these elements:

(1) an initial soil water content independent of depth, z;

(2) the depth-integrated form of the continuity equation;

(3) the depth-integrated form of Darcy's law;

(4) a dynamic soil moisture vertical profile, $\theta(z)$, of lowest extent Z and approximated, until the time $t=t_2$, by a distorted rectangle defined through a shape factor $\beta(\leq 1)$ which depends on $\theta(z=0)=\theta_0$. The functional form adopted for $\theta(z)$ (Corradini et al. 1994) is:

$$\frac{\theta(z)-\theta_i}{\theta_0-\theta_i}=1-exp\left[\frac{\beta z\left(\theta_0-\theta_i\right)-I'}{\left(\beta-\beta^2\right)I'}\right] \tag{2.1}$$

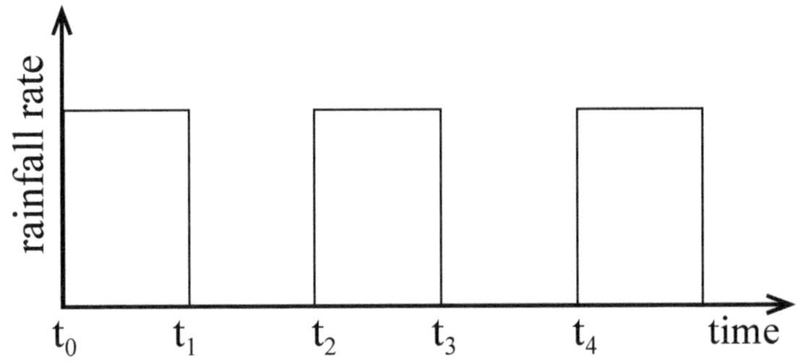

Fig. 2.1. Rainfall pattern used to explain the Corradini et al. (1997; 2000) models.

where I' is the cumulative dynamic infiltration depth ($I'=I-K(\theta_i)t$), with K hydraulic conductivity and the subscript i which henceforth represents quantities at time t=0.

The final equation applicable for time less than that of beginning of the second pulse, $t<t_2$, is:

$$\frac{d\theta_0}{dt} = \frac{(\theta_0-\theta_i)\,\beta(\theta_0)}{I'\left[(\theta_0-\theta_i)\frac{d\beta(\theta_0)}{d\theta_0}+\beta(\theta_0)\right]} \times \left[q_0 - K_0 - \frac{(\theta_0-\theta_i)\,G(\theta_i,\theta_0)\,\beta(\theta_0)\,p\,K_0}{I'}\right] \tag{2.2}$$

where q_0 is the net soil water flux at the surface, $K_0 = K(\theta_0)$ the hydraulic conductivity at the surface, and p is a parameter depending on the profile shape of θ. The quantity $G(\theta_i,\theta_0)$ denotes the net capillary drive at the wetting front, depending on K and the soil water matric potential, Ψ, and in general terms defined as:

$$G(\theta_i,\theta_0) = \frac{1}{K_0}\int_{\Psi(\theta_i)}^{\Psi(\theta_0)} K(\Psi)\,d\Psi \tag{2.3}$$

where the soil hydraulic properties are expressed as (Smith et al. 1993):

$$\Psi(\theta) = \Psi_b \left[\left(\frac{\theta - \theta_r}{\theta_s - \theta_r} \right)^{-\frac{c}{\lambda}} - 1 \right] + d \tag{2.4}$$

$$K(\theta) = K_s \left(\frac{\theta - \theta_r}{\theta_s - \theta_r} \right)^{3+\frac{2}{\lambda}} \tag{2.5}$$

with θ_s surface natural saturation, θ_r residual volumetric water content, Ψ_b air entry head given in tabular form for soil texture classes (Rawls et al. 1983), λ pore size distribution index, c and d empirical coefficients. Equations (2.4) and (2.5) incorporate both the Van Genuchten (1980) and Brooks and Corey (1964) equations as particular cases. The quantities β and p can be determined by expressing β as a linear function g_1 of the scaled water content $\left[\beta(\theta_0) = g_1\left(\frac{\theta_0 - \theta_i}{\theta_s - \theta_r}\right)\right]$ and β_p as an exponential function g_2 of r/K_S.

Equation (2.2) can be solved numerically:

- initially with rainfall over unsaturated soil, q_0=r and I'=[r-K(θ_i)]t, it provides θ_0(t) until surface saturation is reached (see for example point 1 in Fig. 2.2);

- when θ_0 reaches θ_s, $\theta_0=\theta_s$ and $d\theta_0/dt=0$, thus with q_0=r it gives through I' time to ponding, t_{p1} (point 2 in Fig. 2.2);

- for $t_{p1}<t\le t_1$ (see Fig. 2.1), $\theta_0=\theta_s$ and $d\theta_0/dt=0$, thus it provides $q_0=f_c$ (between points 2 and 3 in Fig. 2.2) deduced as:

$$f_c = K_s\left[1 + \frac{(\theta_s - \theta_i)\, G(\theta_i,\theta_s)\, \beta(\theta_s)\, p}{I'}\right] \tag{2.6}$$

where I' in the surface saturation period accumulates from f_c;

- for $t_1<t<t_2$ (see Fig. 2.1), $q_0 = r = 0$ and $d\theta_0/dt<0$, thus it gives θ_0 (t) decreasing with time and elongated vertical profiles.

At the time t_2 starts the second rainfall pulse with $r>K_s$ which leads the surface to saturation at a time t_{p2} (point 5 in Fig. 2.2). However, for $t\geq t_2$, while the structure of Eq. (2.2) remains unchanged, the model includes two alternative application methods depending on the magnitude of r in comparison with the downward redistribution rate, D_F, estimated at $t=t_2$ and expressed through the combination of the elements (1), (2) and (4) by:

$$D_F = \left(-\frac{1}{\beta}\frac{d\beta}{d\theta_0} - \frac{1}{\theta_0 - \theta_i} \right) I_2' \frac{d\theta_0}{dt} - K_i \qquad (2.7)$$

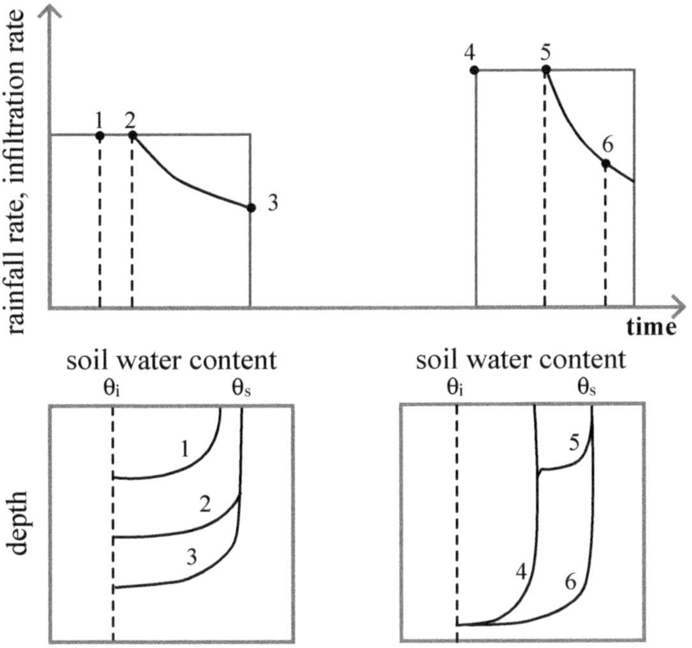

Fig. 2.2. Time evolution of infiltration rate and soil moisture vertical profiles at the times 1-6 above indicated.

Specifically, for $D_F \geq r$, the re-infiltrating water is assumed to be distributed in the whole dynamic profile occurring at $t = t_2$ (see Fig. 2.1). This allows to determine θ_0 (t) in the period of the second pulse directly by Eq. (2.2). On the other hand, for $r>D_F$, the existing profile is considered temporarily invariant and the surface flux creates a superimposed secondary wetting profile (point 5 in Fig. 2.2), thus advancing alongside the pre-existing profile according to Eq. (2.2), with $\theta_i = \theta_0(t_2)$ and I' accumulated after t_2, specifically:

$$\frac{d\theta_0}{dt} = \frac{\left(\theta_0 - \theta_0(t_2)\right)\,\beta(\theta_0)}{\left[I - I(t_2) - K_0(t_2)(t - t_2)\right]\left[\left(\theta_0 - \theta_0(t_2)\right)\frac{d\beta(\theta_0)}{d\theta_0} + \beta(\theta_0)\right]} \times \left[q_0 - K_0 - \frac{(\theta_0 - \theta_i)\,G(\theta_i,\theta_0)\,\beta(\theta_0)\,p\,K_0}{\left[I - I(t_2) - K_0(t_2)(t - t_2)\right]}\right] \quad (2.8)$$

Then, the secondary wetting front advances until its depth approaches the depth of θ_0 (t_2), when the compound profile again becomes a single profile (point 6 in Fig. 2.2) and infiltration becomes again computable directly by Eq. (2.2). If the second pulse produces a secondary wetting profile but the surface does not reach saturation, the successive redistribution stage involves first this profile and then the re-established one after the possible merging of the compound profile. In the case merging has not been reached, if the rainfall rate starting at $t = t_4$ (see Fig. 2.1) is larger than $D_F(t_4)$ associated with the secondary profile a fictitious procedure of consolidation that merges the composite profile existing at $t = t_4$ is applied. The profile consolidation reduces the compound profiles to a maximum of two dynamic components. The model, completed with the procedures of generation of compound and consolidated profiles and above described by using the simple rainfall pattern of Fig. 2.1, can be easily applied to any real time evolution of r.

Corradini et al. (1997) provided also the model calibration using a silty loam soil, and then its testing, through different soil types (denoted as A, B, J, D, E and SL) ranging from fine-textured to sandy loam.

The Richards equation was used as a benchmark. A further example of model testing on Soil B is shown in Fig. 2.3 where most of the model components are involved. During the first rainfall pulse the pre-ponding and post-ponding stages are observed, both simulated by the single profile scheme, which characterizes also the successive redistribution stage. At the beginning of the second pulse there is the formation of a secondary profile, which persists in the new pre-ponding period and later until $t = t_m$ when the merging of the compound profile occurs; then the single profile scheme is re-established. As it can be seen the model is very accurate in terms of infiltration rate as well as of surface water content.

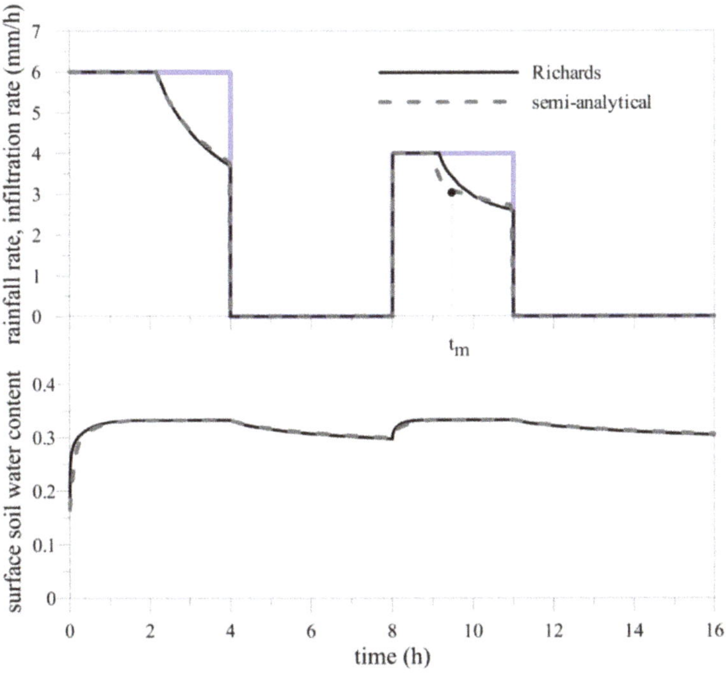

Fig. 2.3. Time evolution of infiltration rate and surface water content for the rainfall pattern shown in the upper diagram on Soil B (Corradini et al. 1997) for initial soil moisture θ_i=0.1659 (for t_m see text).

The model was also checked through laboratory experiments directed to examine the function θ (z,t). Figures 2.4 and 2.5 illustrate a comparison of model and experimental results referred to two experiments of the studies earlier presented by Melone et al. (2006; 2008). As it can be deduced from an overall analysis of the results, the conceptual/semi-analytical model simulates fairly well the observed data even though some uncertainties remain because of the limits forced by the locations of the soil moisture sensors. In any case, for example, Fig. 2.5 shows as the compound profile procedure works well during the re-infiltration stage associated with the rainfall pulse occurring after a long rainfall.

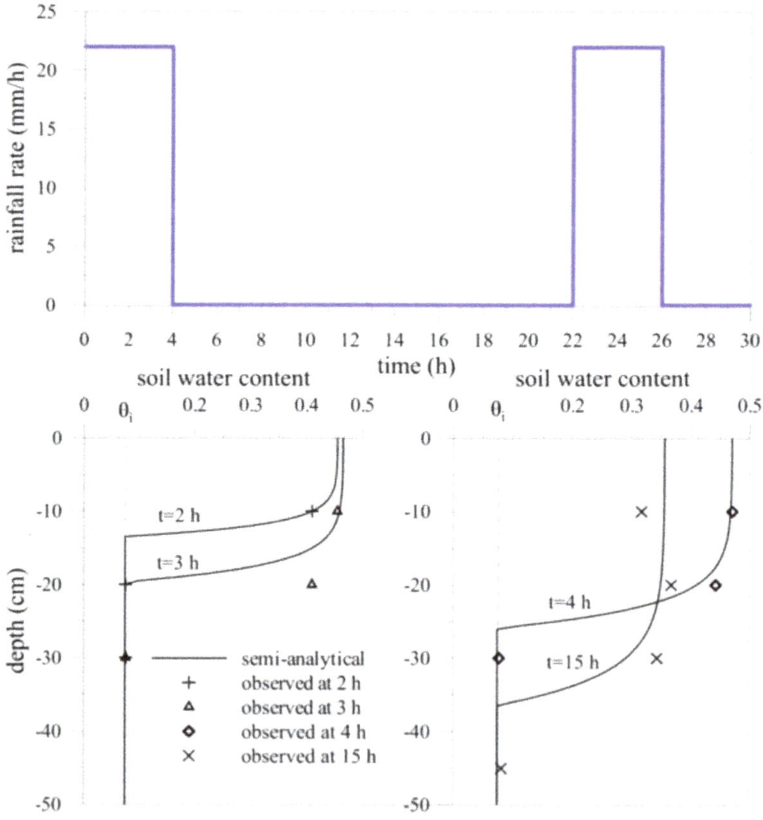

Fig. 2.4. Comparison of soil moisture vertical profiles estimated by the Corradini et al. (1997) model and values observed in a laboratory experiment in a loam soil under the rainfall pattern shown in the upper diagram (Melone et al. 2006).

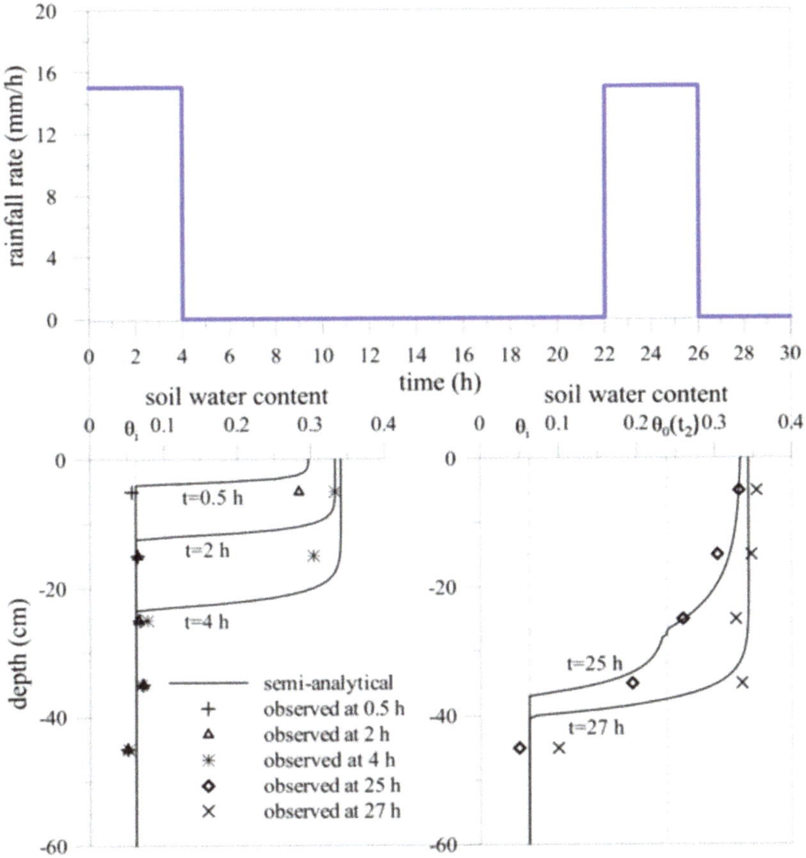

Fig. 2.5. **Comparison of soil moisture vertical profiles estimated by the Corradini et al. (1997) model and values observed in a laboratory experiment in a sandy clay loam soil under the rainfall pattern shown in the upper diagram (Melone et al. 2008).**

2.3 THE MODELING FOR INFILTRATION OF ERRATIC RAINFALLS INTO TWO-LAYERED SOILS

Most natural soils are characterized by a not homogeneous vertical profile that in many cases can be schematized by a two layer approximation (Mualem et al. 1993; Taha et al. 1997) with a thin sealing layer and the underlying parent soil or two layers with the upper one of greater permeability. The possible existence of mechanisms of formation and disruption of a crust has been widely described (Mualem and Assouline,

1989; Mualem et al. 1993; Bullock et al. 1998; Emmerich, 2003; Morbidelli et al. 2011). The basic role of crusted soils, in semi-arid areas, has been recently pointed out by Chen et al. (2013). Profiles with a more permeable upper layer are commonly found for fairly moderate rainfalls over freshly tilled soils or can be artificially realized in the hydrological practice when a fast water drainage is required. Furthermore, a similar fictitious profile can be useful as a first approximation of the infiltration-redistribution cycle in a homogeneous soil with grassy vegetation (Morbidelli et al. 2014).

This short analysis highlights the importance to realize models of infiltration into any two-layered soil under erratic rainfalls. Really, approaches applicable only for crusted soils and under restrictive physical conditions were mainly proposed (for example, Hillel and Gardner, 1970; Moore, 1981; Ahuja, 1983; Smith, 1990; Philip, 1998). A more general model suitable to simulate infiltration of erratic rainfalls into a horizontal two-layered soil, independently of the location of the layer which is less permeable, was proposed by Corradini et al. (2000) by adapting a model earlier formulated by Smith et al. (1999) for crusted soils.

The model by Corradini et al. (2000) is based on the same elements (1)-(4) (Section 2.2) previously used for a homogeneous soil, but adopted here separately in each layer, integrated by the following boundary conditions:

- at the interface between the two layers:

$$\Psi_1(Z_c) = \Psi_2(Z_c) = \Psi_c \qquad t > 0 \tag{2.9}$$

$$K_1\left[\frac{\partial \Psi_1(Z_c)}{\partial z} - 1\right] = K_2\left[\frac{\partial \Psi_2(Z_c)}{\partial z} - 1\right] \qquad t > 0 \tag{2.10}$$

where hereafter the subscripts c, 1 and 2 stand for quantities referred to the interface, the upper layer and the underlying layer, respectively;
- at the lower boundary, Z_{inf}, that is not reached by the dynamic profile:

$$\Psi_2 = \Psi_i \qquad t > 0 \tag{2.11}$$

In addition, the interaction between the two layers is expressed through the flux at the interface, $D_{F,c}$, which is assumed equal to the depth-integrated downward flux in the upper layer:

$$D_{F,c} = K_{1s} \frac{G_1(\Psi_c, \Psi_{10})}{Z_c} + K_{1c} \tag{2.12}$$

The final equations, applicable starting from the time t_c, when the wetting front reaches the interface, are:

$$\frac{d\Psi_{10}}{dt} = \frac{1}{\alpha Z_c C_1(\Psi_{10})}\left[q_{10} - K_{1c} - \frac{K_{1s} G_1(\Psi_c, \Psi_{10})}{Z_c}\right] - \frac{(1-\alpha) C_1(\Psi_c)}{\alpha C_1(\Psi_{10})} \frac{d\Psi_c}{dt} \tag{2.13}$$

$$\frac{d\Psi_c}{dt} = \frac{1}{P_L(\Psi_c, t)}\left[K_{1c} + \frac{K_{1s} G_1(\Psi_c, \Psi_{10})}{Z_c} - K_{2c} - \frac{\beta_2(\theta_{2c}) p_2(\theta_{2c} - \theta_{2i}) K_{2s} G_2(\Psi_i, \Psi_c)}{I_2}\right] \tag{2.14}$$

with $P_L(\cdot)$ and I_2 defined as:

$$P_L(\Psi_c, t) = \left[\beta_2(\theta_{2c}) + \frac{d\beta_2}{d\theta_{2c}}(\theta_{2c} - \theta_{2i})\right] \frac{I_2}{(\theta_{2c} - \theta_{2i})\beta_2(\theta_{2c})} C_2(\Psi_c) \tag{2.15}$$

$$I_2 = I - Z_c\left[\alpha(\theta_{10} - \theta_{1i}) + (1-\alpha)(\theta_{1c} - \theta_{1i})\right] - K_{2i} t \tag{2.16}$$

and

$$C_1(\Psi_{10}) = d\theta_{10}/d\Psi_{10},\ C_1(\Psi_c) = d\theta_{1c}/d\Psi_c,\ C_2(\Psi_c) = d\theta_{2c}/d\Psi_c. \tag{2.17}$$

The quantity α, considered as a constant, is a weighting coefficient for surface wetting and interface wetting in the upper layer. The quantities β_2 and p_2 can be determined by analogy with the functions g_1 and g_2 earlier discussed for homogeneous soils.

For $t \geq t_c$ the system of Eqs. (2.13) and (2.14) may be solved by well-known numerical methods under the same upper boundary conditions previously examined for the solution of Eq. (2.2), that has to be applied for $t < t_c$. We remark that the single Eq. (2.2) or the system Eqs. (2.13)-(2.14) are applicable up to the time t_2 when the 2nd rainfall pulse starts. Then, the two-layered model is applicable, along the lines specified for the homogeneous soil, comparing the relative role of the upper boundary flux in each layer and the corresponding downward flux. Thus, the generation of secondary wetting front and the profile consolidation have to be incorporated even in the use of the differential equation system. For example, a secondary wetting front in the upper layer determines a compound profile, which reduces to a simple profile when its depth approaches Z_c, then a comparison between the downward fluxes of each layer establishes if a secondary wetting front is generated in the underlying soil. In any case, on the basis of the elements aforementioned the model application to rainfall patterns much more complex than that of Fig. 2.1 is straightforward.

Corradini et al. (2000) presented also calibration and testing of the model using the Richards equation as a yardstick. Figure 2.6 shows for a representative event minor errors in the simulated behaviors of both q_0 and θ_0 through the whole cycle infiltration-redistribution-reinfiltration. In addition, Fig. 2.7 illustrates an example of appropriate model application in the simulation of $\theta(z)$ in the stages of infiltration and redistribution for an erratic rainfall distribution over a real soil profile with grassy vegetation, represented as a two-layered soil (Morbidelli et al. 2014). We note as the shape of the simulated curves is fairly well reproduced and the errors in the upper layer are probably due to the assumed fictitious soil profile influenced also by irregular spatial distribution of vegetation.

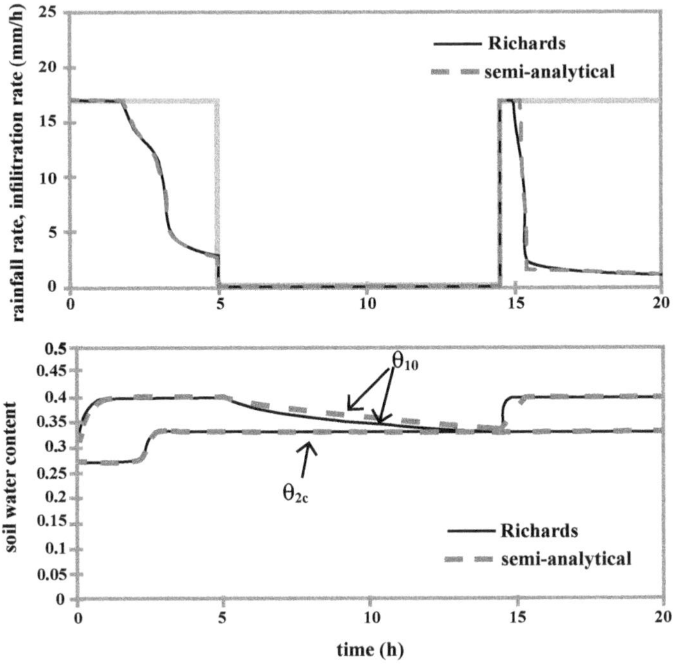

Fig. 2.6. Time evolution of infiltration rate and soil water content, θ_{10} at the surface and θ_{2c} at the interface, for the rainfall pattern shown in the upper diagram. Twenty cm of Soil 2 over Soil 1, as in Corradini et al. (2000), initial soil water matric potential $\Psi_i = -4190$ mm.

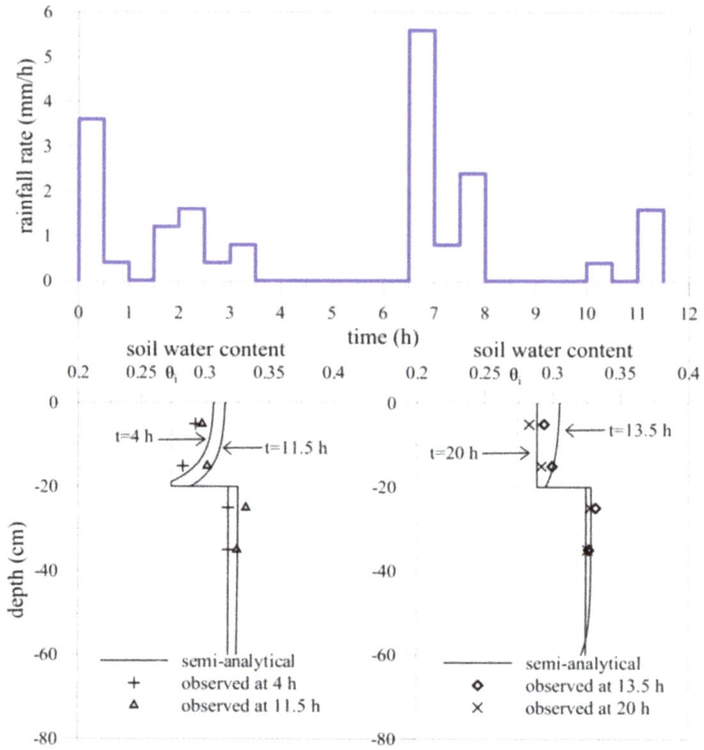

Fig. 2.7. Comparison of soil moisture vertical profiles estimated by the Corradini et al. (1997) model and values observed under the rainfall pattern shown in the upper diagram. Field experiment on Mar 14, 2011 in 20 cm of vegetative more permeable soil over parent loam soil, as in Morbidelli et al. (2014), Ψ_i = - 826 mm.

2.4 CONCLUSIONS

There are important features of a few models considered throughout this review that have to be remarked.

1) To our knowledge the Corradini et al. (2000) model is currently the most general conceptual/semi-analytical approach that enables us to simulate with great accuracy $q_0(t)$ and $\theta_0(z,t)$ under arbitrary rainfall patterns for any two-layered soil combination.

2) The numerical solution of the Richards equation is widely recognized to be appropriate to derive the correct results but that of the above ordinary differential equations requires a much simpler computer code and implies a substantial reduction in computational effort.

3) The Dagan and Bresler (1983) model uses simplified relationships to analyze the spatial variability of moisture content and water flux under conditions of infiltration and redistribution. Specifically, the model uses relationships for the hydraulic properties in the simplest form, a piston flow scheme for the wetting front, a two stages system without reinfiltration. However, it involves important analytical developments even though applicable only in a limited number of cases.

4) The extended Green-Ampt model is obtained from the Corradini et al. (2000) model with the simplification $\beta_p = 1$ but in the field limited by the condition $f_c \leq r$ to be verified with continuity. However, for some applications to problems, which require an averaging procedure over a field its use appears justified. Therefore, considering its very simple formulation the model can be conveniently adopted when areal average infiltration approximations have to be realized (see, for example, Corradini et al. 2011a and 2011b; Govindaraju et al. 2001).

5) Even though not explicitly considered in the model formulation, the water flux at the surface in Eqs. (2.2) and (3.5) can be expressed as a net flux at the soil surface that incorporates rainfall, evapotranspiration and also the run-on (Nahar et al. 2004) expressed in terms of discharge per unit surface.

2.6 ACKNOWLEDGMENTS

Most of this review relies on papers earlier published in cooperation with Dr. Florisa Melone who untimely died leaving us deeply distressed.

2.7 REFERENCES

Ahuja, L. R. 1983. Modeling infiltration into crusted soils by the Green-Ampt approach, *Soil Science Society of America Journal* 47:412-418.

Brooks, R. H. and A. T. Corey. 1964. Hydraulic properties of porous media. *Hydrology Papers 3*, Colorado State University, Fort Collins, Colorado, 27 pp.

Bullock, M. S., Kemper, W. D. and S. D. Nelson. 1988. Soil cohesion as affected by freezing, water content, time and tillage, *Soil Science Society of America Journal* 52: 770–776.

Chen, Li, Sela, S., Svoray T. and S. Assouline. 2013. The roles of soil-surface sealing, microtopography and vegetation patches in rainfall-runoff processes in semiarid areas. *Water Resources Research* 49:1-15.

Chow, V. T., Maidment, D. R. and L. W. Mays. 1988. *Applied Hydrology*. McGraw-Hill, New York.

Corradini, C., Melone, F. and R. E. Smith. 1994. Modeling infiltration during complex rainfall sequences. *Water Resources Research* 30:10:2777-2784.

Corradini, C., Melone F. and R. E. Smith. 1997. A unified model for infiltration and redistribution during complex rainfall patterns. *Journal of Hydrology* 192:104-124.

Corradini, C., Melone, F. and R. E. Smith. 2000. Modeling local infiltration for a two-layered soil under complex rainfall patterns. *Journal of Hydrology* 237:58-73.

Corradini, C., Morbidelli, R., Flammini, A. and R. S. Govindaraju. 2011a. A parameterized model for infiltration in two-layered soils with a more permeable upper layer. *Journal of Hydrology* 396:221-232.

Corradini, C., Flammini, A., Morbidelli, R. and R. S. Govindaraju. 2011b. A conceptual model for infiltration in two-layered soils with a more permeable upper layer: From local to field scale. *Journal of Hydrology* 410:62-72.

Dagan, G. and E. Bresler. 1983. Unsaturated flow in spatially variable fields, 1, Derivation of models of infiltration and redistribution. *Water Resources Research* 19:2:413-420.

Emmerich, W. E. 2003. Season and erosion pavement influence on saturated soil hydraulic conductivity, *Soil Sciences* 168: 637–645.

van Genuchten, M. Th. 1980. A closed-form equation for predicting the hydraulic conductivity of unsaturated soils. *Soil Science Society of America Journal* 44:5:892-898.

Green, W. H. and G. Ampt. 1911. Studies of soil physics, part I – the flow of air and water through soils. *Journal of Agricultural Science* 4:1-24.

Govindaraju, R. S., Morbidelli, R. and C. Corradini. 2001. Areal infiltration modeling over soil with spatially-correlated hydraulic conductivities. *Journal of Hydrologic Engineering*, ASCE 6:150-158.

Hillel, D. and W. R. Gardner. 1970. Transient infiltration into crust topped profiles, *Soc. Am. J.*, 109:2:69-76.

Mein, R. G. and C. L. Larson. 1973. Modeling infiltration during a steady rain. *Water Resources Research* 9:2:384-394.

Melone, F., Corradini, C., Morbidelli, R. and C. Saltalippi. 2006. Laboratory experimental check of a conceptual model for infiltration under complex rainfall patterns. *Hydrological Processes* 20:439–452.

Melone, F., Corradini, C., Morbidelli, R., Saltalippi, C. and A. Flammini. 2008. Comparison of theoretical and experimental soil moisture profiles under complex rainfall patterns. *Journal of Hydrologic Engineering* 13:1170–1176.

Mls, J. 1980. Effective rainfall estimation. *Journal of Hydrology* 45:305-311.

Moore, I. D. 1981. Infiltration equations modified for surface effects. *Journal of Irrigation and Drainage Division* 107:71-86.

Morbidelli, R., Corradini, C., Saltalippi, C., Flammini, A. and E. Rossi. 2011. Infiltration-soil moisture redistribution under natural conditions: experimental evidence as a guideline for realizing simulation models, *Hydrology and Earth System Sciences* 15:2937-2945.

Morbidelli, R., Saltalippi, C., Flammini, A., Rossi, E. and C. Corradini. 2014. Soil water content vertical profiles under natural conditions: matching of experiments and simulations by a conceptual model. *Hydrological Processes*, 28(17), 4732-4742.

Mualem, Y. and S. Assouline. 1989. Modeling soil seal as a non-uniform layer, *Water Resources Research* 25:2101-2108.

Mualem, Y., Assouline, S. and D. Eltahan. 1993. Effect of rainfall induced soil seals on soil water regime: wetting processes, *Water Resources Research* 29:1651-1659.

Nahar, N., Govindaraju, R. S., Corradini, C. and Morbidelli, R. 2004. Role of run-on for describing field-scale infiltration and overland flow over spatially variable soils. *Journal of Hydrology* 286:36-51.

Nielsen, D. R., Biggar, J. W. and K. T. Erh. 1973. Spatial variability of field measured soil-water properties. *Hilgardia* 42:7:215-259.

Parlange, J.-Y., Lisle, I., Braddock, R. D. and R. E. Smith. 1982. The three-parameter infiltration equation. *Soil Science* 133:6:337-341.

Péschke, G. and M. Kutílek. 1982. Infiltration model in simulated hydrographs. *Journal of Hydrology* 56:369-379.

Philip, J. R. 1957. The theory of infiltration, 4, Sorptivity algebraic infiltration equation. *Soil Science* 84:3:257-264.

Philip, J. R. 1969. Theory of infiltration. *Advances in Hydroscience* 5:215-296.

Philip, J. R. 1998. Infiltration into crusted soils. *Water Resources Research* 34:8:1919-1927.

Rawls, W. J., Brakensiek, D. L. and B. Soni. 1983. Agricultural management effects on soil processes: Part I. Soil water retention and Green-Ampt infiltration parameters. *Transactions of the ASAE* 26:6:1747-1752.

Reeves, M. and E. E. Miller. 1975. Estimating infiltration for erratic rainfall. *Water Resources Research* 11:102-110.

Russo, D. and E. Bresler. 1981. Soil hydraulic properties as stochastic processes, 1, Analysis of field spatial variability. *Soil Science Society of America Journal* 45:682-687.

Smith, R. E. 1982. Rational models of infiltration hydrodynamics. Modeling Components of Hydrologic Cycle. *Water Resources Publications, LLC*, Littleton, CO, 107-126.

Smith, R. E. 1990. Analysis of infiltration through a two-layer soil profile. *Soil Science Society of America Journal* 54:5:1219-1227.

Smith, R. E., Corradini, C. and F. Melone. 1993. Modeling infiltration for multistorm runoff events. *Water Resources Research* 29:1:133-144.

Smith, R. E., Corradini C. and F. Melone. 1999. A conceptual model for infiltration and redistribution in crusted soils. *Water Resources Research* 35:5:1385-1393.

Smith, R. E. and J.-Y. Parlange. 1978. A parameter-efficient hydrologic infiltration model. *Water Resources Research* 14:3:533-538.

Smith, R. E. and J.-Y. Parlange. 1989. Comment on "Constant rate rainfall infiltration: A versatile nonlinear model: 2. Applications of solutions" by I. White and P. Broadbridge. *Water Resources Research* 25:5:1051-1053.

Taha, A., Gresillon, J. M. and B. E. Clothier. 1997. Modelling the link between hillslope water movement and stream flow: application to a small Mediterranean forest watershed. *Journal of Hydrology* 203:11-20.

Verma, S. C. 1982. Modified Horton's infiltration equation. *Journal of Hydrology* 58:383-388.

CHAPTER 3

DOCUMENTING CONNECTIVITY OF HYDROLOGIC FLOWPATHS USING TERRESTRIAL DIATOMS AND THERMAL INFRARED IMAGERY

Laurent Pfister[1], Carlos E. Wetzel[1], Núria Martínez-Carreras[1], Jay Frentress[1,2], Luc Ector[1]

[1] Luxembourg Institute of Science and Technology, L-4362 Esch/Alzette, Luxembourg
[2] Oregon State University, Corvallis, Oregon, USA

ABSTRACT

Water sources, flowpaths and transit times have been on the research agendas of experimental hydrologists for decades. Understanding the spatial and temporal variability of hydrologic processes is indeed a key to success for anticipating any catchment's response to climate and/or anthropogenic forcing agents.

The advent of geochemical and isotopic tracers shed new light on the rainfall-runoff transformation process. However, the gained insights remained largely constrained to specific catchments. In addition to contributing to the description of catchment peculiarities, these widespread tracing techniques also have well-known limitations, such as non-stable end-members or incomplete mixing assumptions.

There have been repeated calls for new tracers and high-frequency monitoring strategies to overcome the current status quo in tracer hydrology. Here, we provide examples of two innovative and complementary research avenues in this respect. We currently explore the potential for:

(i) terrestrial diatoms to detect the onset/cessation and connectivity of overland flow in the hillslope-riparian zone-stream (HRS) continuum;
(ii) thermal infrared (IR) imagery to document saturated area dynamics and their role in storm hydrograph generation.

Over the past five years, we have tested these new approaches in a nested catchment set-up in the Attert River basin (Luxembourg). Catchment sizes range from 0.47 to 247 km^2 and their physio-geographic characteristics are very contrasted (geological layers including sandstone, marls, schists and land use varying between forests, grassland and agricultural land).

3.1 INTRODUCTION

Early investigations on streamflow generation relied on measurements of precipitation, discharge, soil moisture and groundwater table fluctuations. They revealed for the first time the contrasted responses of hillslopes and riparian zones to precipitation [Whipkey, 1965; Dunne and Black, 1970], triggered by both surface processes (infiltration excess [Horton, 1933], partial area surface runoff generation [Betson, 1964], variable source area [Cappus, 1960; Hewlett and Hibbert, 1967; Dunne et al. 1975]) and subsurface processes (e.g. subsurface storm flow [Hewlett and Hibbert, 1963], return flow [Dunne, 1978]).

One important limitation of these approaches was their inadequacy for apportioning the relative quantitative and qualitative contributions of distinct landscape units to the storm hydrograph. New momentum was gained in this respect with the advent of geochemical and stable isotope tracers. They helped to substantially improve our understanding of geographic [Sklash and Farvolden, 1979; Hooper and Shoemaker, 1986; Wels et al. 1991] and temporal [Pearce et al. 1986; Sklash et al. 1986; McDonnell, 1990; Burns et al. 2001] sources of stormflow. However, conceptual limitations [Richey et al. 1998; Burns, 2002] and uncertainties [Genereux, 1998; Joerin et al. 2002], inherent to these techniques, were quite rapidly identified. As a consequence, the interpretation of the hydrological, geochemical or isotope signals in storm hydrographs often remains equivocal [Kirchner, 2003]. Equifinality caused by the potential mixing of various water sources or flowpaths, as well as hydrological mechanisms [Kendall et al. 2001], is equally problematic.

New promising avenues have been tested in recent years for overcoming the prevailing status quo in catchment hydrology. The spatial and temporal resolution of available datasets has rapidly increased with new technological developments, such as temperature sensing via fiber optic cables [Selker et al. 2006] and thermal IR imagery [Pfister et al. 2010; Schuetz and Weiler, 2011], or electrical and magnetic geophysical methods [Robinson et al. 2008]. However, the application of these new technologies remains essentially restricted to the point or plot scales. Ultimately, they have not brought the long-awaited progress in terms of the discovery of those macroscale 'hydrologic laws' that had first been advocated by Dooge [1986]. To date, catchment hydrology still has to overcome the limitations of the prevailing concepts dating back to the

1960s by Cappus' [1960] and Hewlett's [1961] Variable Source Area (VSA) concept [critiqued by Ambroise, 2004; McDonnell et al. 2007].

Weak process representations and uncertain model structures [Sivapalan, 2009] eventually are the result of 'uniqueness in field measurements limiting model representations' [Beven, 2000] and the intimately related 'idiosyncrasies' of individual catchment studies [McDonnell, 2003]. In an attempt to grasp the ever-increasing description of catchment heterogeneities, catchment characterization and classification have been suggested as a promising research avenue [McDonnell and Woods, 2004; Kirchner, 2009; Ali et al. 2012]. In this context, concepts and definitions prevailing in ecology, such as connectivity and related threshold measures, were introduced in order to move beyond the obviously simplistic VSA-concept [Ambroise, 2004; Bracken and Croke, 2007; McDonnell et al. 2007].

Pringle [2003] and Freeman et al. [2007] documented the key role of headwater streams in sustaining hydrological connectivity between uplands, riparian areas and streams and thereby structuring ecosystems at regional scales. A holistic view of catchment functionality is expected to emerge from interdisciplinary studies that integrate hydrology and ecology in catchment science [Hannah et al. 2007; Tetzlaff et al. 2007; Wainwright et al. 2011; Booth and Loheide II, 2012].

Connectivity has been recently introduced as a research interface in catchment science. In this respect, it is seen as an emergent property when seen from either the hydrological, geomorphological or ecological perspective. It stands out as a natural backbone to which a multitude of heterogeneous sub-scale processes, functions or habitats map onto [McDonnell et al. 2007; Tetzlaff et al. 2007]. To date, the concept of connectivity still suffers from a multitude of definitions [e.g. Pringle, 2003; Freeman et al. 2007], even though recent attempts have aimed at rendering this usage less ambiguous [Ali and Roy, 2009; Wainwright et al. 2011; Spence, 2010; Spence and Phillips, 2014].

McGlynn and McDonnell [2003] were the first to differentiate and quantify contributions (via hydrometric measurements, isotopic and geochemical data) from dynamic riparian zones and hillslopes throughout a storm hydrograph. Ambroise [2004] insisted on the difference between variable 'active' areas and 'contributing', i.e. temporarily connected areas. Upslope accumulated areas were introduced by Jencso et al. [2009] as a new predictor of hillslope-riparian water table connectivity. Hillslope signatures being often buffered or obscured by riparian zones [Jencso et al. 2010], it is essential to better understand the spatial and temporal dynamics of the connectedness between uplands and the valley bottoms [McGuire and McDonnell, 2010; Phillips et al. 2011].

Given recent calls within the hydrological community for interdisciplinary approaches and exploring new techniques to advance our understanding of runoff mechanisms, we have recently initiated research into eco-hydrological couplings [e.g. Petts et al. 2006; Hannah et al. 2007; Tetzlaff et al. 2007]. More specifically, we have explored the potential for terrestrial diatoms to aid detection of the onset/cessation of surface runoff in the hillslope-riparian-stream (HRS) system [Pfister et al. 2009]. Moreover, we have assessed the potential for thermal IR imagery to gain better insight into the spatial dynamics of water mixing and connectivity of surface saturation in the riparian zone [Pfister et al. 2010].

Here, we develop on recent results obtained on the combined use of diatoms as a biological tracer and thermal IR imagery for detecting the onset/cessation of surface runoff in the hillslope/riparian zone/stream (HRS) continuum. More specifically, we ask whether terrestrial diatom populations exhibit distinct features specific to catchments, as a result of local physiographic catchment characteristics like geology and land cover.

3.2 METHODOLOGY

For studying rapid flowpaths connectivity within the HRS, we relied on the simultaneous use of conventional geochemical and isotopic tracers, biotic tracers (i.e. diatoms) and thermal IR imagery. The monitoring and sampling protocols were carried out in the contrasting Weierbach (0.47 km^2; schists dominated; forested) and Huewelerbach (2.76 km^2; sandstone dominated; forested) experimental catchments, located in the Attert River basin in Luxembourg (Fig. 3.1). Water and terrestrial diatom sampling campaigns were carried out at seasonal and event scales to grasp the spatial variability and temporal dynamics of the terrestrial diatom flushing process.

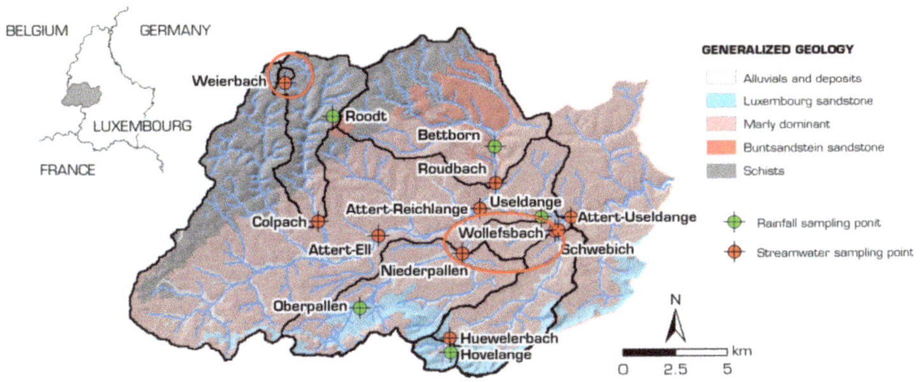

Fig. 3.1. The Attert River basin with nested experimental catchments. The Weierbach (schists) and Wollefsbach (marls) experimental catchments are marked with red circles.

3.2.1 Hydrometric Monitoring

Streamwater samples were collected with automatic samplers (ISCO), coupled to recording stream gauges (Figs. 3.1 and 3.2f). A sequential rainfall sampler (Fig. 3.2a-b), several overland flow collectors (gutters) sealed to the soil surface (Fig. 3.2c-d) and suction cups (Fig. 3.2e) were installed in the Weierbach catchment to collect water and diatom samples. Four 90-mm groundwater wells were instrumented with real-time TD-Divers data loggers (Schlumberger Water Services, Fig. 3.2g) recording water table depth and temperature at 15-min intervals. Conductivity was measured at 30-min intervals using a conductivity meter (WTW). The volumetric water content (VWC) of soils was measured using water content reflectometers (CS616-L, Campbell Scientific, Fig. 3.2h), which use the time-domain measurement method. Five reflectometers were buried parallel to the surface at 10 cm depth and installed along a transect perpendicular to the stream along the HRS continuum. Four probes were installed on the perpendicular to the right bank of the stream (deciduous forest) in the plateau, middle of the hillslope, hillslope foot and in the riparian zone; the fifth reflectometer was installed on the left side of the creek (coniferous forest), in the middle of the slope. Two soil temperature probes were also installed in the same area (right bank, deciduous trees), one on the plateau and another in the riparian zone.

Fig. 3.2. Instrumental set-up in the Weierbach catchment.

3.2.2 Hydrochemical and Stable Isotope Tracers

Collected water samples were analysed in the laboratory for major anions and cations (Cl^-, NO_3^-, SO_4^{2-}, Na^+, K^+, Mg^{2+}, Ca^{2+}), dissolved silica (SiO_2), pH, alkalinity and UV-absorbance (254 nm) after filtration through WHATMAN GF/C glass fiber filters.

Isotopic analyses of $^{18}O/^{16}O$ and D/H were conducted using a LGR Liquid-Water Isotope Analyser (DLT-100, version 908-0008). Streamflow was then separated into (i) event and pre-event water using a traditional two-component hydrograph separation [Sklash and Farvolden, 1979], and into (ii) groundwater, overland flow, soil water and rainwater using an end-member mixing analysis [Christophersen and Hooper, 1992].

3.2.3 Diatom Sampling and Analysis

3.2.3.1 Targeting diatom sampling sites

During baseflow conditions, we have carried out seasonal diatom samplings in the HRS system on various substrates (e.g. in-stream, soils, litter, moss, vegetation), in order to document the spatio-temporal variability of diatom communities.

Infrared imagery was used to select diatom sampling locations *in situ* during storm events. Rainfall or snowmelt sources of water can vary greatly from groundwater temperatures, allowing for the likely source of overland flow to be quickly determined. We used a hand-held infrared camera (FLIR b425) to capture infrared data from areas of overland flow to enhance the diatom sampling. Upwelling groundwater is unlikely to contain diatoms as diatoms are thought to be too large to be transported through the subsurface environment – though this has yet to be tested in the field. However, overland flow, which is largely made up of snowmelt and incident precipitation is more likely to contain diatoms as it has not yet infiltrated into the soil. This technique was used to target diatom sampling in areas where overland flow could potentially contain diatoms. In Fig. 3.3, samples at location A would be more likely to contain diatoms because the temperature at A indicates the overland flow is likely originating from melting snow and incoming rain, as opposed to B where the increased temperature suggests a groundwater source.

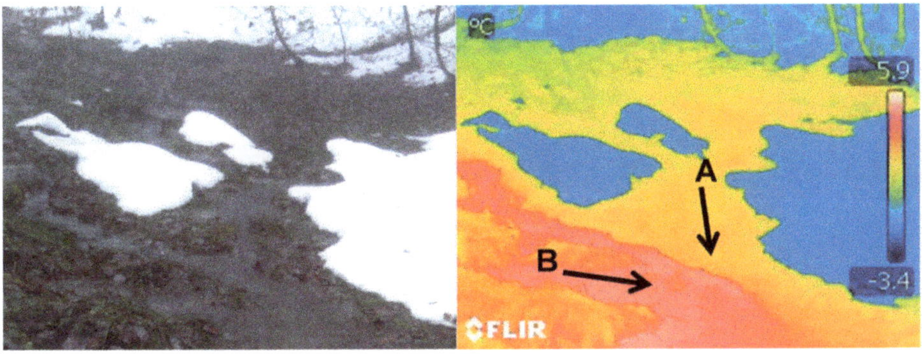

Fig. 3.3. Infrared and digital color images showing overland flow in the same location of the Weierbach catchment during a rain on snow event in January 2011. Location B is largely comprised of upwelling groundwater (~5 °C) while location A is likely snowmelt (~1 °C). Arrows indicate flow direction and mixing of the two sources.

3.2.3.2 Diatom extraction from various organic substrates

In each experimental catchment, the main potential diatom substrates (e.g. leaves, litter, moss and herbaceous vegetation) were identified and sampled using 1-L sampling bottles. Each sample was composed of five sub-samples collected on a 5-m transect parallel to the stream (a subsample was collected every 1 m). Only material on the top surface, well exposed to direct sunlight, was sampled.

Upon arrival at the laboratory, the sampling bottles containing the different diatom-substrate samples were filled with sparkling water, carefully shaken and left in the fridge overnight. The gas bubbles in the sparkling water dislodge diatoms from the substrate. The following day, the sample was passed through a 1-mm screen and rinsed several times with sparkling water until filling a 2-L bottle. The 2-L sample was cooled to 4 °C a minimum of eight hours to allow the diatoms to settle. Next, the supernatant was carefully removed by aspiration using a vacuum pump. The remaining water contained the settled-out diatoms concentrated in a smaller volume (approximately 200 mL).

3.2.3.3 Diatom extraction from open soils

Soil samples were collected using 'soil sampling rings' (Fig. 3.4) in places with erosion/overland flow signs, if possible. Each sample was composed of three sub-samples (i.e. three soil sampling rings per sample).

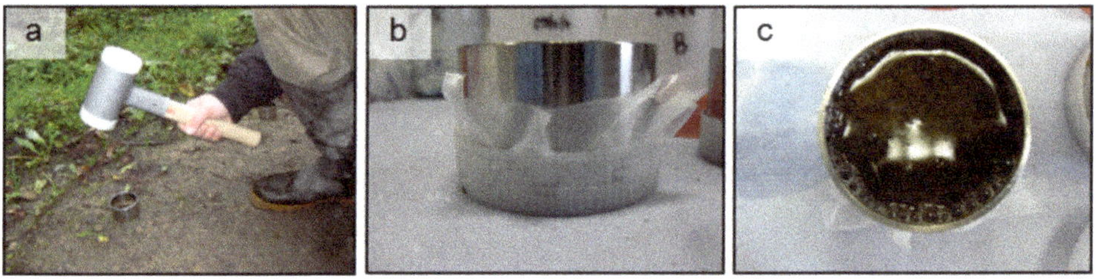

Fig. 3.4. (a) Soil diatom sampling and (b, c) diatom extraction in the laboratory.

Soil-sampling rings were inserted into the ground and slowly removed to retain the soil. One side of the ring was parafilmed to prevent soil loss and sparkling water was poured over the surface of the soil sample to dislodge diatoms. The sparkling water was left in the soil-sampling ring for approximately five seconds before being poured into a clean bottle - each soil ring was rinsed three times with sparkling water.
All collected samples were treated and prepared following standard procedures [European Committee for Standardization, 2003, 2004]. Microscopic slides for light microscopy observations were prepared using a refractive resin (Naphrax®) as mounting medium. Diatom valves were counted (~400 valves were required to adequately

represent species assemblages) and identified on each slide in random transects with a Leica® DMRX light microscope with a 100x oil immersion objective at a magnification of 1000x. The ecological classification was based on the Van Dam et al. [1994] system, refined and updated with the inclusion of information concerning poorly known species [e.g. Wetzel et al. 2013].

3.3 RESULTS

3.3.1 Diatom Species Distribution and Dynamics in the HRS System

For the quantitative analysis of diatom communities, we focused primarily on the Weierbach and Huewelerbach catchments and analysed a total of 273 diatom slides.

3.3.1.1 Diatom species identification

The quantitative analysis of diatom microflora allowed the observation of 324 taxa. In both catchments (Weierbach and Huewelerbach) diatom communities were usually dominated by species from oligotrophic environments [e.g. *Achnanthidium kranzii* (Lange-Bertalot) Round & Bukthiyarova, *Fragilariforma virescens* (Ralfs) D.M. Williams & Round, *Eunotia botuliformis* F. Wild, Nörpel & Lange-Bertalot, *E. minor* (Kützing) Grunow and *Planothidium lanceolatum* (Brébisson) Lange-Bertalot], while the riparian zones and hillslope areas were characterized by the presence of several species that prefer terrestrial habitats, mainly living epiphytically on bryophytes or exposed soils, such as *Chamaepinnularia evanida* (Hustedt) Lange-Bertalot, *Hantzschia abundans* Lange-Bertalot, *Navicula parsura* Hustedt, *Nitzschia harderi* Hustedt, *Orthoseira dendroteres* (Ehrenberg) Round, R.M. Crawford & D.G. Mann, *Pinnularia borealis* Ehrenberg, *P. perirrorata* Krammer, *Stauroneis muriella* J.W.G. Lund, *S. parathermicola* Lange-Bertalot and *S. thermicola* (J.B. Petersen) J.W.G. Lund (Figs. 3.5 and 3.6). The relative percentage of sampled diatoms from terrestrial environments was higher on the hillslope areas and riparian zones, compared to stream samples. This documents the importance of certain areas, acting as diatom sources during storm events, when the diatoms are eventually mobilized from moist or temporarily dry terrestrial habitats into the stream channel.

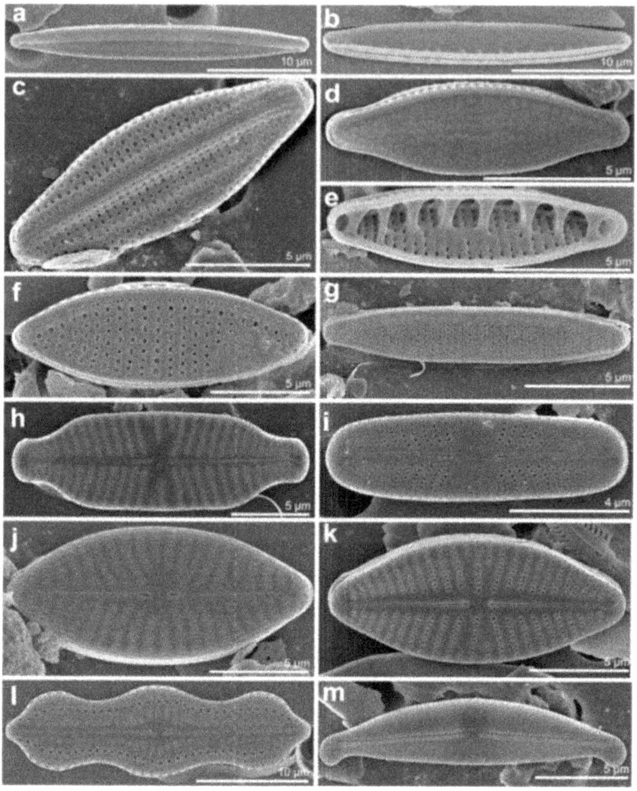

Fig. 3.5. Scanning electron microscope images of diatoms from the Weierbach sub-catchment (a) *Fragilariforma virescens* (Ralfs) D.M. Williams & Round, (b) *Eunotia tenella* (Grunow) Hustedt, (c) *Eunotia nymanniana* Grunow, (d) *Eunotia botuliformis* F. Wild, Nörpel & Lange-Bertalot (e) *Chamaepinnularia obsoleta* (Hustedt) C.E. Wetzel & Ector, (f) *Chamaepinnularia parsura* (Hustedt) C.E. Wetzel & Ector, (g) *Pinnularia perirrorata* Krammer, (h) *Pinnularia borealis* Ehrenberg, (i) *Mayamaea fossalis* (Krasske) Lange-Bertalot, (j) *Orthoseira dendroteres* (Ehrenberg) Round, R.M. Crawford & D.G. Mann, (k) *Planothidium lanceolatum* (Brébisson) Lange-Bertalot and (l) *Stauroneis parathermicola* Lange-Bertalot.

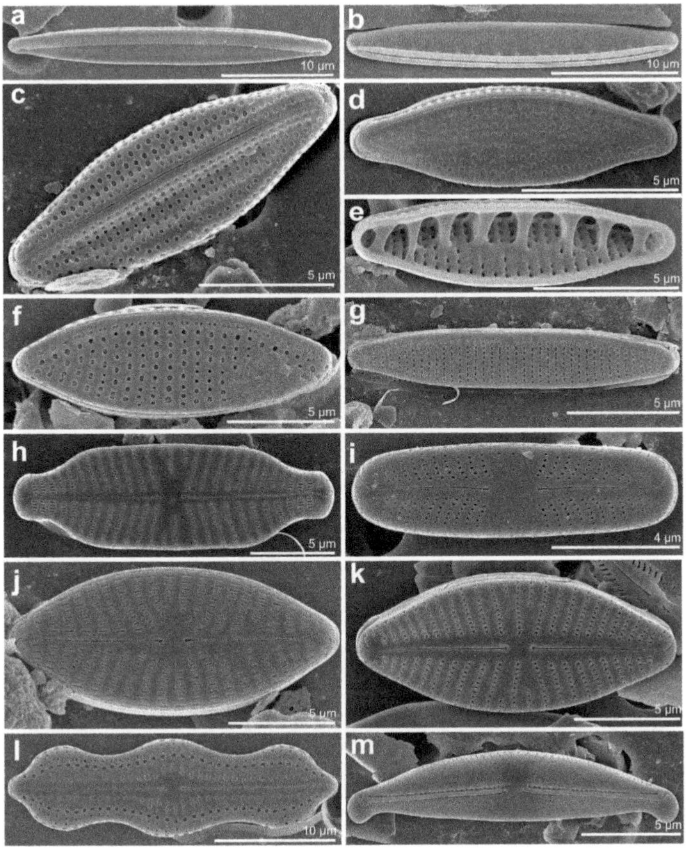

Fig. 3.6. **Scanning electron microscope images of diatoms from the Huewelerbach sub-catchment (a-b) *Nitzschia harderii* Hustedt, (c) *Denticula tenuis* Kützing, (d-e) *Nitzschia solgensis* Cleve-Euler, (f) *Nitzschia amphibia* Grunow, (g) *Nitzschia acidoclinata* Lange-Bertalot, (h) *Geissleria gereckei* Cantonati & Lange-Bertalot, (i) *Eolimna tantula* (Hustedt) Lange-Bertalot, (j) *Navicula antonii* Lange-Bertalot & Rumrich, (k) *Placoneis hambergii* (Hustedt) Bruder, (l) *Biremis motschii* (F. Meister) Lange-Bertalot and (m) *Halamphora montana* (Krasske) Levkov.**

3.3.1.2 Discriminating landscape units (i.e. hillslope, riparian zone, stream) and geological substrates using aquatic and terrestrial diatom communities

A statistical ordination (non-metric multi-dimensional scaling, NMDS) of diatom inventories revealed distinct spatial patterns in the organization of diatom communities along three landscape units (Fig. 3.7): hillslope (mosses), riparian zone (mosses, soil, leaves), stream (epilithon, epipelon / drift at baseflow / drift at storm events). We also noted a high dissimilarity among diatom communities sampled in the riparian zone and hillslope areas of the Weierbach (schists) and Huewelerbach (sandstone) catchments (Fig. 3.8). Among the approximately 400 identified diatom taxa, only a few were common to both geological substrates.

It also is important to note that a large number of identified diatom taxa are considered "endangered", "in regression", "strongly threatened", "very rare" or "under extinction risk" (Lange-Bertalot and Steindorf, 1996; Hofmann et al. 2011). Almost one third of the encountered species (32%) were in one of these categories, while for 47% the risk is either not estimated or data are insufficient. Only 21% of the species were classified as "not threatened" or "widespread species".

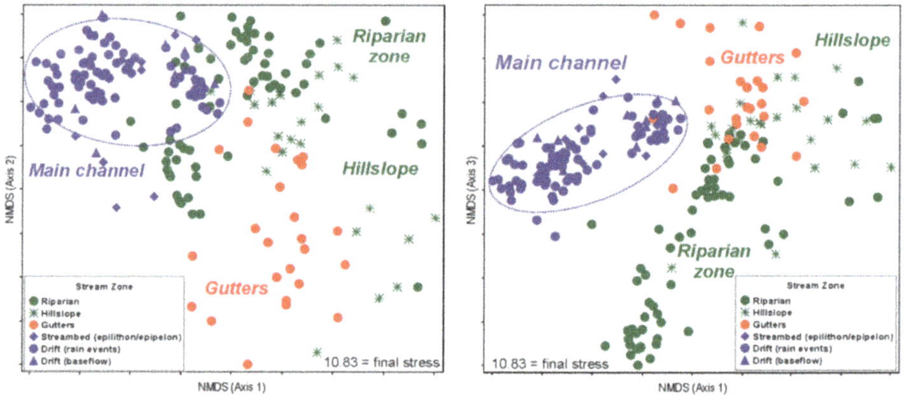

Fig. 3.7. Non-metric multidimensional scaling ordination plots (axis 1 x 2 and 1 x 3) of sites based on 209 diatom assemblages (sites in species space). Inventories from the Weierbach catchment showing distinct landscape units.

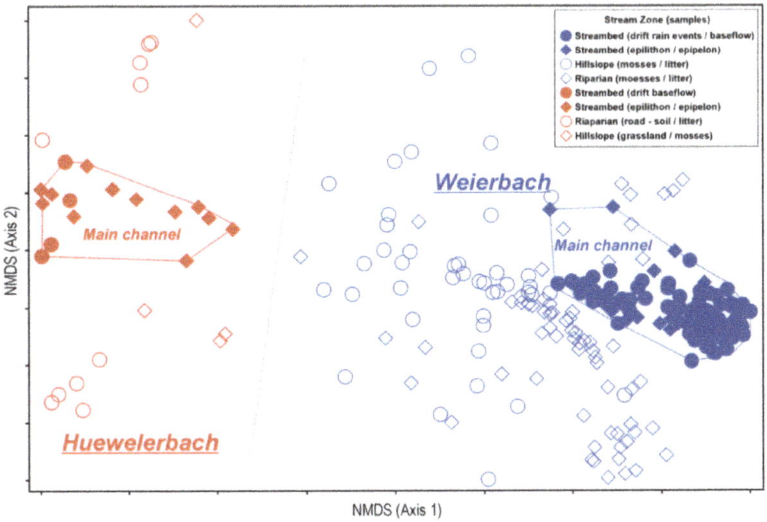

Fig. 3.8. NMDS ordination of 240 diatom assemblages from the Weierbach (n=209) and Huewelerbach (n=31) catchments; full symbols include samples collected from the main channel while empty symbols were collected on distinct terrestrial and aerial habitats.

No diatom frustules were found in groundwater samples, thus indicating that sub-surface transport of diatoms through the soil matrix is unlikely. However, the potential for transport of diatoms through heterogeneous macropore networks remains unexplored.

3.3.2 Connectivity within the HRS System

3.3.2.1 Documenting connectivity by hydrometric measurements

In order to shed light on the linkages between hydrological processes and terrestrial diatom transport through the HRS system, we concentrated our investigations first on the densely monitored Weierbach catchment.

In winter, the Weierbach's response to precipitation is two-peaked, with a first, nearly instantaneous, peak concomitant to rainfall and a second, delayed peak that may occur days later. Stable isotope data analysis indicates that the first peak is generated by event water, while the delayed peak is essentially formed by pre-event water. Different dynamics occur during summer, when the catchment's response to precipitation results in only one sharp peak, which is mainly composed of event water.

We hypothesize that the weathered layers, which tend to be increasingly hydrologically-disconnected nearer to the soil surface, generate this contrasting seasonal behaviour [Juilleret et al. 2011]. We have studied this intermittent connectedness along hillslope-riparian-stream transects via hydrometric observations. Water tables in saprock, bedrock with fractures at its margin, responded significantly to storm events and changes in water levels correlate to precipitation. While the riparian zone remained saturated throughout the year, soil volumetric water content decreased with distance upslope and the magnitude of soil moisture response to rainfall diminished with distance upslope as well. Soil moisture indicated the onset of connectivity between the hillslope and the riparian zone during high flows, especially in winter when lateral subsurface flow and two-peaked events occur. We hypothesize that during summer direct precipitation on the stream and the riparian zone generates a single sharp peak of the hydrograph and the lower soil moisture and water table levels inhibit connectivity between the hillslope and the riparian zone, resulting in smaller runoff response.

3.3.2.2 Documenting connectivity by terrestrial diatom tracing

Stream water samples taken during base flow, as well as the rising and falling limbs of storm hydrographs have shown a systematic flushing of terrestrial diatoms to the stream in response to incident precipitation and increasing discharge [Pfister et al. 2009].

A storm event of June 2011 illustrates the connectedness within the HRS system, with a flush of terrestrial diatoms occurring in the Weierbach immediately after incident precipitation (Fig. 3.9). This observation strongly supports the hypothesis that terrestrial diatoms can serve as biotic tracers of the onset of surface runoff connected to the stream.

However, conventional geochemical tracing applied to that same event suggests a completely different hydrograph generation process, with major groundwater and especially soil-water contributions (Fig. 3.10).

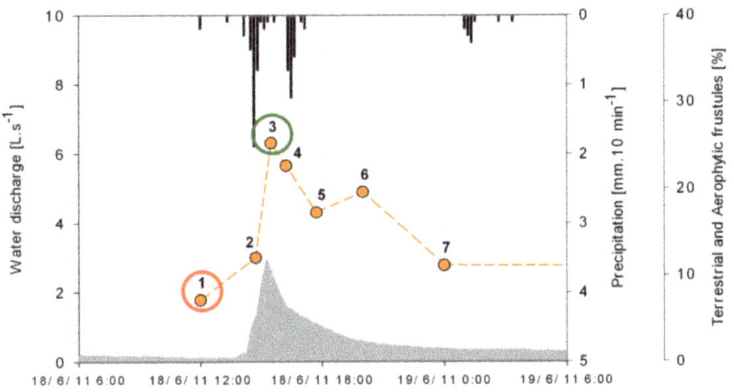

Fig. 3.9. **Example of terrestrial and aerophytic diatom percentage flushing in the Weierbach during a storm event in June 2011 (red circle: base flow; green circle: peak flow).**

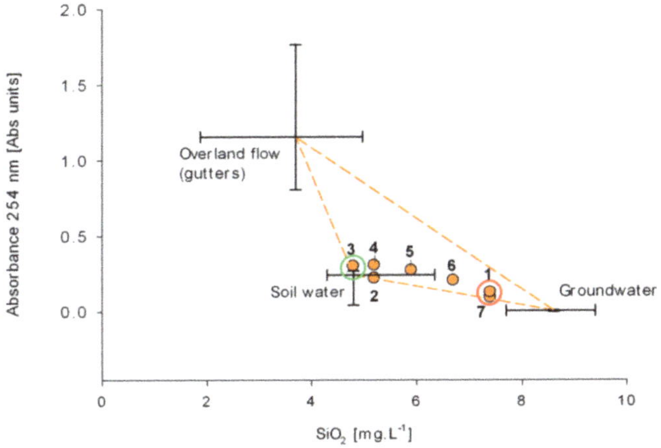

Fig. 3.10. **Mixing diagram for a storm event in the Weierbach in June 2011 (red circle: base flow; green circle: peak flow).**

While groundwater seems to dominate the storm hydrograph, terrestrial diatoms are somehow transported to the stream despite the inability of relatively big diatom species to move through the subsurface [Tauro et al. submitted]. Simulated rainfall experiments have demonstrated strong diatom depletion from the soil surface, supporting the possibility that diatoms are transported via overland flow, though further research into diatom transport pathways is necessary [Coles et al. submitted]. While diatom results also indicated a strong catchment-specific species assemblage – and likely the potential to act as catchment signatures – we do not yet know at what scale species assemblages will continue to be unique and discernible.

A surprising and important result from our diatom research to date has been that while terrestrial diatoms are mobilized rapidly to the stream during rainfall events, thus reflecting surface flowpaths and connectivity, the geochemical and isotopic evidence suggests that groundwater contributions dominate the hydrograph response. There appears to be a paradox in that groundwater, stored in the catchment during previous events, makes up most of the stream response during storms – though the relative abundance of terrestrial diatom species in the stream, which move along surface pathways and do not appear capable of traveling through the subsurface, quickly increases during events. Unraveling this paradox is important to understand the mechanisms supporting streamflow response and controls on stream chemistry during rainfall events.

3.4 DISCUSSION

The vast majority of hydrologic research over the past 50 years has conceived of catchments as black boxes where we relate inputs and outputs but ignore the internal mechanisms controlling catchment behaviour and thus never try to answer fundamental questions like 'how does the watershed function?' [Black, 1997], or 'where the water goes when it rains, the flowpaths it takes to the stream and how long it resides in the catchments' [McDonnell, 2003]. Black [1997] summarized the operating of a watershed into three hydrological functions (water collection, storage and discharge), two ecological functions (chemical reactions along sites and pathways, providing habitats for the flora and fauna), and two integrative responses (attenuation of the precipitation input signal and flushing of water and dissolved/suspended elements/matter). In the light of this operational framework for defining watershed function, recent literature has increasingly advocated interdisciplinarity, and more specifically coupled hydro-ecological approaches, as promising new avenues for advancing our understanding of hydrological processes through investigations on linkages between hydrological processes and catchment (ecological) functions [Pringle, 2003; Bracken and Croke, 2007; McDonnell et al. 2007; Tetzlaff et al. 2007; Cammeraat and Kooijman, 2009].

3.4.1 Prospecting Hydrological Connectivity: Where do we go from here?

Our overarching research question for this work focuses on how hydrological connectivity is defined in the HRS system and across catchments. Consequently, we aim to understand the spatio-temporal dynamics of saturated areas, the onset/cessation of their connectedness, and eventually their contribution to the storm hydrographs via rapid surface or sub-surface flowpaths.

We see potential for embedding this innovative topic under a larger umbrella of fundamental research on hydrological (water sources and flowpaths) and ecological (dynamics of terrestrial diatoms) functioning of watersheds.

To achieve this goal, we will pursue the following objectives:

(i) Further investigate the spatial variability of terrestrial diatom communities, by relating them to physiogeographical characteristics (e.g. geology, land cover, soil types, soil moisture, etc.).

(ii) Explore the potential depletion of the terrestrial diatom 'reservoir' across several rainfall events over a short period of time.

(iii) Determine whether terrestrial diatoms (soil and bryophytic diatoms) are mobilized during precipitation events and flushed to the stream through a subsurface network of macropores in the soils, as opposed to the prevailing assumption that only surface overland flow is capable of transporting diatoms to the stream. This subsurface, preferential transport pathway has been recently documented by Fishkis et al. [2010] for phytolith (amorphous silicon dioxide minerals generated via silicification of plant cell walls, similar in size, abundance and material to diatoms). We now must target a more specific investigation of diatom transport pathways – either surface (e.g. washing out of diatoms living on bryophytes and surface of soils), subsurface through the shallow soil layers (e.g. macropores: roots, macrofaunal burrows, cracks, etc.), or deep percolation (even though no diatoms have yet been found in groundwater samples, we need to investigate the hypothetical percolation of small and large diatoms through soils).

(iv) Document and constrain our perceptual models of the onset of connectivity inside the HRS system through a multi-tracer approach, combining geochemical elements, stable isotopes and diatoms.

3.4.2 Further Development Opportunities for Terrestrial Diatom Tracing

A positive side-effect of the continuation of this interdisciplinary research is related to the potential for terrestrial diatoms to assess soil quality with a view to sustainable soil quality management. Deformed diatom cells can indeed be indicators of biological state [Van Gestel and Van Brummelen, 1996; Falasco et al. 2009a, b], and the sensitivity of diatoms to pesticides is known to be very species dependent [e.g. Metting, 1981; Debenest et al. 2008, 2009]. This aspect may well be of interest in the context of a potential future European Framework Directive on soils [see Heger et al. 2012a, b and references therein].

3.5 CONCLUSION

We have documented significant heterogeneity in terrestrial diatom communities as well as their transport to the stream during storm events. We have also highlighted the potential for this biological tracer, combined with thermal IR imagery, to address the question of hydrological connectivity. However, we have also identified a number of open questions that need to be addressed before this new approach may become a standard tool for fundamental research on hydrological (water sources and flowpaths) and ecological (dynamics of terrestrial diatoms) functioning of watersheds.

We expect these investigations to open new avenues for tracing water origin and flowpath across multiple scales as well as assessing soil quality. These investigations will also add new, fundamental knowledge regarding diatom ecology. Most importantly perhaps is that such knowledge will provide new model targets for constraining predictive models of runoff generation – something currently lacking due to the small number of additional criteria that can be incorporated to test model structure and output.

3.6 REFERENCES

Ali, G., Tetzlaff, D., Soulsby, C., McDonnell, J.J., and Capell, R. 2012. A comparison of similarity indices for catchment classification using a cross-regional dataset. *Advances in Water Resources*, 40: 11-22.

Ali, G., and Roy, A.G. 2009. Revisiting hydrologic sampling strategies for an accurate assessment of hydrologic connectivity in humid temperate systems. *Geography Compass*, 3: 350-374.

Ambroise, B. 2004. Variable 'active' versus 'contributing' areas or periods: a necessary distinction. *Hydrological Processes*, 18: 1149-1155.

Betson, R.P. 1964. What is watershed runoff? *Journal of Geophysical Research*, 69: 1541-1552.

Beven, K. 2000. Uniqueness of place and process representations in hydrological modelling. *Hydrology and Earth System Sciences*, 4: 203-213.

Black, P.E. 1997. Watershed functions. *Journal of the American Water Resources Association*, 33: 1-11.

Booth, E.G., and Loheide II, S.P. 2012. Comparing surface effective saturation and depth-to-water-level as predictors of plant composition in a restored riparian wetland. *Ecohydrology*, 5: 637-647.

Bracken, L.J., and Croke, J. 2007. The concept of hydrological connectivity and its contribution to understanding runoff-dominated geomorphic systems. *Hydrological Processes*, 21: 1749-1763.

Burns, D.A. 2002. Stormflow hydrograph separation based on isotopes: the thrill is gone – what's next? *Hydrological Processes*, 16: 1515-1517.

Burns, D.A., McDonnell, J.J., Hooper, R.P., Peters, N.E., Freer, J.E., Kendall, C., and Beven, K. 2001. Quantifying contributions to storm runoff through end-member mixing analyses and hydrologic measurements at the Panola Mountain Research Watershed (Georgia, USA). *Hydrological Processes*, 15: 1903-1924.

Cammeraat, L.H., and Kooijman, A.M. 2009. Biological control of pedological and hydro-geomorphological processes in a deciduous forest ecosystem. *Biologia*, 64: 428-432.

Cappus, P. 1960. Bassin expérimental d'Alrance. Étude des lois de l'écoulement. Application au calcul et à la prévision des débits. *La Houille Blanche A*: 493-520.

Christophersen, N., and Hooper, R.P. 1992. Multivariate analyses of stream water chemical data: the use of principal components analyses for the end-member mixing problem. *Water Resources Research*, 28: 99-107.

Coles, A.E., Wetzel, C.E., Martínez-Carreras, N., Frentress, J., Ector, L., Hoffmann, L., McDonnell, J.J., and Pfister, L. in review. Diatoms as a tracer of hydrological connectivity: are they of infinite supply? *Ecohydrology*.

Debenest, T., Pinelli, E., Coste, M., Silvestre, J., Mazzella, N., Madigou, C., and Delmas, F. 2009. Sensitivity of freshwater periphytic diatoms to agricultural herbicides. *Aquatic Toxicology*, 93: 11-17.

Debenest, T., Silvestre, J., Coste, M., Delmas, F., and Pinelli, E. 2008. Herbicide effects on freshwater benthic diatoms: Induction of nucleus alterations and silica cell wall abnormalities. *Aquatic Toxicology*, 88: 88-94.

Dooge, J.C.I. 1986. Looking for hydrologic laws. *Water Resources Research*, 22: 46S-58S.

Dunne, T., and Black, R.D. 1970. Partial area contributions to storm runoff in a small New England watershed. *Water Resources Research*, 6: 1296-1311.

Dunne, T. 1978. *Field studies of hillslope flow processes*. In: Kirkby, M.J. (Ed.), Hillslope Hydrology. Wiley, New York: 227-293.

Dunne, T., Moore, T.R., and Taylor, C.H. 1975. Recognition and prediction of runoff-producing areas in humid regions. *Hydrological Sciences Bulletin*, 20: 305-327.

European Committee for Standardization. 2004. *Water quality – guidance standard for the identification, enumeration and interpretation of benthic diatom samples from running waters*. EN 14407:2004. European Committee for Standardization, Brussels.

European Committee for Standardization. 2003. *Water quality – guidance standard for the routine sampling and pretreatment of benthic diatoms from rivers*. EN 13946:2003. European Committee for Standardization, Brussels.

Falasco, E., Bona, F., Badino, G., Hoffmann, L., and Ector, L. 2009a. Diatom teratological forms and environmental alterations: a review. *Hydrobiologia*, 623: 1-35.

Falasco, E., Bona, F., Ginepro, M., Hlúbiková, D., Hoffmann, L., and Ector, L. 2009b. Morphological abnormalities of diatom silica walls in relation to heavy metal contamination and artificial growth conditions. *Water SA*, 35: 595-606.

Fishkis, O., Ingwersen, J., Lamers, M., Denysenko, D., and Streck, T. 2010. Phytolith transport in soil: A field study using fluorescent labelling. *Geoderma*, 157: 27-36.

Freeman, M.C., Pringle, C.M., and Jackson, C.R. 2007. Hydrologic connectivity and the contribution of stream headwaters to ecological integrity at regional scales. *Journal of the American Water Resources Association*, 43: 5-14.

Genereux, D. 1998. Quantifying uncertainty in tracer-based hydrograph separations. *Water Resources Research*, 34: 915-919.

Hannah, D.M., Sadler, J.P., and Wood, J.P. 2007. Hydroecology and ecohydrology: a potential route forward? *Hydrological Processes*, 21: 3385-3390.

Heger, T.J., Imfeld, G., and Mitchell, E.A.D. 2012a. Special issue on "Bioindication in soil ecosystems": Editorial note. *European Journal of Soil Biology*, 49: 1-4.

Heger, T.J., Straub, F., and Mitchell, E.A.D. 2012b. Impact of farming practices on soil diatoms and testate amoebae: A pilot study in DOK-trial at Therwin, Switzerland. *European Journal of Soil Biology*, 49: 31-36.

Hewlett, J.D., and Hibbert, A.R. 1967. *Factors affecting the response of small watersheds to precipitation in humid areas*. In: Sopper WE, and Lull HW (Eds), International Symposium on Forest Ecology. Pergamon, Oxford: 275-290.

Hewlett, J.D., and Hibbert, A.R. 1963. Moisture and energy conditions within a sloping soil mass during drainage. *Journal of Geophysical Research*, 68: 1081-1087.

Hewlett, J.D. 1961. *Watershed Management*. In: Annual Report 1961. USDA Forest Service, Southeastern Forest Experiment Station, Asheville (NC, USA), 61-66.

Hofmann, G., Werum, M., and Lange-Bertalot, H. 2011. *Diatomeen im Süsswasser-Benthos von Mitteleuropa. Bestimmungsflora Kieselalgen für die ökologische Praxis über 700 der häufigsten Arten und ihre Ökologie*. H. Lange-Bertalot (Ed.), A.R.G. Gantner Verlag K.G., Rugell, 908 p.

Hooper, R.P., and Shoemaker, C.A. 1986. A comparison of chemical and isotopic hydrograph separation. *Water Resources Research*, 22: 1444-1454.

Horton, R.E. 1933. The role of infiltration in the hydrologic cycle. *Transactions of the American Geophysical Union*, 14: 446-460.

Jencso, K.G., McGlynn, B.L., Gooseff, M.N., Bencala, K.E., and Wondzell, S.M. 2010. Hillslope hydrologic connectivity controls riparian groundwater turnover: Implications of catchment structure for riparian buffering and stream water sources. *Water Resources Research*, 46: W10524, doi:10.1029/2009WR008818.

Jencso, K.G., McGlynn, B.L., Gooseff, M.N., Wondzell, S.M., Bencala, K.E., and Marshall, L.A. 2009. Hydrologic connectivity between landscapes and streams: Transferring reach- and plot-scale understanding to the catchment scale. *Water Resources Research*, 45: W04428, doi:10.1029/2008WR007225.

Joerin, C., Beven, K.J., Iorgulescu, I., and Musy, A. 2002. Uncertainty in hydrograph separations based on geochemical mixing models. *Journal of Hydrology*, 255: 90-106.

Juilleret, J., Iffly, J.F., Pfister, L., and Hissler, C. 2011. Remarkable Pleistocene periglacial slope deposits in Luxembourg (Oesling): pedological implications and geosite potential. *Bulletin de la Société des naturalistes luxembourgeois*, 112: 125-130.

Kendall, C., McDonnell, J.J., and Gu, W. 2001. A look inside 'black box' hydrograph separation models: A study at the Hydrohill catchment. *Hydrological Processes*, 15: 1877-1902.

Kirchner, J.W. 2009. Catchments as simple dynamical systems: Catchment characterization, rainfall-runoff modeling, and doing hydrology backwards. *Water Resources Research*, 45: W02429, doi:10.1029/2008WR006912.

Kirchner, J.W. 2003. A double paradox in catchment hydrology and geochemistry. *Hydrological Processes*, 17: 871-874.

Lange-Bertalot, H., and Steindorf, A. 1996. Rote Liste der limnischen Kieselalgen (Bacillariophyceae) Deutschlands. *Schriftenreihe für Vegetationskunde*, 28: 633-678.

McDonnell, J.J., Sivapalan, M., Vaché, K., Dunn, S., Grant, G., Haggerty, R., Hinz, C., Hooper, R., Kirchner, J., Roderick, M.L., Selker, J., and Weiler, M. 2007. Moving beyond heterogeneity and process complexity: A new vision for watershed hydrology. *Water Resources Research*, 43: W07301, doi:10.1029/2006WR005467.

McDonnell, J.J., and Woods, R. 2004. On the need for catchment classification. *Journal of Hydrology*, 299: 2-3.

McDonnell, J.J. 2003. Where does the water go when it rains? Moving beyond the variable source area concept of rainfall-runoff response. *Hydrological Processes*, 17: 1869-1875.

McDonnell, J.J. 1990. A rationale for old water discharge through macropores in a steep, humid catchment. *Water Resources Research*, 26: 2821-2832.

McGlynn, B.L., and McDonnell, J.J. 2003. Quantifying the relative contributions of riparian and hillslope zones to catchment runoff. *Water Resources Research*, 39: 1310, doi:10.1029/2003WR002091.

McGuire, K., and McDonnell, J.J. 2010. Hydrological connectivity of hillslopes and stream: Characteristic timescales and non-linearities. *Water Resources Research*, 46: W10543, doi:10.1029/2010WR009341.

Metting, B. 1981. The systematics and ecology of soil algae. *Botanical Review*, 47: 195-312.

Pearce, A.J., Stewart, M.K., and Sklash, M.G. 1986. Storm runoff generation in humid headwater catchments: 1. Where does the water come from? *Water Resources Research*, 22: 1263-1272.

Petts, G., Morales, Y., and Sadler, J. 2006. Linking hydrology and biology to assess the water needs of river ecosystems. *Hydrological Processes*, 20: 2247-2251.

Pfister, L., McDonnell, J.J., Hissler, C., and Hoffmann, L. 2010. Ground-based thermal imagery as a simple, practical tool for mapping saturated area connectivity and dynamics. *Hydrological Processes*, 24: 3123-3132.

Pfister, L., McDonnell, J.J., Wrede, S., Hlúbiková, D., Matgen, P., Fenicia, F., Ector, L., and Hoffmann, L. 2009. The rivers are alive: on the potential for diatoms as a tracer of water source and hydrological connectivity. *Hydrological Processes*, 23: 2841-2845.

Phillips, R.W., Spence, C., and Pomeroy, J.W. 2011. Connectivity and runoff dynamics in heterogeneous basins. *Hydrological Processes*, 25: 3061-3075.

Pringle, C. 2003. The need for a more predictive understanding of hydrologic connectivity. *Aquatic Conservation: Marine and Freshwater Ecosystems*, 13: 467-471.

Richey, D.G., McDonnell, J.J., Erbe, M.W., and Hurd, T.M. 1998. Hydrograph separations based on chemical and isotopic concentrations: a critical appraisal of published studies from New Zealand, North America and Europe. *Journal of Hydrology*, 37: 95-111.

Robinson, D.A., Binley, A., Crook, N., Day-Lewis, F.D., Ferré, T.P.A., Grauch, V.J.S., Knight, R., Knoll, M., Lakshmi, V., Miller, R., Nyquist, J., Pellerin, L., Singha, K., and Slater, L. 2008. Advancing process-based watershed hydrological research using near-surface geophysics: a vision for, and review of, electrical and magnetic geophysical methods. *Hydrological Processes*, 22: 3604-3635.

Schuetz, T., and Weiler, M. 2011. Quantification of localized groundwater inflow into streams using ground-based infrared thermography. *Geophysical Research Letters*, 38: L03401, doi:10.1029/2010GL046198.

Selker, J.S., Thévenaz, L., Huwald, H., Mallet, A., Luxemburg, W., van de Giesen, N., Stejskal, M., Zeman, J., Westhoff, M., and Parlange, M.B. 2006. Distributed fiber-optic temperature sensing for hydrologic systems. *Water Resources Research*, 42: W12202, doi:10.1029/2006WR005326.

Sivapalan, M. 2009. The secret to 'doing better hydrological science': change the question! *Hydrological Processes*, 23: 1391-1396.

Sklash, M.G., Stewart, M.K., and Pearce A.J. 1986. Storm runoff generation in humid headwater catchments: 2. A case study of hillslope and low-order stream response. *Water Resources Research*, 22: 1273-1282.

Sklash, M.G., and Farvolden, R.N. 1979. The role of groundwater in storm runoff. *Journal of Hydrology*, 43: 45-65.

Spence, C., and Phillips, R.W. 2014. Refining understanding of hydrological connectivity in a boreal catchment. *Hydrological Processes*, doi: 10.1002/hyp.10270

Spence, C. 2010. A paradigm shift in hydrology: Storage thresholds across scales influence catchment runoff generation. *Geography Compass*, 4: 819-833.

Tauro, T., Martínez-Carreras, N., Barnich, F., Juilleret, J., Wetzel, C.E., Ector, L., Hissler, C., and Pfister, L. (in review). Diatom percolation through soils: a proof of concept laboratory experiment. *Ecohydrology*.

Tetzlaff, D., Soulsby, C., Bacon, P.J., Youngson, A.F., Gibbins, C., and Malcolm, I.A. 2007. Connectivity between landscapes and riverscapes – a unifying theme in integrating hydrology and ecology in catchment science. *Hydrological Processes*, 21: 1385-1389.

Van Dam, H., Mertens, A., and Sinkeldam, J. 1994. A coded checklist and ecological indicator values of freshwater diatoms from The Netherlands. *Netherlands Journal of Aquatic Ecology*, 28: 117-133.

Van Gestel, C.A., and Van Brummelen, T.C. 1996. Incorporation of the biomarker concept in ecotoxicology calls for a redefinition of terms. *Ecotoxicology*, 5: 217-225.

Wainwright, J., Turnbull, L., Ibrahim, T.G., Lexartza-Artza, I., Thornton, S.F., and Brazier, R.E. 2011. Linking environmental regimes, space and time: Interpretations of structural and functional connectivity. *Geomorphology*, 126: 387-404.

Wels, C., Cornett, R.J., and Lazerte, B.D. 1991. Hydrograph separation: a comparison of geochemical and isotopic tracers. *Journal of Hydrology*, 122: 253-274.

Wetzel, C.E., Martínez-Carreras, N., Hlúbiková, D., Hoffmann, L., Pfister, L., and Ector, L. 2013. New combinations and type analyses of *Chamaepinnularia* species from aerial habitats in Central Europe. *Cryptogamie Algologie*, 34: 149-168.

Whipkey, R.Z. 1965. Subsurface stormflow from forested slopes. *International Association of Hydrological Sciences Bulletin*, 10: 74-85.

CHAPTER 4

SATELLITE REMOTE SENSING FOR CLOUDS AND PRECIPITATIONS: A REVIEW OF SATELLITE TECHNIQUES DEVELOPED AT THE INSTITUTE OF METHODOLOGIES FOR ENVIRONMENTAL ANALYSIS-NATIONAL RESEARCH COUNCIL

Elisabetta Ricciardelli, Francesco Di Paola, Domenico Cimini, Filomena Romano, and Mariassunta Viggiano

Institute of Methodologies for Environmental Analysis,
National Research Council (IMAA/CNR), Potenza, Italy

ABSTRACT

Satellite remote sensing is very useful for the monitoring and characterization of cloudy areas and for the monitoring of precipitation events. In particular, satellite monitoring of extreme precipitation events is particularly attractive for real-time operational purposes. The detection and classification of cloudy areas is important for monitoring convective and stratiform clouds to which heavy and moderate rain are respectively associated. In this paper, two techniques - developed at the *National Research Council* of Italy *Institute of Methodologies for Environmental Analysis* (CNR-IMAA) - are reviewed. These were developed for the recognition of convective clouds as well as the monitoring of extreme precipitation events: the *Precipitation Evolving Technique* (PET) and the *Rain Class Evaluation from Infrared and Visible* observations (RainCEIV). PET uses Infrared (IR) channels from the Meteosat Second Generation-Spinning Enhanced Visible and Infrared Imager (MSG-SEVIRI) coupled with Rain Rate (RR) obtained from Passive-Microwave-based algorithm to provide near real time RR maps. RainCEIV uses Visible (VIS) and IR observations from MSG-SEVIRI to determine near real-time cloud classification and rain class evaluation. RainCEIV classifies SEVIRI pixels as *non-rainy*, *light-to-moderate rain,* or *heavy-to-very heavy rain.* In addition, two cloud classification schemes developed at CNR-IMAA, namely MACSP (Cloud Mask Coupling of statistical and Physical Methods) and C_MACSP (Classification Cloud Mask Coupling of Statistical and Physical Methods), are also described.

4.1 INTRODUCTION

Precipitation measurements are essential for short term hydrological and long term climate studies. Operational networks of rain gauges and weather radars provide fairly accurate rain rate measurements, but they leave large areas uncovered. Because of this, satellite remote sensing becomes the essential tool for the detection and characterization of the rainy areas in regions where this information is missing. Passive-Microwave (PMW) observations from Low Earth Orbiting (LEO) satellites are exploited for precipitation retrieval even if with coarse spatial and temporal resolution. Visible and Infrared (VIS/IR) observations from Geosynchronous (GEO) satellites are crucial for the characterization of cloudy and rainy areas with better spatial and temporal resolution than PMW-LEO. For these reasons, PMW and VIS/IR appear complementary and their combination enhances the continuous monitoring of extreme events.

Over the past decades several rain rate retrieval methods based on passive imager observations have been developed (Kucera et al. 2013). LEO satellite PMW observations provide more direct and performing techniques for the retrieval of precipitation when compared with the IR/VIS techniques. The geostationary satellite IR/VIS techniques work better over areas where rainfall originates from deep convection than in areas where it originates from stratiform systems. However, the low spatial and temporal resolutions of PMW observations make them less attractive for monitoring extreme events, characterized by small scale and high temporal variability. Operational networks of weather radars provide rain rate measurements characterized by high temporal and spatial resolution, but leaves large areas of the Earth uncovered, where information on the occurrence and intensity of rainfall would be missing. Rain rate estimates from passive imagers on geostationary satellite may bridge PMW-LEO and weather radar gaps. Many techniques based on geostationary satellite observations have been developed in order to estimate rain rate values or confidences. Moreover, the geostationary satellite observations are useful to identify and to monitor the spatial and temporal evolution of the convective cells. Kidd and Levizzani (2011) provide a broad overview of the recent techniques developed in order to estimate Rain Rate values starting from the geostationary satellite IR/VIS observations. In this paper, two techniques - developed at the *National Research Council* of Italy *Institute of Methodologies for Environmental*

Analysis (CNR-IMAA) - are reviewed. Besides, considering that the cloud detection and classification constitute an essential starting point for the recognition of the raining areas, two cloud detection and classification algorithms developed at CNR-IMAA are also described. In particular, Section 4.2 shows the characteristics of the instruments whose observations have been used for the implementation of the described techniques, Section 4.3 describes cloud detection and classification (MACSP and C_MACSP) algorithms, while Sections 4.4 and 4.5 summarize the PET and the RainCEIV techniques, respectively. Finally, in Section 4.6 the advantages of the described techniques are discussed.

4.2 INSTRUMENT AND DATA DESCRIPTION

Both PET and RainCEIV, as well as the cloud detection and classification algorithms, use observations from the Spinning Enhanced Visible and InfraRed Imager (SEVIRI) on board the Meteosat Second Generation (MSG) (Schmetz et al. 2002). Meteosat satellites are spin-stabilized and the instruments aboard them have been designed so to provide permanent Visible and Infrared imaging of the Earth. The first Meteosat Second Generation (MSG-1 also named Meteosat-8) was launched on 28 August 2002 and became operational on 29 January 2004. MSG-2 (Meteosat-9) was launched on 21 December 2005 and MSG-3 (Meteosat-10) was launched on 5 July 2012 and future MSG-4 is planned for launch in 2014-2015. The MSG system provides accurate weather monitoring data through SEVIRI, which observes the Earth in 12 spectral channels. These provide precise data throughout the atmosphere, giving improved quality to the initial conditions for Numerical Weather Prediction models. The SEVIRI is a 50-cm-diameter aperture line-by-line scanning radiometer. It observes the Earth-atmosphere system with a spatial sampling of 3 km at the sub-satellite point in 11 channels, while the remaining High Resolution Visible (HRV) channel covers half the full disk with a 1-km spatial sampling at the sub-satellite point. The actual instantaneous field of view is about 4.8 km at the sub-satellite point for all channels except 1.67 km for the HRV channel. The major improvements with respect to previous generation are its enhanced spectral characteristics, its higher temporal resolution (15 min), the improved signal-to-noise ratio, and the higher precision of data storing, going from 8 bits (256 levels) on Meteosat-7 to 10 bits (1024 levels) on Meteosat-8 (Schmetz et al. 2002).

The rain rate maps derived from Advanced Microwave Sounding Units (AMSU-B) and Microwave Humidity Sounder (MHS) have been used both for the training phase of RainCEIV and the initialization of PET. AMSU-B is on board the National Oceanic and Atmospheric Administration (NOAA) satellites while MHS is on board the European Polar Satellite MetOp-A. AMSU-B and MHS are cross-track, line scanning microwave radiometer measuring radiances in five channels in the frequency range from 89 GHz to 190 GHz. AMSU-B and MHS exploit similar channels: center frequencies for the AMSU-B channels are 89, 150, 183 ± 1, 183 ± 3, and 183 ± 7 GHz; while for the MHS channels are 89, 157, 183 ± 1, 183 ± 3, and 190 GHz. Therefore, the two instruments differ just at channel number 2 (150 vs. 157 GHz) and 5 (183 ± 7 vs. 190 GHz); though these two channel duplets show the same fundamental features. The AMSU-B and MHS field of view (FOV) have a circular shape (diameter of about 16 km at nadir) that becomes ellipsoidal away from the nadir and the dimension of the axes reaches 51 km for the cross-track direction and 25 km for the along-track direction at the maximum scanning angle (Bennartz, 2000). The purpose of the instrument is to receive and measure radiation from a number of different layers of the atmosphere information on humidity profiles.

Regarding the cloud masks that will be described in Section 4.3, the cloud mask from MODIS (the Moderate-Resolution Imaging Spectroradiometer) observations and that from CPR (Cloud Profiling Radar)/CALIOP (Cloud–Aerosol Lidar with Orthogonal Polarization) observations were used to validate the cloud masks developed at CNR-IMAA. Thus, we consider worthwhile to briefly describe MODIS and CPR/CALIOP instrument characteristics.

MODIS, an instrument aboard the Terra and Aqua - Earth Observing System (EOS) – satellites, provides global coverage every two days and measures radiances at 36 wavelengths (from 0.4 μm to 14.5 μm) at 250 m spatial resolution in two visible bands, 500 m resolution in five visible bands and 1000 m resolution in the infrared bands (Ackerman et al. 1998).

The CloudSat and CALIPSO (Cloud-Aerosol Lidar and Infrared Pathfinder Satellite Observation) mission were launched together in April 2006. . CloudSat maintains a close formation with Aqua and CALIPSO, providing near-simultaneous and co-located observations also taken with the instruments of these two platforms (Stephens at al. 2002). The Cloud Profiling Radar (CPR), a 94 GHz near nadir pointing millimetrewavelength radar, is the primary instrument on CloudSat. It measures the power backscattered by clouds as a function of the distance from the radar and it provides vertically-resolved information on cloud position, cloud ice and liquid water content, precipitation, heating rates as well as radiative fluxes. The vertical resolution is 500 m and the horizontal resolution is about 1.4 km (cross-track) by 2.5 km (along-track). The CPR operates in a short-pulse mode and yields measurements at a minimum detectable sensitivity of -28 dBz (Im et al. 2005; Stephens et al. 2002). CALIPSO consists of three nadir viewing instruments: CALIOP, the Imaging Infrared Radiometer (IIR) and the Wide Field Camera (WFC). This instrument synergy enables the retrieval of a wide range of aerosol and cloud characteristics (Winker et al. 2007). The CALIOP horizontal resolution is 70 m sampled at 333 m intervals along-track, and the vertical resolution is 30 m. CALIPSO flies at about 20 seconds behind CloudSat.

4.3 CLOUD DETECTION AND CLASSIFICATION ALGORITHM: AN OVERVIEW

The algorithms developed at the CNR-IMAA for the cloud detection and classification are the Cloud Mask Coupling of Statistical and Physical Methods (MACSP, Ricciardelli et al. 2008) and the Cloud Classification Mask Coupling of Statistical and Physical Methods (C_MACSP, Ricciardelli et al. 2014, Di Paola et al. 2014). MACSP flags each SEVIRI pixels as *cloudy* or *cloud-free*, while C_MACSP classifies each SEVIRI pixel as *cloud-free*, *low-middle cloud*, *high thin cloud*, *high thick cloud* or *convective cloud*.

MACSP (C-MACSP) uses physical, statistical and temporal approaches in order to obtain a good performance in cloud detection (and classification). In particular, a non-parametric classifier, the K-Nearest Neighbour (K-NN) is used both in the MACSP/C_MACSP statistical and temporal modules. . Non-parametric classifiers do not assume an *a priori* known parametric form for the probabilities, but they estimate them directly from the design samples. They implement the decision rule locally and the probabilities are estimated for each sample to be classified. In order to assign the MSG-SEVIRI pixel to a clear or cloudy class, the K-NN classifier uses textural and spectral features estimated in 3-by-3 pixel boxes. Textural and spectral features characterizing each pixel are extracted from infrared and visible MSG-SEVIRI images acquired at 3.9 μm, 8.7 μm, 10.8 μm, 12 μm, 0.6 μm, 0.8 μm and 1.6 μm.

The features used in classification algorithm have been selected by applying the Fisher criterion selection. It is crucial here to describe the spectral and textural features that have been used as components of the feature vector in the techniques summarized in this paper. Spectral and textural features are important in order to identify objects in an image. Spectral features account the average tonal variations in the visible or in the infrared portion of the spectrum, whereas the textural features give information about the spatial distribution of tonal variations within a spectral band. Tone and texture are always present in an image, but one property can sometimes dominate the other. When a small area of an image has a slight variation of features of discrete grey tone, its dominant area property is tone (spectral features). On the contrary, when a small area has a wide variation of features of discrete grey tone, the dominant area property is texture. The size of the small area-patch and the number of distinguishable discrete features are crucial in this distinction. As the number of distinguishable discrete features decreases, the tonal properties predominate. In fact, when the small area is one resolution cell, the only property present is tone. As the number of distinct features of discrete grey tone increases within the small area, the texture properties prevail. Both in MACSP and C_MACSP statistical and temporal algorithm (and also in the RainCEIV statistical algorithm that will be summarized in Section 4.5), a 3-by-3 pixel box has been chosen as grid element in order to have significant values for textural and tonal features representative for each pixel. The pixel to be classified is the central point of the grid element. The statistical algorithm gives in output the probabilities that a pixel belongs to the cloud-free class and to the cloudy classes.

The physical algorithm computes the clear/cloudy probabilities for every pixels. The MACSP/C_MACSP physical algorithm does not use real time ancillary data. However, the land cover and elevation map is the only information used in defining threshold. This piece of information is not provided in real time but through monthly maps. The physical algorithm is based on a set of multi spectral threshold tests applied to each pixel of the image. Its crucial point is to determine optimal thresholds that have been derived using both a measured and a simulated clear datasets. The mixed dataset is representative for different areas characterized by different land cover, at different times and weather conditions. The algorithm selects the thresholds on the basis of the satellite and solar zenith angle, land cover, and according to the brightness temperature ranges of the image. The brightness temperature tests, the brightness temperature differences tests as well as the reflectance tests have been chosen on the basis of the different behaviour of the clouds at the different wavelengths. These are listed in Table 2 of Ricciardelli et al. (2008). The C_MACSP (Ricciardelli et al. 2014, Di Paola et al. 2014) physical algorithm has been updated by adding a new test that considers the brightness temperature difference between the window channel centered at 10.8 μm and water vapour channel centered at 6.2 μm. Moreover, the training dataset for the C_MACSP statistical algorithm as well as the threshold values for the C_MACSP physical tests have been updated considering the new class for *convective clouds*. In order to obtain reliable values both for the training dataset and threshold values, a double matching of MSG-SEVIRI pixels with the rain rate values inferred from the PMW observations and those derived from the Italian Weather Radar Network (coordinated by the Italian Department of Civil Protection, DPC). The surface rain intensity from PMW observations have been obtained from the Precipitation Estimation at Microwave Frequencies (PEMW, by Di Tomaso et al. 2010).

The statistical and physical algorithms run separately and their outputs (probabilities that the pixel belongs to a cloud-free or a cloudy classes) are compared. When they agree the pixel is classified accordingly, otherwise the statistical temporal algorithm is applied. This test is also applied to those pixels having low confidence to be clear or cloudy. The temporal test takes advantage of the best MSG-SEVIRI temporal resolution and it applies the K-Nearest Neighbour classifier to the spectral and textural features calculated in "temporal" 3-by-3 pixel boxes, defined "temporal" because their elements belong to the three subsequent MSG-SEVIRI images, acquired at the time of interest and 15 and 30 minutes before. The MACSP algorithm was validated against the MODIS cloud mask and compared with CPR (Cloud Profiling Radar) and MPEF (**M**eteorological **P**roduct **E**xtraction **F**acility, software package from Eumetsat (Eumetsat, 1996)) cloud masks results. The outcomes show that the MACSP detects 91.8% of the total number of the pixels used for validation against MODIS cloud mask correctly, while the MPEF cloud mask detects 89.2% of them correctly. In particular, the MACSP classifies as cloudy 8.8% of the pixels classified by the MODIS cloud mask as clear, while the MPEF cloud mask classifies as cloudy 12.1% of them. The MACSP detects 91.2% of the cloudy CPR pixels and 90.8% of the cloud free CPR pixels, considered for comparison, correctly. On the other hand, the MPEF and CPR cloud masks agree in the detection of 90.7% and 90.2% of the cloudy pixels and of the cloud-free pixels, respectively.

C_MACSP was validated against an independent test dataset of MSG-SEVIRI pixels manually classified as belonging to the cloud-free and cloudy classes. The manual classification has been made by carefully observing the SEVIRI RGB compositions of MSG-SEVIRI channels properly chosen. Moreover, in order to detect correctly the convective clouds, the rain rate values derived from the radar network have been considered. The analysis of the test samples (300 samples for the cloud-free class and for each cloudy class) results that C_MACSP classifies *high thick* clouds and *convective* clouds with an accuracy higher than 95%. It classifies *low/middle* clouds with an accuracy higher than 91%. The *high thin* clouds are detected with an accuracy of 87.6% during daytime and nighttime over sea, and with an accuracy of 85% and 84% during daytime and nighttime, respectively, over land.

In validating MACSP results against MOD35/MYD35 MODIS cloud mask, the pixel classified as uncertain by the MODIS cloud mask were not considered for validation purposes, but they have been considered in a subsequent study (Ricciardelli et al. 2010). This study proposes a technique that uses the information from the MSG-SEVIRI to inquire the areas classified as uncertain by the operative MOD35/MYD35 MODIS cloud mask algorithm. This technique applies a K-NN classifier to the spectral and textural features and IR brightness temperature differences calculated from a time-series of MSG-SEVIRI images. This technique was used to re-classify a set of uncertain MOD35 and MYD35 pixels. Regarding MOD35/MYD35 uncertain pixels analysed over land, 52.0% of them were classified as cloudy when using HRV, 48.7% of them were classified as cloudy when using IR features, and 37.0% of them were classified as cloudy when using IR/VIS features. Regarding MOD35/MYD35 uncertain pixels analysed over sea, 54.5%, 45.4% and 49.7% were classified as cloud-free, when HRV, IR, and IR/VIS features are used as inputs, respectively. 39.8%, 31.8%, and 37.3% of the pixels analysed over land, and 40.7%, 47.4% and 38.0% of the pixels analysed over sea were classified as cloudy when using HRV, IR, and IR/VIS features as inputs, respectively. Some pixels remain unclassified due to the low cloud-free/cloudy confidence that the K-NN classifier associates with them. After visual analyses, most of these pixels are found to be very thin clouds. A set of comparisons was made with CPR/CALIOP 2B-Geoprof-Lidar results. During the daytime, the 2B-Geoprof-Lidar results were compared with the results obtained by the K-NN classifier when using both IR/VIS and HRV textural and spectral features. The K-NN classifier agree with the 2B-Geoprof-Lidar in the detection of about 74% of the uncertain MYD35 pixels, examined both over land and sea, as cloud-free. During the nighttime, the K-NN classifier tallies with the 2B-Geoprof-Lidar algorithm in the detection of 80.9% and 79.7% of the uncertain MYD35 pixels, respectively examined over land and sea, as cloud-free. Most of these discrepancies are due to different times of overpass, pixel size, and characteristics of the sensors.

4.4 PET OVERVIEW

The Precipitation Evolving Technique, PET (Di Paola et al. 2012), is a multi-sensor approach for convective rain cell continuous monitoring that uses IR brightness temperature (TB) maps of water vapor (6.2 µm) and thermal-IR (10.8 µm) channels from SEVIRI radiometer to propagate forward in time and space the last available Rain Rate (RR) maps derived from AMSU and MHS PMW observations. This technique is based on two different modules, the first for morphing and tracking rain cells and the second for dynamic calibration IR-Rain Rate. A combination of the morphing and calibration outputs provides rainfall map at IR-SEVIRI space and time scale, and the whole procedure is reiterated by using the last PET-RR map output until a new PMW-based rainfall map is available.

On the basis of the *Global Convective Diagnostic* (GCD) approach, PET uses the brightness temperature differences between water vapour channel and the thermal-IR channel ($TB^{6.2\ \mu m}$ - $TB^{10.8\ \mu m}$) (ΔTB^{GCD}) maps to recognize convective areas in a better way than the use of a single IR channel, highly reducing the number of false alarms such as those caused by the cirrus or others non-precipitating clouds. PET consists of two modules:

- Morphing module – This module drives precipitating cells of the latest rain rate map by modifying their positions and shapes, using a multi-scale and multi-threshold pattern recognition approach on two consecutive GCD maps.
- Calibration module – This module computes a dynamic RR-GCD relationship using a probabilistic histogram matching method to calculate a $RR[t_0+n\Delta t]$ – $\Delta TB^{GCD}[t_0+n\Delta t]$ lookup table (LUT) to be applied to $TB^{GCD}[t_0+(n+1)\Delta t]$ to obtain $RR[t_0+(n+1)\Delta t]$.

Final RR map is obtained as a linear combination of the RR maps obtained by the Morphing module and the RR map obtained by the Calibration module. This linear combination is computed pixel-by-pixel by evaluating the goodness of the Morphing procedure. In detail, at the first iteration, by using $RR[t_0]$ and $\Delta TB^{GCD}[t_0]$ and $\Delta TB^{GCD}[t_{0+15'}]$ maps, the Morphing and Calibration modules produce two new RR maps that, when combined, produce a final $RR[t_{0+15'}]$ result on the SEVIRI grid. When a new SEVIRI observation is available at $t=t_0+30'$, a new iteration is performed, in order to produce a new RR map. Generally speaking, for each time step $t_n=t_0+n\Delta t$ with Δt equal to SEVIRI temporal sampling, Morphing and Calibration modules use $\Delta TB^{GCD}[t_0+n\Delta t]$ and $\Delta TB^{GCD}[t_0+(n+1)\Delta t]$, to propagate $RR[t_0+n\Delta t]$ forward in time and space to obtain $RR[t_0+(n+1)\Delta t]$, until a new AMSU/MHS overpass is available and new estimates are produced.

The PET express purpose is not to estimate the precipitation but to propagate the starting PM-RR map in time and space, so that, different PM-based RR estimates can produce different results. For this reason, PET evaluation is possible only by comparing its results with the closest PM-based RR estimates obtained using the same algorithm and the same PW sensor used for PET initialization. In the light of this, PET performance has been evaluated by using some case studies with two different PM algorithms developed at *Massachusetts Institute of Technology* (MIT) (Surussavadee and Staelin, 2008) and at CNR-IMAA (Di Tomaso et al. 2009).

The validation of PET against the rainfall rates estimated by the above-mentioned algorithms has been performed through the assessment of both dichotomous and continuous scores, used to assess quantitatively the accuracy of rain detection and estimation, respectively. Dichotomous assessment has been calculated on different RR classes using as rain/no-rain the thresholds 0, 0.2, 1, 2, 5 and 10 mm/h. The results show a POD of 60%-70%, FAR 20%-40%, and Bias and Accuracy close to 1, showing an overall ability to distinguish rainy from no-rainy areas. Continuous assessment shows that the binned scatterplot (Figure 7 of Di Paola et al. 2012) is close to the bisector with a reasonably good correlation between AMSU and PET RR (R^2=0.98, RMSE=0.4 mm/h). The small overestimate of low RR bins and the small underestimate of high RR bins is a consequence of the RR distribution.

To sum up, PET shows a good ability to propagate RR depending on the propagation time primarily, as well as on the nature of the storm under consideration. While its performance is acceptable for 2÷3 hours propagation time, it may become unsatisfactory after 4÷5 hours past the last AMSU/MHS overpass.

4.5 RAINCEIV OVERVIEW

The RainCEIV (Rain Class Evaluation from Infrared and Visible observations) (Ricciardelli et al. 2014) uses VIS/IR observations from SEVIRI to provide near real time cloud classification and rain classes maps. It consists of two modules:

- a cloud classification algorithm - the C_MACSP cloud detection and classification algorithm described in Section 4.3;
- a statistical algorithm that evaluates the rain class for each pixel classified as *low/middle cloud*, *high thick cloud* or *convective cloud* by the first module (C_MACSP).

RainCEIV uses the k-Nearest Neighbour Mean (k-NNM) classifier to associate each SEVIRI pixel to the:

- *non-rainy* (rain rate<0.5 mm/h) class;
- *light-to-moderate rain* (0.5=<rain rate<=4 mm/h) class;
- *heavy-to-very-heavy rain* (rain rate> 4 mm/h) class.

The k-NNM classifier uses as input the spectral and textural features determined in IR/VIS SEVIRI images. For each class, it finds the k nearest neighbours of the pixel to be classified, and it determines the mean value of the distances between the pixel and the k-nearest neighbours. The lowest mean distance principle is used to determine the class to which the pixel belongs. The spectral and textural features used as inputs in the k-NNM classifier have been chosen by using the Fisher criterion so as to select the features characterized by the highest discriminatory power. Subsequently, the *k* parameter and the number of the features to be included in the vector of features have been chosen on the basis of the results obtained by applying RainCEIV to a test dataset.

The training dataset has been built by double matching a set of SEVIRI observations with PEMW-derived and radar-derived co-located

RR values. PEMW (Precipitation Estimation at Microwave frequencies, Di Tomaso et al. 2009) is an algorithm developed at the CNR-IMAA. PEMW results has also been used to get RR maps for PET and then to validate PET results. PEMW operative version (OPEMW) is operationally run at CNR-IMAA. The OPEMW has been validated against radar-derived RR values and rain gauge surface rain intensity (Cimini et al. 2013). OPEMW accuracy in identifying rainy/non-rainy areas is 98%, Heidke Skill Scores are 45% and 42% with respect to radar-derived and rain gauge rain rate values, respectively.

RainCEIV was validated on the basis of the rain rate observations from the operational weather radar network coordinated by DPC. The dichotomous statistical scores indicate that RainCEIV detects correctly as rainy or non-rainy about 96% of the pixel examined. The Bias score (1.36 for daytime validation and 1.58 for nighttime validation) indicates the RainCEIV tendency to overestimate rainy pixels, especially during nighttime. However, the POD (Probability of Detection) (81% both for daytime and nighttime validation) indicates good skills for RainCEIV to detect rainy areas. The fact that RainCEIV overestimates rainy areas is due principally to the misclassification of *non-rainy* pixels to *light-to-moderate* rainy pixels. Moreover, it is important to bear in mind that the statistical scores interpretation could be negatively affected by the different spatial resolution and the radar-SEVIRI temporal distance.

4.6 CONCLUSIONS

This paper is a review of techniques developed at the CNR-IMAA for continuous monitoring of rainy events, especially extreme precipitation events. The identification of rainy areas requires, as a starting point, the detection and the characterization of cloudy areas. Because of this, the algorithms for the cloud detection and classification developed at the CNR-IMAA are also presented. Both the cloud detection/classification algorithm and the precipitation estimation techniques summarized in this paper use the Infrared and Visible observations from the Meteosat geostationary satellite. Infrared and Visible observations are very useful for cloud mask and classification purposes but they are limited in determining rain rate values when compared with the PMW observations. Nevertheless their higher temporal and spatial resolutions make them very useful for the continuous monitoring of precipitation events, especially extreme precipitation events. An example of the PET and

C_MACSP (RainCEIV) usefulness for the near-real time monitoring of extreme events is proposed in Di Paola et al. (2014), where the extreme convective event that occurred on 21 February 2013 on the east coast of Sicily, causing flash floods over Catania, is analyzed. This event was characterized by extraordinarily intense rainfall over a short period, discharging more than 50 mm of rain in an hour (between 15:00 and 16:00 UTC). This intense precipitation was not forecast by Numerical Weather Prediction (NWP) models. The case study was analysed by applying C_MACSP and PET, and it is worth considering C_MACSP (RainCEIV) and PET complementarity for near real-time monitoring of extreme events. When PET rain rate maps are available, C_MACSP gives further information about the cloud cover and classification. When PET is not available, C_MACSP and RainCEIV give information about the presence of convective clouds and evaluate the rainfall class (the RainCEIV results related to the Catania flashflood are shown in Ricciardelli et al. 2014). These algorithms are very useful for monitoring purposes; though of course they do not provide forecast, they can contribute in observing weather events not correctly forecast by NWP models.

4.7 REFERENCES

Ackerman, S. A., Strabala, K., Menzel, W. P., Frey, R. A., Moeller, C. C., Gumley, L. E. 1998. Discriminating clear-sky from clouds with MODIS, *Journal of Geophysical Research*, 103: 141-157.

Bennartz, R. 2002. Optimal Convolution of AMSU-B to AMSU-A. J, *Atmos. Oceanic Technol.*, 17: 1215–1225.

Cimini, D., Romano F., Ricciardelli, E., Di Paola, F., Viggiano, M., Marzano, F. S., Colaiuda, V., Picciotti, E., Vulpiani, G., and Cuomo, V. 2013. Validation of satellite OPEMW precipitation product with ground-based weather radar and rain gauge networks, *Atmos. Meas. Tech. Discuss.*, 6: 4279-4312, doi:10.5194/amtd-6-4279-2013.

Di Paola, F., Casella, D., Dietrich, S., Mugnai, A., Ricciardelli, E., Romano, F., and Sanò, P. 2012. Combined MW-IR Precipitation Evolving Technique (PET) of convective rain fields, *Nat. Hazards Earth Syst. Sci.*, 12: 3557-3570.

Di Paola, F., Ricciardelli, E., Cimini, D., Romano, F., Viggiano, M., and Cuomo, V. 2014 Analysis of Catania Flash Flood Case Study by Using Combined Microwave and Infrared Technique, *Journal of Hydrometeorology*, doi: 10.1175/JHM-D-13-092.1, in press.

Di Tomaso, E., Romano, F., and Cuomo, V. 2009. Rainfall estimation from satellite passive microwave observations in the range 89 GHz to 190 GHz, *Journal of Geophysical Research*, 114: D18203, doi:10.1029/2009JD011746.

Kidd, C. and Levizzani, V. 2011. Status of satellite precipitation retrievals. *Hydrology and Earth System Sciences*, 15: 1109-1116. doi: http://dx.doi.org/10.5194/hess-15-1109-2011.

Kucera P. A., E. E. Ebert, F. J. Turk, V. Levizzani, D. Kirschbaum, F. J. Tapiador, A. Loew, and M. Borsche, PRECIPITATION FROM SPACE - *Advancing Earth System Science, Bull. Amer. Met. Soc.*, 365-375, March, doi:10.1175/BAMS-D-11-00171.1, 2013.

EUMETSAT, 1996. Meteorological Products Extraction Facility (MPEF), *Algorithm Specification Document*, Doc. No. MTP.SPE030, EUMETSAT.

Im, E., Durden S. L., Wu, C. 2005. Cloud profiling radar for the CloudSat mission, IEEE *Aerospace and Electronic Systems Magazine*, 20: 15-18.

Ricciardelli, E., Romano, F., Cuomo,V. 2008. Physical and statistical approaches for cloud identification using Meteosat Second Generation-Spinning Enhanced Visibile and Infrared Imager Data, *Remote Sensing of Environment*, 112:2741-2760.

Ricciardelli, E., Romano, F., Cuomo, V. 2008. Corrigendum to "Physical and statistical approaches for cloud identification using Meteosat Second Generation-Spinning Enhanced Visible and Infrared Imager Data, Remote Sensing of Environment 112: 2741-2760, (2008), *Remote Sensing of Environment*, 112: 3991-3991.

Ricciardelli, E., Romano, F., Cuomo, V. 2010. A technique for classifying uncertain MOD35/MYD35 pixels through Meteosat second generation-spinning enhanced visible and infrared imager observations. IEEE *Transactions on Geoscience and Remote Sensing*, 48: 2137–2149.

Ricciardelli, E., Cimini, D., Di Paola, F., Romano, F., and Viggiano, M. 2014. A statistical approach for rain intensity differentiation using Meteosat Second Generation–Spinning Enhanced Visible and InfraRed Imager observations, *Hydrol. Earth Syst. Sci.*, 18: 2559-2576, doi:10.5194/hess-18-2559-2014.

Schmetz, J., Pili, P., Tjemkes, S., Just, D., Kermann, J., Rota, S., et al. 2002. An introduction to Meteosat Second Generation (MSG), *BAMS*, 977–992.

Stephens, G. L. , Vane, D., Roain, R. J., Mace, G., Sassen, K., Wang, Z., et al. 2002. The CloudSat mission and the A-Train: a new dimension of space based observations of clouds and precipitation, *BAMS*, 83:04_Ricciardelli.doc1771–1790.

Tapakis, R., Charalambides, A. G. 2012. Equipment and methodologies for cloud detection and classification: A review, *Solar Energy*, 95: 392-430.

Winker, D., Hunt, W., McGill, M. 2007. Initial performance assessment of CALIOP. *Geophysical Research Letter*, 34, L19803, doi:10.1029/2007GL030135.

CHAPTER 5

A SATELLITE MULTI-SENSOR APPROACH FOR FLOODED AREAS DETECTION AND MONITORING

Teodosio Lacava[1,2*], Emanuele Ciancia[2], Irina Coviello[2], Carmine Di Polito[2], Mariapia Faruolo[1], Nicola Pergola[1,2], Valeria Satriano[2], Valerio Tramutoli[1,2]

[1] Institute of Methodologies for Environmental Analysis (IMAA), National Research Council - C.da S. Loja – 85050 Tito Scalo (Italy)

[2] School of Engineering, University of Basilicata – Via dell'Ateneo Lucano, 10 – 85100 Potenza (Italy)

*Corresponding author: teodosio.lacava@imaa.cnr.it

ABSTRACT

Timely, frequently updated and accurate information about the area affected by flood events are crucial for local authorities and decision makers to promptly react to the disaster and to put in place effective countermeasures devoted to limit damages and human live losses. Remote sensing data can provide a useful support to the full flood management cycle, showing, in particular, a high potential in performing a rapid and cost-effective flood mapping during the emergency phase. In this situation, an integrated use of satellite data, acquired by both optical and passive or active microwave instruments, can provide adequate and frequently updated information on soil conditions. In this framework, long series data coming from AMSU (Advanced Microwave Sounding Unit) and AVHRR (Advanced Very High Resolution Radiometer) sensors, both onboard of several Meteorological Operational (MetOp) and National Oceanic and Atmospheric Administration (NOAA) satellites, were processed and jointly analyzed for studying the August 2002 Elbe river flood occurred in Germany. A general methodology for multi-temporal satellite data analysis, the Robust Satellite Techniques (RST) approach, has been implemented on both data sets, trying to assess the potential of such an integrated system for a near real time monitoring of soil conditions, in terms of moisture variation and/or water presence, during the analyzed event.

5.1 INTRODUCTION

Among natural disasters, floods are one of the most significant, causing serious economic and social losses and in some instances human casualties (Wierzbicki et al. 2013). In the future, such a scenario is expected to get worse because of climate change effects at global scale: the global exposure to floods would increase with a variable intensity depending on the degree of warming (Hirabayashi et al. 2013).

Satellite data acquired in different region of the electromagnetic spectrum, from visible to microwave (MW), both from active and passive sensors, have been largely used to support the full flood management cycle, providing, depending on the specific characteristics of the used sensor/satellite system, near real time and reliable information on the processes involving soils (O'Grady et al. 2014; Pekel et al. 2014). Information about precipitation and rain rate estimation, snow depth and soil moisture maps as well as river water level and flood maps can then be produced (Lacava et al. 2010a; Matgen et al. 2011, Brocca et al. 2011, Tarpanelli et al. 2013, 2014) and integrated with the outputs of other systems (i.e. hydrological models). Concerning the emergency phase, the timely delivery of this information, soon after and during the event, is crucial for a better coordination of the emergency chain as well as for helping disaster managers in taking adequate decisions devoted to reduce the social and economic impact of such a phenomenon (Faruolo et al. 2013).

The possible contribution of remote sensing during the flood emergency response phase mainly depends on the spatial/spectral/ temporal features of the used data as well as on the accuracy of the methods used to generate relevant information from satellite observations. Considering advantages and drawbacks of active and passive satellite systems, a combined use of optical and MW sensors is preferred to provide reliable and timely information under any weather conditions and during both day and night. Optical sensors onboard meteorological satellites, both on polar and geostationary orbit, allow for a temporal resolution up to few minutes and hence a prompt overpass over the affected areas, with a medium-low spatial resolution (from 0.25 to 5 km). On the other hand, the main advantage of MW data is their availability even in the presence of clouds and, considering active system, their high spatial resolution. The main limit of Synthetic Aperture Radar (SAR) sensors is their coarse temporal resolution, often inadequate to follow flood dynamics. Indeed, passive microwave sensors acquire data at a high temporal frequency (1-2 times per day) on a near-global scale, even if with a low spatial resolution. In addition, passive microwave and optical sensors are often carried out on the same platform, allowing for a natural co-location and a simultaneous acquisition. In this way, an integration of optical and microwave data can allow for a better investigation (both in space and in time) of the flood events. In this paper, we assess the potential of such a strategy, using passive microwave and optical data acquired by two sensors onboard Meteorological Operational (MetOp) and National Oceanic and Atmospheric Administration (NOAA) satellites. Specifically, we investigated the complementary role of the Advanced Microwave Sounding Unit (AMSU) and Advanced Very High Resolution Radiometer (AVHRR) data. A multi-sensor approach, based on a general change detection scheme, named Robust Satellite Techniques (RST – Tramutoli 2005, 2007), was applied to automatically identify water in/on the soil. AMSU data were used for a continuous and frequently updated monitoring of soil conditions aimed at an effective early warning of floods while AVHRR ones were employed to map, with a higher detail than AMSU, the extent of areas affected by the flood. To this aim, multi-year time series of such data were processed and analyzed for studying the Elbe river flood event occurred in the South of Germany in the second half of August 2002 (Toothill 2002; RMS 2003).

5.2 THE ROBUST SATELLITES TECHNIQUES (RST) APPROACH

The RST approach (Tramutoli 2005, 2007) is a multi-temporal scheme of satellite data analysis, which automatically identifies, at pixel level, statistically significant anomalies of the investigated signal, by implementing the ALICE (*Absolutely Llocal Index of Change of the Environment*) index:

$$\otimes_V\ (x,y,t) = \frac{V(x,y,t) - \mu_V\ (x,y)}{\sigma_V\ (x,y)} \tag{5.1}$$

where, $V(x,y,t)$ is the investigated observable, selected according to the type of phenomenon to be studied; $\mu_V(x,y)$ and $\sigma_V(x,y)$ are, respectively, its monthly temporal mean and its natural variability (i.e. standard deviation) computed by collecting and processing a multi-year data-set of co-located, cloud-free and unperturbed imagery. The ALICE index is a standardized variable, characterized by a Gaussian behavior (i.e. with mean equal to zero and standard deviation equal to 1). This means that the probability of occurrence of values higher/lower than 2 standard deviations is around 2% and it becomes lower than 0.2 for values higher/lower than 3 (Tramutoli 1998).

When implemented on AMSU data to monitor soil moisture variations, the investigated signal (i.e. the $V(x,y,t)$ observable in Eq. 5.2) is the Soil Wetness Index (SWI=Bt89-Bt23) defined as the difference between the brightness temperature (Bt) achieved in two AMSU window channels (i.e. channel 15, at 89 GHz and channel 1, at 23.8 GHz) differently sensitive to the presence of water in the soil (Songyan et al. 2000). An increase of the soil water content or of the water fraction affecting pixel, should generate an increase of SWI. In previous works (Lacava et al. 2005; Lacava et al. 2010b; Manfreda et al. 2011), SWI has been standardized following the RST prescriptions and computing the Soil Wetness Variation Index (SWVI), defined as:

$$SWVI(x,y,t) = \frac{SWI(x,y,t) - \mu_{SWI(x,y)}}{\sigma_{SWI(x,y)}} \tag{5.2}$$

The used AMSU images have an original spatial resolution of 48 km, but during the processing they have been re-sampled at 25 km. High values of *SWVI* in correspondence of a relative and significant increase of soil wetness and/or of the water affected areas should be measured (Lacava et al. 2005).

When implemented on AVHRR data, the soil and water spectral behaviors in visible (VIS) and near infrared (NIR) regions of the electromagnetic spectrum were exploited to discriminate flooded pixels over the scene. As previous studies showed (Lacava et al. 2010a; Faruolo et al. 2013), the NIR and VIS reflectances difference, R_{NIR}-R_{VIS}, represents the signal to be investigated to map flood affected areas, being it more sensitive to water presence than the ratio R_{NIR}/R_{VIS}. The ALICE index, implemented on daytime AVHRR images at 1km of spatial resolution, has the following formulation:

$$\otimes_{NIR-VIS}(x,y,t)=\frac{R_{NIR-VIS}(x,y,t)-\mu_{NIR-VIS}(x,y)}{\sigma_{NIR-VIS}(x,y)} \tag{5.3}$$

In correspondence of flooded areas, negative $\otimes_{NIR-VIS}(x,y,t)$ values should be observed as described in previous works (Lacava et al. 2010a; Faruolo et al. 2013).

5.3 RESULTS AND DISCUSSION

The investigated test case is the Elbe River flood which hit the South of Germany between August 6 and 13, 2002. Such an extreme event caused severe damages and losses of human lives (Toothill 2002; RMS 2003). According to RST prescriptions, all AMSU and AVHRR images acquired during the months of August from 2002 to 2010 were collected and processed, in order to compute the two proposed ALICE indices (Eq. (5.2) and (5.3)).

In Fig. 5.1 shows an example of the results achieved by implementing SWVI for the AMSU data of 13 August 2002 at 0619 UTC. In the image pixels showing SWVI value statistically different from the expected one for that specific conditions of observation (i.e. August, morning NOAA 15 data) are shown in different blue tones. Large areas of Central Europe were characterized by conditions wetter than normal as consequence of the extreme precipitations which affected the area in the previous days as well as during the preceding hours (the brown areas identify the raining clouds present over the scene) (Ulbrich et al. 2003a, 2003b). In particular, we focused our attention on the areas within the red box just in the West of Germany, where the most affected area along the Elbe river (NASA 2002; ZKI 2002; RMS 2003; ESA 2005; EUROMAP 2006), located between Wittenberg and Torgau cities, are present.

In Fig. 5.2 the trend of SWVI averaged over the pixels within the considered box for August 2002 is reported. Missing values refer to rainy or no-data (i.e. the considered area is outside of the satellite swath). The black dashed line marks the value SWVI = 2 representing a statistical significant critical level to discriminate anomalous values (i.e. SWVI > 2), referred to days in which soils were likely to be wetter than expected.

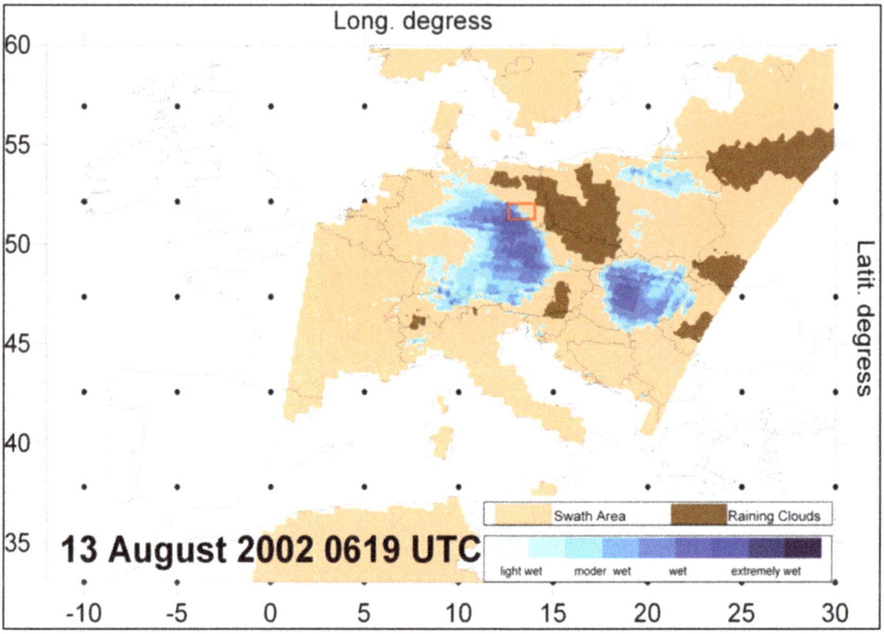

Fig. 5.1. SWVI map of 13 August 2002 at 0619 UTC; red box refers to the analysed area.

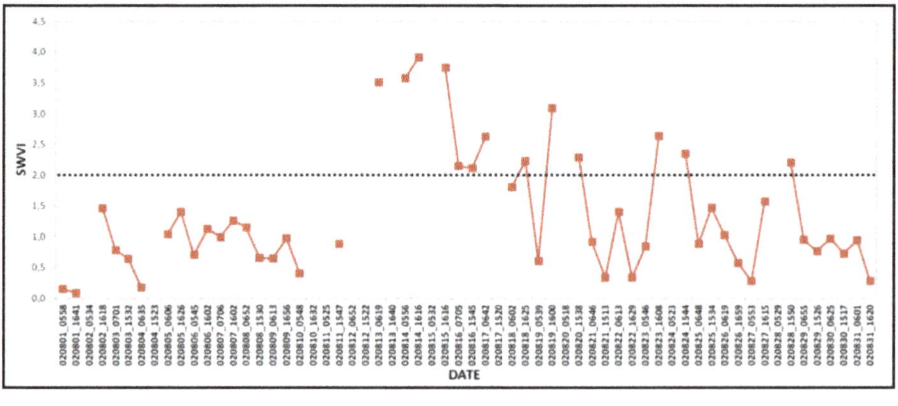

Fig. 5.2. SWVI temporal pattern for August 2002 for the area within the red box in Fig. 5.1; the black dashed line marks SWVI values > 2.

The SWVI temporal pattern is well in agreement with the evolution of the phenomenon in the investigated area, as documented by independent sources (RMS 2003). SWVI values within almost the natural variability have been detected during the first ten days of the period indicating that the first phase of the event has not directly affected the area within the red box in terms of soil water content variation. Between the morning of the 10 and 13 of August, instead, the area has been highly interested by the precipitation whose first direct consequence on the surface has been already shown in Fig. 5.1. The presence of heavy rains over the target area is moreover confirmed by the missing SWVI values for this period. Then, in the following days, SWVI reached its maximum (i.e. SWVI ~ 4) confirming the probable presence of very wet soils. In this second phase of the event, the information carried out by SWVI is mostly related to the increase of the soil water content over the analyzed areas. Such an indication is fundamental in the flood cycle management, because it highlights areas potentially at higher flood risk. It should be stressed that the detected pattern is independent from the time (morning or afternoon) of the AMSU pass, confirming, on one side, the consistency of the detected anomalous signal and, on the other side, the robustness of the proposed approach. After the high SWVI peak on 14 and 15 of August, its values start to decrease, indicating a sort of restoring of the normal conditions of soils in terms of water content. This decrease continues until 19 August when a new abrupt increase of the index is recorded, in correspondence of the beginning of the last phase of the flood. Starting from that day, in fact, the Elbe wave flood hit the investigated areas, overcoming the riverbanks reaching its maximum extension between 19 and 20 of August (Toothill 2002; RMS 2003; Ulbrich et al. 2003a, 2003b). In this period the SWVI response is mainly related to the increase of the soil fraction covered by water. In this phase, differently from the previous days, the anomalous SWVI values, although generally decreasing in relative intensity, still appear for several days, until August 28, as a consequence of two isolate but quite strong rain episodes which affected the area on 21 (about 10 mm of rain) and 27 (25 mm) of August 2002. After this date, SWVI definitively returns to values (i.e. SWVI $\leq$ 1) revealing a complete restoration of the normal conditions of the soils.

The area within the red box was also investigated by implementing the RST approach on AVHRR data for the considered period. The first almost cloud free AVHRR image showing the presence of "flooded" pixels, in terms of $\otimes_{NIR\text{-}VIS}$, is dated 16 of August at 11:37 GMT (Faruolo et al. 2013). In particular, for this image, concerning the area within the red box already analyzed by AMSU, 66 pixels (i.e. 0.15% of the considered scene) were identified as flooded (Fig. 5.3a) notwithstanding a large portion of the area (about 20%) was still covered by clouds. After that, a progressive increase of the flood-affected area was documented and confirmed by the results also depicted in Fig. 5.3b, where a total flood extent map is shown. Such a map, referred to the most flood affected soils (i.e. the ones between Wittenberg and Torgau) within

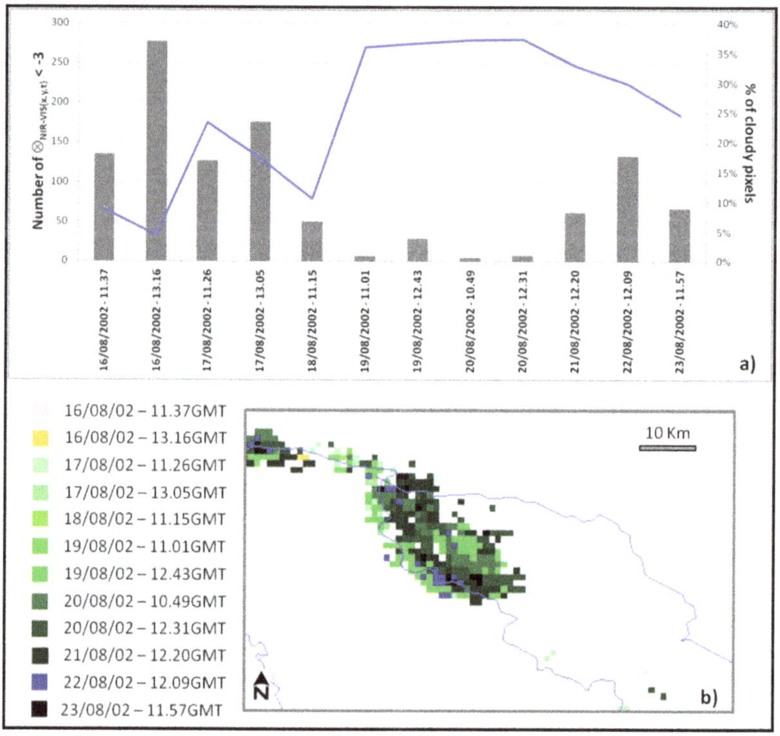

Fig. 5.3. **a) Temporal trend of number of pixels (blue line) detected as flooded by the AVHRR based ALICE index ($\otimes_{NIR\text{-}VIS} < -3$) on the area within the red box in Fig 3.1 for the AVHRR images from 16 to 23 August 2002, coupled with the percentage of cloudy pixels for each day; b) flood progression map obtained by implementing the proposed approach on the 12 considered AVHRR images acquired over the same target area, the map is referred to the most affected flooded area within the AVHRR scene.**

the investigated scene, was generated combining pixels detected as flooded at the most reliable level of confidence of the index ($\otimes_{NIR-VIS} < -3$) from 16 to 23 August, 2002 by analyzing 12 AVHRR images. As expected, water firstly started to inundate the areas close to the riverbanks, then it moved far from them, reaching its maximum extensions after the 19 of August, confirming the results already achieved by SWVI.

The results shown in the map highlight the potential of the optical-based RST approach in detecting the presence of flooded pixel in absence of clouds (Faruolo et al. 2013).

The obtained achievements clearly state the feasibility of the proposed approach: passive microwave data can be used to provide a first warning about the presence of soil wetter than normal and then to follow its evolution, also in presence of clouds. Optical data, instead, may help in "downscaling" at a local level such as information, showing, with a higher detail, the areas mostly affected by the event. It should be stressed that in order to assure an almost continuous (day and night) monitoring of the investigated areas, the RST approach could be easily exported also on thermal infrared data. Such an attempt has been already done using Moderate Resolution Imaging Spectroradiometer (MODIS) data (Faruolo et al. 2010), confirming the results previously described.

Finally it should be highlighted that AMSU and AVHRR, in particular the AMSU-A module, have started to fly together on the same NOAA platform in 1998 (i.e. NOAA 15). Since then, they have been simultaneous working on several NOAA and MetOp satellites (Table 5.1). Such constellations allow for an increase of the sensors temporal resolution, up to a few hours. As above said, in a general system devoted to soil conditions monitoring, this could represent an effective add-on, allowing for continuous and frequently updated observations, provided that a satellite based methodology able to give reliable information is used.

In the recent past multi-spectral features like the NOAA-MetOp ones were guaranteed by the Earth Observing Satellite (EOS) Aqua satellite, where MODIS and the Advanced Microwave Sounding Unit (AMSR-E) have been flying together for about ten years (Aug 2002 – October 2011, when AMSR-E stopped to work). The integration of data coming from the two sensors furnished good results for flooded area detection and monitoring (Lacava et al. 2014). Moreover, MODIS VIS data are available at higher spatial resolution than AVHRR (i.e. 0.25 km instead of

Table 5.1. Satellites having both AMSU and AVHRR onboard (adapted by http://www.oso.noaa.gov/poesstatus/).

Satellite	Operational since	Operational up to	Local Time Ascending Node
NOAA 15	15/12/1998	Today	16:51:13
NOAA 16	20/03/2001	09/06/2014	21:01:39
NOAA 17	15/10/2002	10/04/2013	19:02:32
NOAA 18	30/08/2005	Today	15:23:30
MetOp-A	21/05/2007	Today	21:29:05
NOAA 19	02/06/2009	Today	13:38:39
MetOp-B	04/24/2013	Today	21:31:45

1.1 km) thus they may help in better downscaling the information, contributing to map flooded areas with improved details (Faruolo et al, 2013). In October 2011 the NPP-SUOMI was launched, carrying onboard the Advanced Technology Microwave Sounder (ATMS) and the Visible Infrared Imaging Radiometer Suite (VIIRS). ATMS is the advanced version of AMSU and VIIRS originates from AVHRR and MODIS, so that their integration will provide the capability of detecting and monitoring flood event. Also data, which will be acquired by the next ESA Sentinel satellites (especially Sentinel 2 and 3), will have similar potential, provided that adequate satellite data analysis techniques, able to assure a high trade-off between accuracy and sensitivity regardless the considered signal/sensor, such as RST, will be used.

5.4 CONCLUSION

Integrating passive microwave and optical satellite data may support the flood management cycle, especially during the emergency phase, when timely information are essential for local authorities and decision makers to quickly identify the most affected areas where first reliefs have to be addressed. Results shown in this work, where AMSU and AVHRR data have been jointly used to investigate the flood event, which hit Central

Europe during August 2002, confirm the potential of such a strategy. In particular, nine years of these data have been analyzed following the Robust Satellite Techniques approach philosophy, implementing two different indices, one based on microwave data, able to provide information about both variation in soil water content and flooded area presence, and another one, based on optical data, helpful in mapping water affected areas with a higher detail. Achieved results confirmed the added value, in terms of early detection, continuous monitoring and mapping capability, that such an integrated satellite-based scheme might provide, also as a possible decision support system during crisis, especially considering the high temporal resolution (up to few hours) at which these information can currently be provided.

5.5 REFERENCES

Brocca, L., Hasenauer, S., Lacava, T., Melone, F., Moramarco, T., Wagner, W. Dorigo, W., Matgen, P., Martínez-Fernández, J., Llorens, P., Latron, J. Martin, C. Bittelli, M. 2011. Soil moisture estimation through ASCAT and AMSR-E sensors: An intercomparison and validation study across Europe. *Remote Sensing of Environment* 115: 3390-3408, doi:10.1016/j.rse.2011.08.003

ESA, 2005. Satellite survey of Elbe flood helps Swiss re insure for disaster. Available on line at: http://www.esa.int/esaEO/SEMR8C6Y3EE_index_1.html.

EUROMAP, 2006. Elbe Flood 2002. Available on line at: http://www.euromap.de/samples/exam_080.html.

Faruolo, M., Coviello, I., Lacava, T., Pergola, N. and Tramutoli, V. 2010. On the potential of Robust Satellite Techniques (RST) approach for flooded areas detection and monitoring using thermal infrared data. *Proceedings of the 2010 IEEE International Geoscience and Remote Sensing Symposium*, ISBN on IEEE Xplore entry: 978-1-4244-9566-5, doi: 10.1109/IGARSS.2010.5653900, pp. 914-917.

Faruolo, M., Coviello, I., Lacava, T., Pergola, N. and Tramutoli, V. 2013. A multi-sensor exportable approach for automatic flooded areas detection and monitoring by a composite satellite constellation. *IEEE Transactions on Geoscience and Remote Sensing* 51-4-1: 2136 – 2149, doi: 10.1109/TGRS.2012.2236336.

Hirabayashi, Y., Mahendran, R., Koirala, S., Konoshima, L., Yamazaki, D., Watanabe, S., Kim H. and Kanae S. 2013. Global flood risk under climate change. *Nature Climate Change 3*: 816–821 doi:10.1038/nclimate1911, doi:10.1038/nclimate1911.

Lacava, T. Cuomo, V, Di Leo, E. V., Pergola, N., Romano, F., and Tramutoli, V. 2005. Improving soil wetness variations monitoring from passive microwave satellite data: the case of April 2000 Hungary flood. *Remote Sensing of Environment* 96-2: 135-148, doi: 10.1016/j.rse.2005.01.015.

Lacava, T., Filizzola, C., Pergola, N., Sannazzaro, F. and Tramutoli, V. 2010a. Improving flood monitoring by the Robust AVHRR Technique (RAT) approach: the case of the April 2000 Hungary flood. *International Journal of Remote Sensing* 31-8: 2043 – 2062, doi:10.1080/01431160902942902.

Lacava, T. Brocca, L., Calice, G., Melone, F., Moramarco, T., Pergola, N. and Tramutoli, V. 2010b. Soil moisture variations monitoring by AMSU-based soil wetness indices: a long-term inter-comparison with ground measurements. *Remote Sensing of Environment* 114-10: 2317-2325, doi:10.1016/j.rse.2010.05.008.

Lacava, T., Brocca, L., Coviello, I., Faruolo, M., Pergola, N., Tramutoli, V. 2014. Integration of optical and passive microwave satellite data for flooded area detection and monitoring. In Engineering Geology for Society and Territory - Volume 3, pp. 631-635, *Springer International Publishing*. doi 10.1007/978-3-319-09054-2_126. http://dx.doi.org/10.1007/978-3-319-09054-2_126.

Manfreda, S. Lacava, T. Onorati, B. Pergola, N. Di Leo, M., Margiotta, M. R. and Tramutoli, V. 2011. On the use of AMSU-based products for the description of soil water content at basin scale. *Hydrology and Earth System Sciences* 15: 2839-2852, doi:10.5194/hess-15-2839-2011.

Matgen, R; Hostache1, G; Schumann, L; Pfister, L; Hoffmann, H. H. G.; Savenije. 2011. Towards an automated SAR-based flood monitoring system: lessons learned from two case studies. *Physics and Chemistry of the Earth* 36: 241-252, doi: 10.1016/j.pce.2010.12.009.

NASA, 2002. Flooding on Elbe River. Available on line at: http://earthobservatory.nasa.gov/NaturalHazards/view.php?id=10055.

O'Grady, D., Leblanc, M., and Bass, A. 2014. The use of radar satellite data from multiple incidence angles improves surface water mapping. *Remote Sensing of Environment*, 140. pp. 652-664, doi: 10.1016/j.rse.2013.10.006.

Pekel, J.-F., Vancutsem, C., Bastin, L., Clerici, M., Vanbogaert, E., Bartholomé, E. and Defourny P. 2014. A near real-time water surface detection method based on HSV transformation of MODIS multi-spectral time series data. *Remote Sensing of Environment* 140: 704–716, doi: 10.1016/j.rse.2013.10.008.

RMS, Risk Management Solutions. 2003. Central Europe Flooding, August 2002, Event Report. Available on line at:
http://loritoye.com/Media/Central_Europe_Flood_2002.pdf.

Songyan, G., Huilin, G., Yuanjing, Bolin Z., Wenjian, Z., Hong, Q., and Maonong, R. 2000. Flood detection and Monitoring by using of AMSU-A microwave data. *Proceedings of the 20th International ATOVS Study Conference*, Budapest, Hungary, 20-26 September 2000.

Tarpanelli, A., Brocca, L., Lacava, T., Melone, F., Moramarco, T., Faruolo, M., Pergola, N., and Tramutoli, V. 2013. Toward the estimation of river discharge variations using MODIS data in ungauged basins, *Remote Sensing of Environment* 136:47-55, doi: 10.1016/j.rse.2013.04.010.

Tarpanelli, A., Brocca, L., Barbetta, S., Faruolo, M., Lacava, T. and Moramarco, T. 2014. Coupling MODIS and radar altimetry data for discharge estimation in poorly gauged river basins. *IEEE Journal of Selected Topics in Applied Earth Observations and Remote Sensing (JSTAR)*, doi: 10.1109/JSTARS.2014.2320582. 8-1: 141-148.

Toothill, J. 2002. EQECAT Technical Report. Central European Flooding August 2002, ABS Consulting, Available on line at: http://www.absconsulting.com/resources/catastrophe_reports/flood_rept.pdf.

Tramutoli, V. 1998. Robust AVHRR Techniques (RAT) for Environmental Monitoring: theory and applications. Earth Surface Remote Sensing II, Giovanna Cecchi, Eugenio Zilioli, Editors, *Proceedings of SPIE* Vol. 3496, 101-113.

Tramutoli, V. 2005. Robust Satellite Techniques (RST) for natural and environmental hazards monitoring and mitigation: ten years of successful applications. The 9th International Symposium on Physical Measurements and Signatures in Remote Sensing, S. Liang, J. Liu, X.n Li, R. Liu, M. Schaepman Editors, Beijing (China), *ISPRS*, Vol. XXXVI (7/W20), 792-795. ISSN 1682-1750.

Tramutoli, V. 2007. Robust Satellite Techniques (RST) for Natural and Environmental Hazards Monitoring and Mitigation: Theory and Applications. Fourth International Workshop on the Analysis of Multitemporal Remote Sensing Images. 18-20 July, 2007, Louven, Belgium, doi: 10.1109/MULTITEMP.2007.4293057.

Ulbrich, U., Brücher, T., Fink, A.H., Leckebusch, G.C., Krüger, A. and Pinto, J.G. 2003a. The central European floods of August 2002: Part 1 ± Rainfall periods and flood development. *Weather* 58: 371□377, doi: 10.1256/wea.61.03A.

Ulbrich, U., Brücher, T., Fink, A.H., Leckebusch, G.C., Krüger, A. and Pinto, J.G. 2003b. The central European floods of August 2002: Part 2 – Synoptic causes and considerations with respect to climatic change. *Weather* 58: 434□442, doi: 10.1256/wea.61.03B.

Wierzbicki, G.; Ostrowski, P.; Mazgajski, M.; Bujakowski, F. 2013. Using VHR multispectral remote sensing and LIDAR data to determine the geomorphological effects of overbank flow on a floodplain (the Vistula River, Poland). *Geomorphology* 183: 73-81, doi: 10.1016/j.geomorph.2012.06.020.

ZKI, 2002. Crisis management during the Elbe flood in August 2002. Available on line at: http://www.zki.dlr.de/article/1047.

CHAPTER 6

A NEW APPROACH TO ESTIMATE THE ENTROPIC PARAMETER FOR DISCHARGES MEASUREMENTS AT UNGAUGED RIVER SITES

(CASE OF THE COASTAL ALGIERS WATERSHED)

Abdelhadi Ammari[1], Tommaso Moramarco[2]

[1] Ecole Nationale Supérieure d'Hydraulique. Blida. Algeria.
a.ammari@ensh.dz

[2] IRPI, Consiglio Nazionale delle Ricerche. Perugia. Italy. T.Moramarco@irpi.cnr.it

ABSTRACT

In river monitoring, it is quite difficult to sample velocity points during high floods, especially in the lower portion of flow area, due to the danger that operators might face during measurement as well. In this context an important contribution is provided by the entropy theory, which identifies a linear relationship between the mean flow velocity and the maximum flow velocity. This theory can also be easily sampled during high floods for its position in the upper portion of the flow area. The entropic relationship is robust and is based on the estimation of a sole parameter M. Considering the ungauged river sites, the M estimation is not simple, however if one was able to estimate the dependence of M on hydraulic river characteristics and/or morphological basin ones, then it would be possible to assess the mean flow velocity and, hence, the discharge just by sampling the maximum velocity through, for instance, non-contact radar sensors. Therefore, based on the above insights, this work attempts, on one hand, to identify a likely dependence between hydraulic river characteristics (Manning's roughness, hydraulic slope, etc.) and the entropy parameter and, on the other hand, determine simple formulations between the entropy parameter M and morphological characteristics of the basin, subtended from the river site of interest, in terms of drainage density, drainage area and perimeter. Algerian gauged river sites are used as case study.

Validation was made on four hydrometric stations in order to check the accuracy of the obtained model results. The results showed that the estimation of M based on watershed morphological characteristics is quite satisfactory with an error, which doesn't exceed 5%, for all investigated river sites except one site, wherein the error does not exceed 12% anyhow.

Key words: entropy, ungauged sites, discharge measurements, Coastal Algiers watershed

6.1 INTRODUCTION

Usually in hydrological practice, discharges measurements are generally addressed using standard methods based especially on the velocity field exploration. The velocity profiles in different verticals are obtained by sampling one-three or more velocity points, but sometimes during high floods it is very complicated to carry out the sampling velocity points especially in the lower portion of the flow area (Ardiclioglu et al, 2007). These approaches are based on the turbulent boundary layer theory (Fenton, 2002), which considers the logarithmic law of the velocity distribution, or the Prandlt Von Karman law (Chow, 1959). These sampling difficulties during high floods considerably reduce the number of data for high flood and also the accuracy of the discharge measurements (Chiu and Chen, 1999). For these reasons, the entropy method developed by Chiu (1987) and based on the estimation of a simple parameter M, was tested on some Wades of Algerian Watershed (Ammari & Remini, 2010), and it gave a good results in terms of discharges estimation and also wetted area calculation. Therefore, in order to address the estimation of M for ungauged river sites it is of interest to seek a relationship between M and morphological characteristics of basin. A first attempt was done by Moramarco, et al. (2008) in order to describe the M parameter in function of Rosgen river classification.

The paper purpose is toinvestigate the correlation, on one hand, between M and hydraulic parameters of the river gauged sites, like hydraulic radius, hydraulic slope and manning roughness in order to extrapolate rating curves at river sites and, on the other hand, which is of considerable interest for ungauged sites, between M and some morphological watershed characteristics.

6.2 MEAN VELOCITY ESTIMATION

Based on the concept of entropy, Chiu developed a simple expression, which can describe a possible velocity profile (Chiu 1987, 1988, 1989):

$$u = \frac{U_{max}}{M} ln\left[1 + \left(e^{M} - 1\right)\frac{\xi - \xi_0}{\xi_{max} - \xi_0}\right] \tag{6.1}$$

The appealing aspect behind this theory is that by integrating Eq. (6.1) over the flow area, the mean flow velocity, u_m, can be easily assessed (Chiu, 1988) as:

$$u_m = \Phi(M) u_{max} \tag{6.2}$$

Where Φ(M) is a function of the entropy parameter M and is inferred after the entropy maximization (Chiu and Hsu 2006), yielding:

$$\Phi(M) = \frac{U_m}{U_{max}} = \frac{e^M}{e^M - 1} - \frac{1}{M} \tag{6.3}$$

Therefore, if Φ(M) is constant for the same gauged site, M can be inferred by Eq. (6.3) and used to describe the velocity profiles according to the Eq. (6.1)

6.3 THE BENEFIT IN USING THE ENTROPY PARAMETER M: AND ITS ESTIMATION AT UNGAUGED SITES

Let's consider the well-known Manning Equation:

$$Q = \frac{\sqrt{S_f}}{n} AR^{2/3} = \alpha AR^{2/3} \tag{6.4}$$

with Sf, energy grade line slope, n; Manning's roughness; A, flow area and R, hydraulic radius.

Using the entropy method we can turn the last equation into:

$$\alpha = \frac{\sqrt{S_f}}{n} = \frac{Q}{AR^{2/3}} = \frac{u_m A}{AR^{2/3}} = \frac{u_m}{R^{2/3}} = \frac{\Phi(M)u_{max}}{R^{2/3}} = \frac{\Phi(M)u_{max}}{h_m^{2/3}} \qquad (6.5)$$

For large flow depths, the hydraulic radius approaches the hydraulic depth (h_m=A/T), where T is the top width of the channel.

Therefore, if a good topographical survey of the river site is available (Ammari and Remmini, 2010), it will be theoretically easy to define the ratio between the square root of the energy line slope and the manning roughness by exploiting M in Eq. (6.5), once u_{max} and h_m is sampled. Therefore, once α is estimated by M, and considering that α is a constant for high stage, Eq. (6.4) would allow to extrapolate the rating curve just by knowing the conveyance $AR^{2/3}$ at river site. As a consequence, it is of interest to estimate M at ungauged river sites and to this end a simple multi-linear expression is proposed here, wherein M is a function of morphological characteristics of basin.

Therefore, considering the importance of M parameter, for addressing its estimation at ungauged river sites where no streamflow data are available, a simple multi-linear expression is investigated by assuming M as a function of morphological characteristics of basin. Specifically, the M values found at river sites were correlated to known characteristics of the basin of each site in terms of basin area (Ar), perimeter (Per), mean altitude (H_m), slope index (I), drainage density (D_d), concentration time (T_c) and torrentiality coefficient (C_t), yielding:

$$M = K_1 Ar + K_2\,Per + K_3 D_d + K_4 C_t + K_5 I + K_6 H_m + K_7 T_c \qquad (6.6)$$

$K_{1,2.....7}$: Multi-linear regression coefficients.

6.4 STUDY ZONE

The coastal Algiers watershed is an extended watershed from the East (Bejaia) to the West (Mostaganem) of Algeria covering an area of 11972 km² with an average wide of 24 km and a 500km long, this wide extension gives it a great climatic, hydrologic, geomorphologic and so on. Diversity. As shown in Fig. 6.1, the watershed is limited to the North by

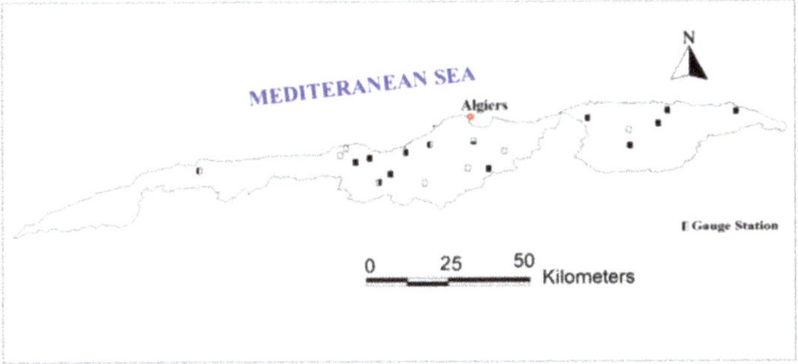

Fig. 6.1. Distribution of the gauge stations over the Coastal Algiers watershed.

the Mediterranean Sea, in the South by the Soummam, Isser and Cheliff watershed, which also borders it from the West. In the East we find the Constantinois watershed.

Topography of the watershed is characterized my an important mountainous chains which are considered as a southern border, called Tellian Atlas where altitudes exceed 1000 m easily, like Chrea (1550m at the refuge), and Djurdjura (2308m at Lalla Khedidja). All wadies took their sources from those mountains. Eighteen gauged river sites sections for the first and second step were considered for the analysis. The gauge stations over the whole watershed are divided into three parts: East one (Sebaou), Center (El Harrach, Mazafran and Cherchell) and the West (Tenes). The watershed is well drained with many wadies (small rivers) especially in the East over the Sebaou Sub-watershed. Most part of those wadies are controlled by fixed or mobile gauge stations, but the data collected presents a large blank periods, even during high floods, where the flow is so important that it is difficult to conduct traditional measuring methods like the exploration of the velocity fields by current meters. So the National Hydraulic resources agency uses other techniques like rating curves and floating bodies, but that methods proved weakness to measure high discharges during floods where the volume of water is more than 60% of the water drained by the wadies and more than 80% of the sediments transported, where the necessity to develop new methods to optimize time and money in order to get more data with a good accuracy especially during high floods.

6.5 FIELD DATA

The collected gauge data is widely different in terms of periods, water levels and discharges (see Table 6.1). In addition, data don't cover a wide range of high discharges, showing the difficulties to carry out a lot of velocity measurements under high floods conditions.

Table 6.1. Data Characteristics: N, Q, and D stand for Number of events, Discharge and Maximum flow depth

Sub-watershed	Station	N	Q (m3/s)	D (m)	Period
OuedSebaou	Baghlia	121	0.01 - 173.12	0.1 - 6.65	1990 - 2001
OuedAissi	RN 30	185	0.01 - 13.82	0.1 - 1.78	1989 - 2008
OuedSebaou	Freha	74	0.022 - 2.82	0.1 - 0.7	1989 - 2006
OuedEl	Baraki	125	0.6 - 9.71	0.15-0.55	1993 - 2006
OuedEl	Gorges de larbaa	141	0.06 - 7.56	0.1- 0.65	1987 - 1996
OuedEl	RocherdesPigeons	300	0.08 - 14.05	0.1 -0.98	1989 - 2007
OuedMazafran	Fer à Cheval	317	0.014 - 349.60	0.1 - 7.8	1985 - 2005
OuedChiffa	AmontdesGorges	166	0.015 - 10.92	0.1 - 1.1	1990 - 2004
OuedDjer	Attatba	117	0.011 - 4.26	0.1 - 0.85	1990 - 2007
OuedDjer	El Affroun	98	0.01 - 3.17	0.1 - 0.85	1990 - 2007
OuedDjer	Boumedfaa	63	0.035 - 21.3	0.1 - 0.85	1996 - 2005
CôtiersCherch	Hadjout	37	0.01-1.11	0.1-0.93	2001-2003
Côtier Cap	Ouled Ali	128	0.03 - 6.57	0.1- 0.7	1988 - 2007
CôtiersCherch	Pont RN11	53	0.08 - 5.30	0.13 - 2.04	1996 - 2006
CôtiersCherch	BorjGhobrini	52	0.036 - 15.96	0.1- 0.95	1988 - 1994
CôtiersCherch	Mesdour	111	0.2- 3.08	0.13-0.45	1996-2007
CôtierTenes	Sidi Akacha	115	0.03 - 7.8	0.1 - 0.85	1988- 2007

6.6 DATA ANALYSIS AND DISCUSSION

A good relationship between the mean and maximum velocities was found, where the ratio $\Phi(M)$ between the two velocities can be considered constant for all the gauge stations investigated; confirming the results by Chiu (1989, 1995, 1999, 2006) and Moramarco (2004, 2010, 2011). As an example, in Fig. 6.2 a comparison Um-Umax is shown for three gauged sites.

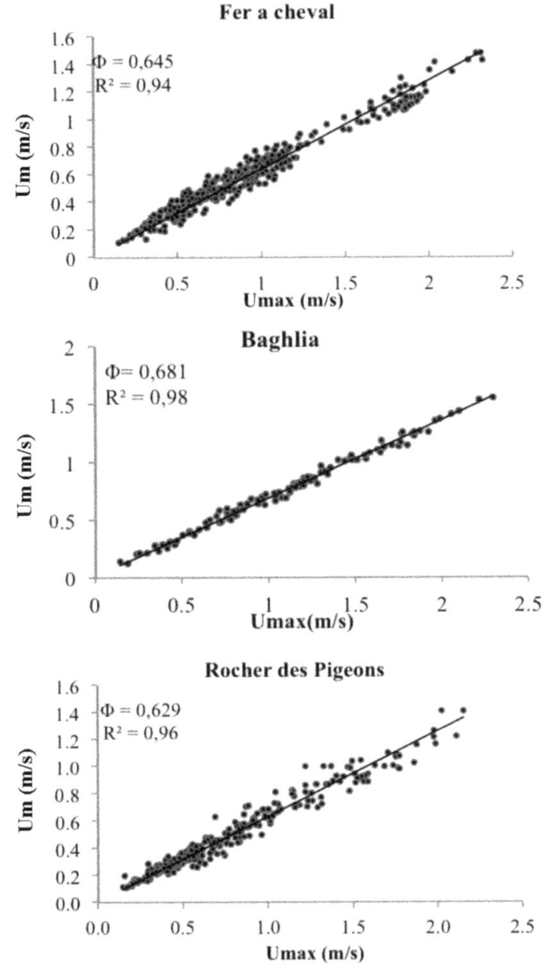

Fig. 6.2. Mean / Max velocity relationship for three gauge stations.

All results are summarized in Table 6.2.

Table 6.2 Entropy parameter M and the ratio Φ(M); mean/maximum velocities (Ammari, 2012).

Gauge Station	River	Φ (M)	M	R^2
Baghlia	Sebaou	0.681	2.36	0.98
Hadjout	Bourkika	0.636	1.71	0.94
RN30	Aissi	0.641	1.78	0.97
Freha	Dis	0.662	2.08	0.95
Gorges de Larbaa	Djemmaa	0.627	1.59	0.98
BARAKI	El harrach	0.627	1.59	0.96
Rocher des pigeons	El harrach	0.629	1.61	0.96
Fer à cheval	Mazafran	0.645	1.84	0.94
Amont des gorges	Chiffa	0.641	1.79	0.97
Attatba	Bouroumi	0.613	1.4	0.93
El affroun	Djer	0.588	1.07	0.92
Boumedfaa	Djer	0.587	1.06	0.96
Azeffoun (RN24)	M'letat	0.673	2.24	0.94
Ouled Ali	Barek	0.619	1.48	0.97
Pont RN11	Belah	0.659	2.03	0.97
Bordj ghobrini	El hachem	0.627	1.58	0.95
Mesdour	Boukdir	0.640	1.77	0.94
Sidi Akacha	Allalah	0.648	1.88	0.94
El Ababsa	HAREZZA	0.645	1.97	0.96
PONT D9	HAMIZ	0.620	1.17	0.94
REGHAIA	REGHAIA	O.634	1.69	0.82
ALTAIRAC	SEMAR	0.677	2.25	0.95

Assuming that the entropy function Φ(M) is constant for each gauged site; the reliability of Eq. (6.5) is tested, i.e., the ratio between energy line slope (S_f) and Manning roughness (n) is checked as a function of parameters which can be measured easily such Maximum velocity and hydraulic radius (or even the hydraulic depth), and Entropy parameter M. Figure 6.3 shows for three gauge sites the good relationship between real values of α parameter and estimated ones using the entropy parameter, and it tends to unity slope line.

From that, we conclude that it is possible to estimate the ratio between the energy line slope and the Manning roughness using the

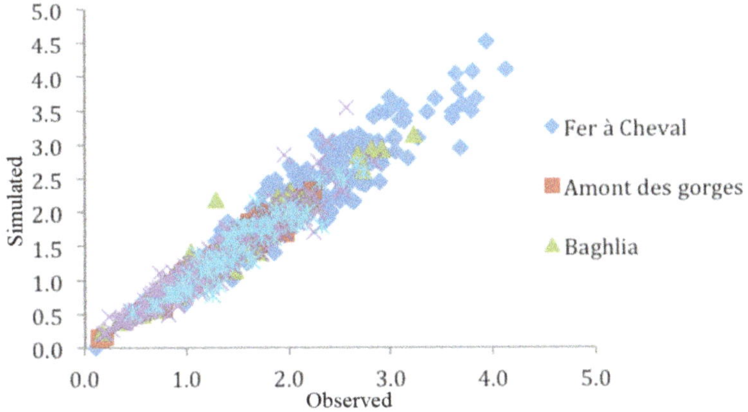

Fig. 6.3. Real /Entropy estimated values of α parameter.

entropy parameter, which, on one hand, increases the importance of the latter as a characteristic parameter of a considered river site and, on the other hand, fosters the use of Eq. (6.4) to extrapolate the rating curve at river site, considering that α is quite constant for high flows.

Therefore, it is of considerable interest to investigate the way to assess the entropy parameter for ungauged river sites. Based on that, the analysis based on Eq. (6.6) would allow to identify a simple model able to evaluate the entropy parameter M (and hence the entropy function $\Phi(M)$) for un ungauged site starting from some characteristics of the draining basin. Over the investigated watersheds we have many important ungauged river sites, like Damous Wadi on which a Dam of $125Mm^3$ is in construction, this example of unavailable streamflow data have a real impact on the management of the water resources especially in regions of water scarcity.

Results are plotted in Fig. 6.4 and as can be seen, the Mean Altitude, slope index and concentration time don't have a great influence on the entropy parameter estimation, so these morphological characteristics were discarded and a new correlation is investigated by using the remaining ones as shown in Fig 6.5. The Standardized coefficients (also known as β-coefficients), are the estimates resulting from an analysis carried out on independent variables:

http://en.wikipedia.org/wiki/Dependent_and_independent_variables
that have been standardized so that their variances are used to have satisfactory indications on what parameters are most significant.

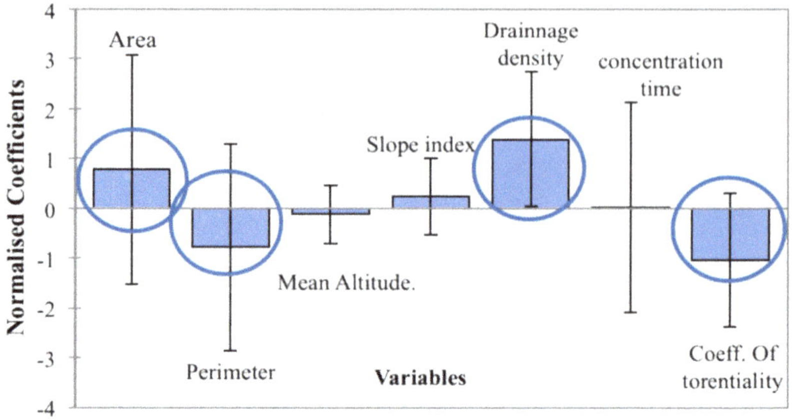

Fig. 6.4. 1st Run; Normalized coefficients (95% confidence interval).

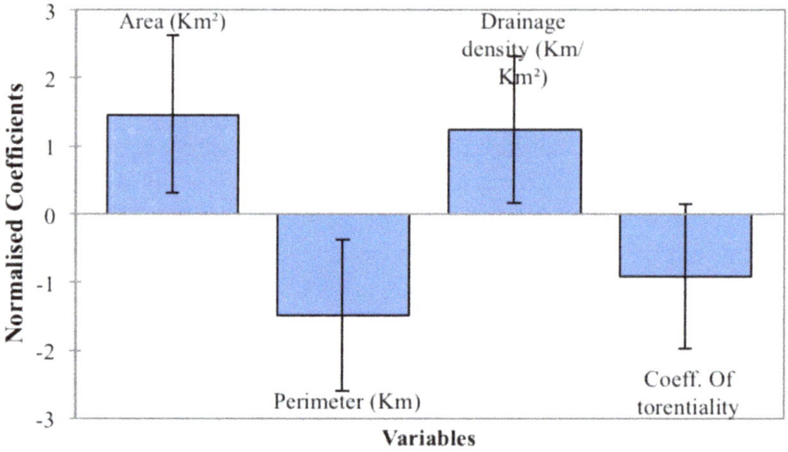

Fig. 6.5. 2nd Run, Normalized coefficients (95% confidence interval).

The proposed correlated model is a function of four drained basin characteristics with a satisfactory correlation according to the traditional statistical tests (Student and Fisher).

$$M = 1.21 + 7.75E^{-4}Ar - 9.87E^{-3}Per + 4.05E^{-1}D_d - 1.61E^{-2}C_t \tag{6.7}$$

The relationship between actual and predicted M and Φ(M) values are illustrated in the Fig. 6.6 and Fig. 6.7, respectively.

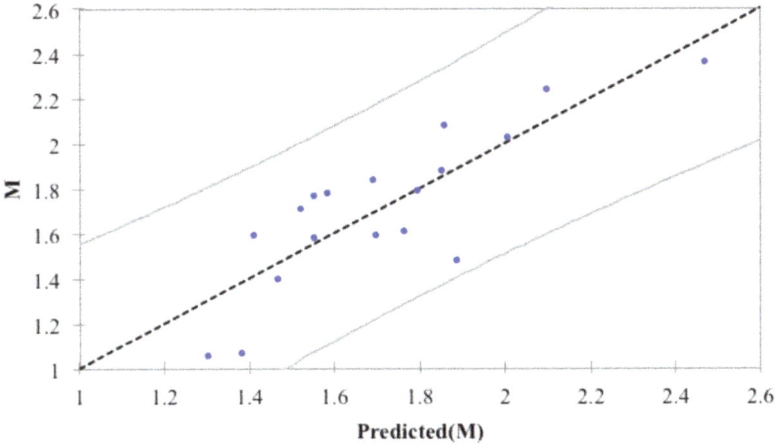

Fig. 6.6. Real /Predicted M values (95% confidence interval).

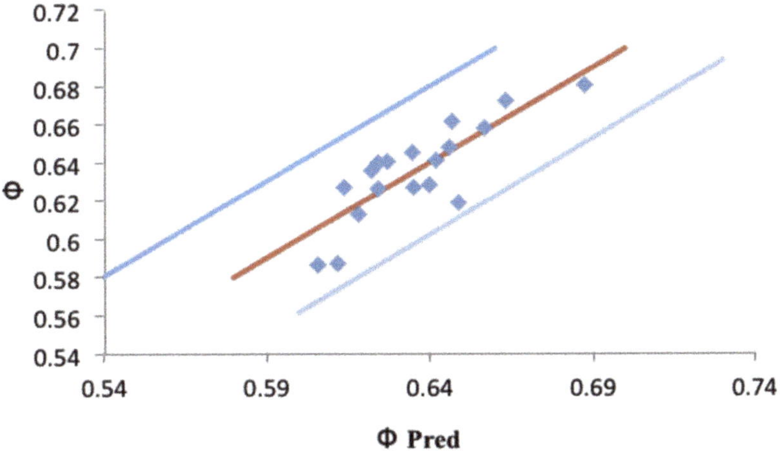

Fig 6.7. Real/Predicted Φ values (95% confidence interval).

As can be seen, all points are close to the straight line (1:1) and into the confidence interval (95%).

In order to validate that model, the four neighboring gauged stations of Altairac, Ababsa, Reghaia and Pont D9 are considered (See Table 6.2).

We observe that only ALTAIRAC is out of the confidence interval (see Fig. 6.8), while the other three stations are inside.

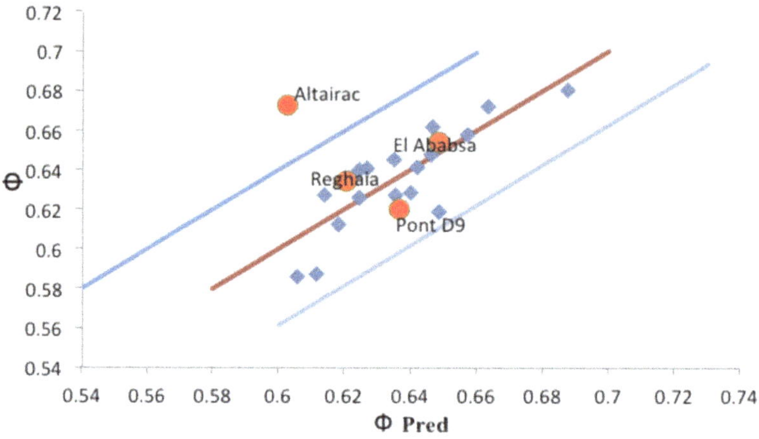

Fig 6.8. Real/Predicted Φ values (95% confidence interval). Validation.

Figure 6.9 shows the maximum error in estimating M value for all river sites and, as can be seen, only for Altairac gauged site the error exceeds 10%, that however is not so bad if one things that no streamflow information was used to estimate it.

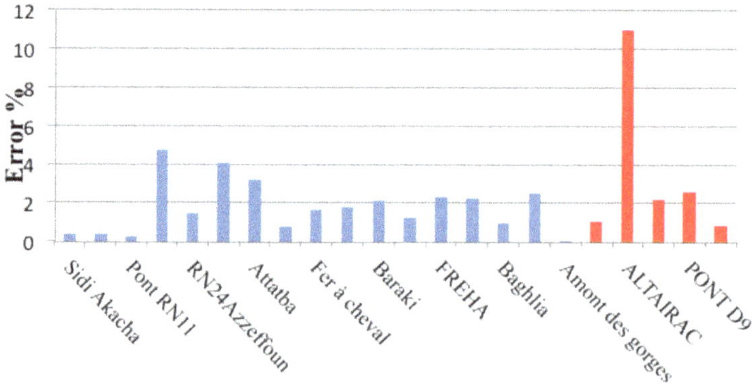

Fig 6.9. Φ (M) estimated errors (black bars for validation sites).

6.7 CONCLUSION

The entropy model was shown by many papers to be a sound tool for overcoming the problem of streamflow measurements for high flows allowing to achieve a reliable estimation of discharge just sampling the maximum flow velocity and knowing the entropy parameter M.

Besides that, this work also shows the robustness of M parameter in estimating some hydraulic parameters linked to the energy slope and the Manning's roughness and of interest for the conventional streamflow monitoring techniques.

For ungauged river sites, a statistical model to estimate M as a function of geomorphological characteristics of the site's subtended basin is proposed here. The approach is based on a simple correlation equation between the entropy parameter M and four drained basin characteristics identified through a multi-linear statistical procedure. For the validation of the model, streamflow data of four gauged sites not used in testing procedure are used. The application of the developed model shows a good performance in terms of $\Phi(M)$ estimation with an error less than 5% and only for one river site the error exceeds 10%. Overall the application of the model to ungauged sites may be very appealing for the flow monitoring considering that no information in terms of streamflow data is used. Future steps will be the application of the statistical model to other watersheds with different geomorphological characteristics of the ones investigated here.

6.8 REFERENCES

Ammari, A. 2012. *Vulnérabilité à l'envasement des Barrages (cas du bassin hydrographique des côtiers Algérois*). Doctorate of Science thesis. Biskra University. Algeria.

Ammari, A., and Remini, B. 2010. Estimation of Algerian rivers discharges based on Chiu's equation. *Arab Journal of Geosciences*. 3:1:59-65.

Ardiclioglu, M., Atabay, S., and Omran, M. 2007. Estimation of discharge and shear stress distribution in natural channels based on entropy concept. *32nd IAHR Congress, Venise*. Italy.

Chow, V. T. 1959. *Open-Channel Hydraulics*. McGraw-Hill, NY.

Chiu, C.L. 1987. Entropy and probability concepts in hydraulics. *Journal Hydraulic Engineering*. ASCE. 133:5:583–600.

Chiu, C.L. 1988. Entropy and 2-D velocity distribution in open channels. *Journal Hydraulic Engineering*. ASCE. 114:7:738–756.

Chiu, C.L. 1989. Velocity distribution in open channel flow. *Journal Hydraulic Engineering*. ASCE 115:5:576–594.

Chiu C.L., and Chen, Y.C. 1999. Efficient methods of measuring discharges and reservoir sediment inflow. *The international workshop of risk analysis in dam safety assessment*, Taipei, Taiwan: 97–116.

Chiu C.L., and Hsu, S.M. 2006. Probabilistic approach to modeling of velocity distributions in fluid flows. *Journal of Hydrolology*. 316:28–42.

Chiu, C.L., and Said, C.A.A. 1995. Maximum and mean velocities and entropy in open channel flow. *Journal of Hydraulic Engineering*. ASCE. 121:1:26–35.

Fenton, J.D. 2002. The application of numerical methods and mathematics to hydrography. *Proc. of the 11thAustralian Hydrograhic Conference*. Sydney. July 2002.

Moramarco, T., Ammari, A., Burnelli, A., Mirauda, D. and Pascale, V. 2008. Entropy Theory Application for Flow Monitoring in Natural Channels. iEMSs 2008: *International Congress on Environmental Modelling and Software*.

Moramarco, T., Saltalippi, C., Singh, V.P. 2004. Estimation of mean velocity in natural channels based on Chiu's velocity distribution equation. *Journal of Hydraulic Engineering*. ASCE.9:1:42-50.

CHAPTER 7

INFLUENCE OF HYDRAULIC GEOMETRY RATIOS ON THE ENTROPY PARAMETER IN OPEN CHANNEL FLOW

Michele Greco
School of Engineering, University of Basilicata, Potenza, Viale dell'Ateneo Lucano 10, 85100, Italy, tel.: +39 0971 205199, fax.: +39 178 2277113,
e-mail: michele.greco@unibas.it.

ABSTRACT

The way to evaluate the entropy parameter M through the ratio between medium and maximum velocity, $\Phi(M)$, still represents a relevant issue feeding a reach discussion among researchers. The invariance of $\Phi(M)$ for sections along the same river only at high flow seems to be consistent while for low stage $\Phi(M)$ can be affected by the influence of roughness, through the relative submergence as well as the aspect ratio related to the cross section geometry. Moving from these bases, using laboratory and field data, the classical hydraulic relationships on entropy velocity profile, uniform flow and regime theory, a predictor for entropy parameter is proposed for open channel flow. The work proposes general relationship existing between the parameter $\Phi(M)$ and the relative submergence taking into account the aspect ratio of the flow. The relationship has been applied to a set of experimental velocity data collected both in laboratory and in field, showing a good response of the theoretical model but selecting different behaviour dependent on the roughness scale. In fact, $\Phi(M)$ is strongly influenced by the ratio depth/roughness for values less than 4 when large and intermediate roughness scale occurs, while it might be assumed almost constant for small roughness scale.

Keywords: entropy velocity profile, submergence, roughness, aspect ratio, water discharge.

7.1 INTRODUCTION

The knowledge of flow velocity distribution is an essential requirement in dealing with stage–discharge relationships, sediments transport processes and prediction of morphological behaviour in alluvial streams, design of stable channels, flood control works and mathematical and physical modelling of flows.
Knowing, even to predict, discharge is a necessity for water resources management, hydraulic design, hydrologic analysis, drought and flood forecasting as well as water quality monitoring.

Measurements of discharge at gaging stations require information about the mean velocity in a number of sub-sections across the river/channel and knowledge of cross-section geometry at measuring locations. Multiple depth and velocity measurements are taken by current meter across the channel to be able to calculate the total stream discharge at a given moment. Such discharge values are used to define the cross-section rating curve, which represents an essential operative tool for river flow management during flood events especially.

In natural rivers, velocity distribution is affected by channel geometry, vegetation and roughness. Due to the limitations of classical hydraulic methods, Chiu (1987) derived the velocity distribution law basing on the concept of informational entropy introduced by Shannon (1948) and Tsallis (1988). Such velocity profile has been widely employed in many different flow cases and improved by relevant and meaningful both theoretical and applied contributions derived from robust experimental knowledge. Main aspect of such model is related to the need of one parameter M said as entropy parameter. Such parameter is depending on the ratio between the mean cross section velocity over maximum velocity, $\Phi(M)$. This velocity distribution has been widely employed in many different flow cases and enriched by relevant and meaningful both theoretical and applied contributions derived from robust experimental knowledge (Chiu 1989, 1991; Chiu et al. 1995; De Araújo et al. 1998; Greco, 1999, 2000, 2013; Greco and Mirauda, 2004, 2014; Greco et al. 2014; Moramarco et al. 2004; Mirauda et al. 2011; Chiu and Hsu, 2006; Marini et al. 2011). Main aspect of such a model is related to the need of just one parameter M known as entropy parameter. Such parameter is depending on the ratio between the mean cross section average velocity and the maximum one, $\Phi(M)$.

The assessment of $\Phi(M)$, in order to apply the entropy velocity profile, still represents a relevant issue feeding a reach debate among researchers mainly addressed to the reasonable ground about the invariance of $\Phi(M)$, thus M, for sections along the same river (Xia, 1997; Ardiclioglu et al. 2007; Moramarco and Singh, 2010). Therefore M should be assumed as a peculiar characteristic not only of the monitored site but also of the river reach where sites are located. This result might be very interesting to derive the rating curve to be employed or river monitoring and control at all the stages. In fact, the knowledge of M and, hence, of the entropic relationship can be crucial to accomplish the item that velocity measurements during high floods can be made only in the upper portion of flow area where the maximum velocity occurs.

On the other hand, the same experimental results which ensure to assume constant the entropy parameter when high stages or floods occur, allow us to achieve a possible uncertainty concerning the variability of the ratio between the mean and maximum velocities during shallow flows there where the role played by bed roughness on the flow asset is relevant.

Based on this last item, the paper deals with the analytic-theoretical derivation of the relationship between $\Phi(M)$ and the relative submergence, here defined as the ratio D/d, using classical open channel hydraulics equations. The $[\Phi(M)\text{-}(D/d)]$ relationship has been applied to experimental velocity data collected both in laboratory and in field, outlining a good response of the proposed model enlightening different response dependent on the roughness scale. In fact, $\Phi(M)$ is strongly influenced by the ratio depth/roughness for values of D/d less than 4 when large and intermediate roughness scale occur (Bathurst et al. 1981, 1985), while it might be assumed almost constant to 0.66 for small roughness scale ($D/d>4$) according to those observed during high stages and floods (Moramarco and Singh, 2010; Greco, 2013).

Furthermore, the relationship $\Phi(M)$-B/D, where B is the flow width, seems to be conditioned by whether or not the flow is confined, like artificial channel, instead of natural cross section. The comparison between the two set of data, laboratory versus field, enlighten the effect of the aspect ratio which is strongly related to $\Phi(M)$ for flume velocity data while it results not dependent on $\Phi(M)$ for river measurements. Further, even this last issue enforce the difference between the $\Phi(M)$ ratio behaviours for flow in presence of high roughness and low roughness, remaking the value of *D/d=4* as operative threshold for natural open channel flow.

This result is very interesting not only from a scientific point of view but, indeed, improves the suitability of the entropy velocity distribution at high stage in shallow water flows as well, enforcing the model as a strongly operative tool.

7.2 ENTROPY VELOCITY DISTRIBUTION AND RELATIVE SUBMERGENCE

Based on the informational entropy concept introduced by Shannon (1948), Chiu (1987) proposes the entropy velocity distribution in a free surface flow as:

$$u = \frac{U_{max}}{M} ln \left[1 + (e^M - 1) \frac{\xi - \xi_0}{\xi_m ax - \xi_0} \right] \quad (7.1)$$

where u is the time-averaged and, therefore, time invariant velocity on an iso-velocity with the assigned value ξ, in which u is 0 at ξ_0, corresponding to the channel boundary, and u reaches the maximum value, U_{max}, at ξ_{max}, which may occur at or below the water surface. M is the dimensionless entropy parameter introduced in the entropy-based derivation (Chiu, 1988; Chiu and Said, 1995; Chiu and Tung 2002; Luo and Singh, 2011; Cui and Singh, 2013) and can be used as a measure of uniformity of probability of the velocity distributions. The value of M can be determined by the mean, U_m, and maximum velocity values derived from the following equation:

$$\Phi(M) = \frac{U_m}{U_{max}} = \left(\frac{e^M}{e^M - 1} - \frac{1}{M} \right) \quad (7.2)$$

This parameter has proved to be really powerful for characterizing and comparing various patterns of velocity distributions and the status of the flow system. The mean velocity value, the location of such a velocity (Chiu and Said, 1995; Chiu and Tung, 2002) and the energy coefficient (Chiu, 1992) can be derived through the knowledge of M. Further, a linear relation between mean and maximum velocities was observed by

Xia (1997) collecting the velocity data in some cross-sections of the Mississippi River and later worldwide accepted by the hydraulic researcher community (De Araújo et al. 1998; Greco, 1999, 2000, 2013; Greco and Mirauda, 2014; Moramarco et al. 2004; Mirauda et al. 2011; Chiu and Hsu, 2006; Moramarco and Singh, 2010; Marini et al. 2011; Luo and Singh, 2011; Cui and Singh, 2013; Greco et al. 2014).

Equation (7.2), indeed, signifies the fundamental relationship, in terms of operational procedure, of the entropy velocity distribution and results extremely fruitfully tool to evaluate the entropy parameter passing through the knowledge of the ratio between mean and maximum velocities, $\Phi(M)$.

Therefore, by the way of classical hydraulic and open channel flow equations, (Rouse, 1950; Streeter, 1961) an easily explaining of the relationship among mean velocity, maximum velocity, dynamic and geometric characteristics of the flow can be addressed. That is, considering the Chezy equation, the mean velocity might be derived as follow:

$$U_m = C\sqrt{gR_iS_o} = Cu_* \tag{7.3}$$

in which $u_* = \sqrt{gR_iS_o}$ is the shear velocity, R_i is the hydraulic radius, S_0 is the energy slope, g is the gravity acceleration and C is the dimensionless resistance coefficient.

Moreover, C can be expressed through the Colebrook equation, or similar, extended to the open channel flow in fully developed turbulent flown like:

$$C = \frac{1}{k} ln\left(\frac{13.3\varphi R_i}{\varepsilon}\right) \tag{7.4}$$

where k is the Von Karman constant, ε is the equivalent bottom roughness and φ is the shape coefficient. Thus, assuming that $R_i \propto D$ and $\varepsilon \propto d$, where D is the water depth and d is the characteristic bottom roughness height (i.e. d_{50} or d_{84}), thus Eq. (7.4) can be written as:

$$C = \frac{1}{k} ln\frac{D}{d} + \frac{1}{k} lnC_0 \tag{7.5}$$

with C_0 is dimensionless coefficient introduced in order to express C as function of D/d. The Eq. (7.3) becomes:

$$\frac{u_m}{u_*} = \frac{1}{k} ln \frac{D}{d} + \frac{1}{k} ln C_0 \tag{7.6}$$

Equation (7.2) shows that the parameter M can be estimated, for the investigated cross section, on basis of pairs (U_m; U_{max}) of available data from measurements sampling. It is necessary to point out that U_{max} is unknown, but it can be considered as the maximum value in the data set of velocity points sampled during the velocity measurements. The maximum velocity prompts an important message about channel flow as it defines the range of velocity distribution. The location of maximum velocity is of interest, as the maximum velocity does not always occur at the water surface, but might be located below it, allowing the dip phenomenon due to secondary currents (Nezu and Nakagawa 1993). Thus, in open channel flow the secondary motion might transport the low momentum fluid from the near bank to the middle of flow and the high-momentum fluid from the free surface toward the bed. The resulting velocity distribution has been widely investigated by Yang et al. (2004) who propose the following dip-modified log-law for open channel flow:

$$\frac{u(y)}{u_*} = \frac{1}{k} ln \frac{y}{y_0} + \frac{\alpha}{k} ln \left(1 - \frac{y}{D}\right) \tag{7.7}$$

in which y is the distance from the bottom and α is the dip-velocity ($\alpha = D/y_{max} - 1$ with y_{max} the distance from the bottom at which U_{max} occurs).

Thus, combining Eqs. (7.6) and (7.7) results:

$$\frac{u_{max}}{u_*} = \frac{1}{k} ln \left(\frac{D}{y_0 (1+\alpha)}\right) + \frac{\alpha}{k} ln \left(\frac{\alpha}{1+\alpha}\right) \tag{7.8}$$

where y_0 is the location where the velocity profile assumes the zero value, generally adopted to be proportional to the bottom roughness height, d, as suggested by Ferro (2003):

$$y_0 = C_\xi d \tag{7.9}$$

in which C_ξ is empirical parameter, therefore:

$$\frac{u_{max}}{u_*} = \frac{1}{k} ln\left(\frac{D}{d}\right) + \frac{1}{k} ln\left(\frac{\alpha^\alpha}{C_\xi (1+\alpha)^{1+\alpha}}\right) \tag{7.10}$$

Lastly, the ratio between Eq. (7.5) and Eq. (7.10) explicitly outlines Φ(M) as log-dependent on the relative submergence, *D/d*:

$$\Phi(M) = A_\Phi \, ln \frac{D}{d} + B_\Phi \tag{7.11}$$

where A_Φ and B_Φ are numerical coefficients.

Equation (7.11) clearly explains the possible effect of bed roughness on the entropy velocity distribution in open channel flow.

7.3 EXPERIMENTAL DATA ANALYSIS

The above mentioned relationship between the velocity ratio, Φ(M), and the relative submergence, *D/d*, has been applied to a wide volume of data, collected both in laboratory and in field. Such database is of considerable interest as it covers a wide interval of relative submergence, ranging from 1 up to 1.9. This is of importance for analyses addressed to the flow resistance estimation, such as pointed out by Bathurst (1985) who suggested different flow resistance types as a function of the relative submergence.

The data set, indeed, can be synthetically resumed in:

- *field data* collected on several monitoring cross sections along different rivers in Southern Italy: slope in the range 0.2-0.8%, water discharge in the range 3.7-9 mc/sec, mean sediment diameter, d_{50}, in the range 3-8.6 cm and relative submergence (here assumed $D/d = D/d_{50}$), in the range 5-17;

- *laboratory data* sampled on a rectangular flume with a regular bed roughness (wooden sphere, *d*=3.5 cm), slope in the range 0.05-1%, water discharge in the range 7-72 lts/sec and relative submergence (*D/d*), in the range 2-7 (Mirauda et al. 2011; Greco and Mirauda, 2014).

The set (U_m; U_{max}) is actually affected by the scaling effect due to the different physical domain existing between flume and river, as shown in

Fig. 7.1. In the same figure, the linear regressions differentiated among the two data set are reported and showing quite similar value of Φ(M).

Further, while the pairs (U_m; U_{max}) referred to field data seem to be distributed around the linear regression showing a uniform trend of spreading, the laboratory data allow us to observe a general tendency to overestimate the real ratio Φ(M) when the velocities decrease indicating a possible dependence of the velocity ratio on local flow conditions. Thus, assuming the relative submergence as influencing factor the variation in the value of Φ(M) and plotting this ratio versus the corresponding relative submergence *D/d*, it is possible to observe different behaviour acting upon the roughness scale (Fig. 7.2).

The velocity ratio Φ(M) is strappingly depending on the relative submergence for values of *D/d* less than 4 when large and intermediate roughness scale occurs (Bathurst et al. 1981; Bathurst, 1985), while the ratio Φ(M) might be assumed almost invariable for small roughness scale (*D/d*>4). This last issue agrees with that observed during high stages and floods by Moramarco and Singh (2010) which assume constant the ratio between mean and maximum velocities and close to 0.6.

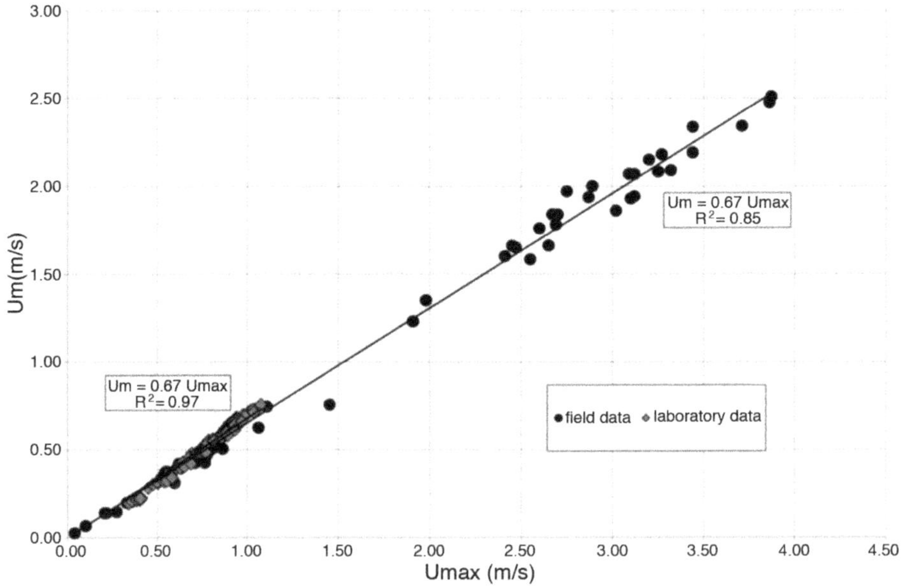

Fig. 7.1. Mean and maximum velocities observed in laboratory and field.

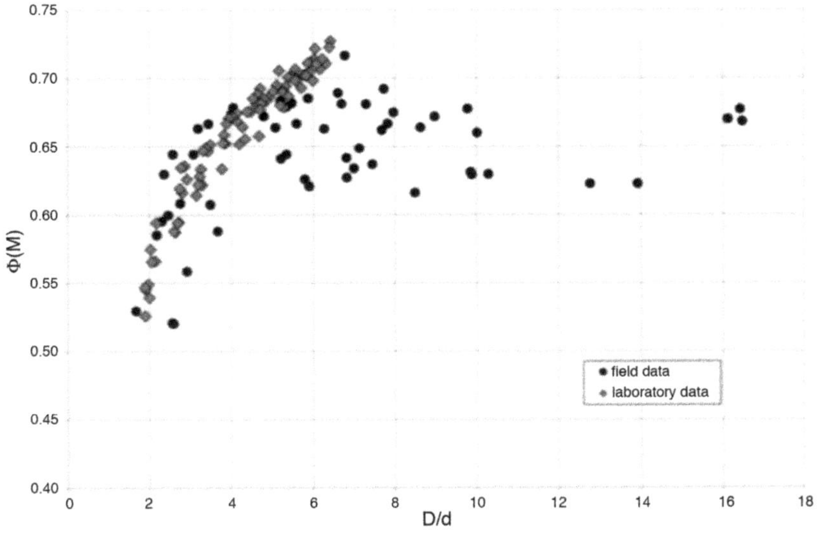

Fig. 7.2. Mean and maximum velocities ratio, Φ(M), versus relative submergence, D/d, for laboratory and field data set.

Thus Fig. 7.2 can be resumed in Fig. 7.3, in which two sub-domain might be demarcate fixing the value of the relative submergence equal to 4:

- large-intermediate roughness flow: the velocity ratio is dependent on the relative submergence both in cases of artificial and natural open channel flow;

- small roughness flow: the velocity ratio does not strictly depend on the relative submergence and can be assumed constant for natural open channel flow while maintains a certain dependence on the ratio *D/d* in case of artificial channel flow.

Furthermore, relevant inference arises from the experimental data analysis observing the mutual influence between relative submergence and aspect ratio.

The Fig. 7.4 reports such variables differentiated among field and laboratory data. How it is possible to point out, the two sets of data are covering two different scenarios with respect to the aspect ratio, *B/D*. In fact, as well known in literature (Ferro, 2003), the dip-velocity phenomenon is strongly influenced by the value of the aspect ratio ($B/D<6$) and tends to decrease as *B/D* increases.

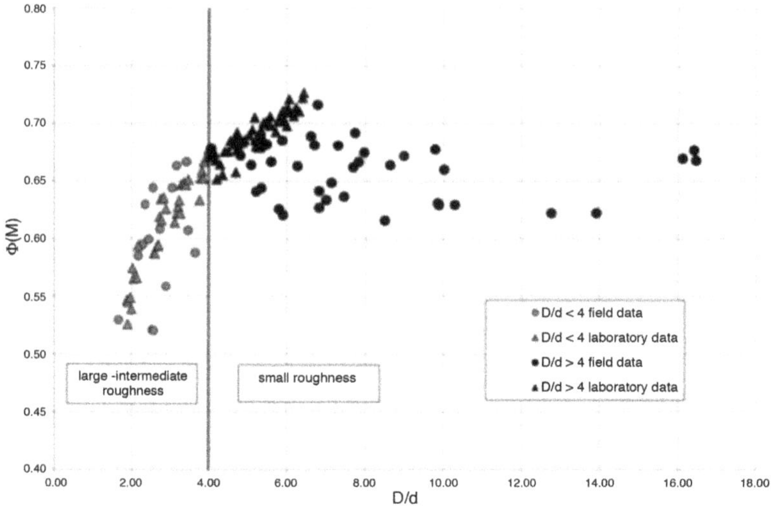

Fig. 7.3. **Velocity ratio and relative submergence among field and laboratory data selected by roughness scale.**

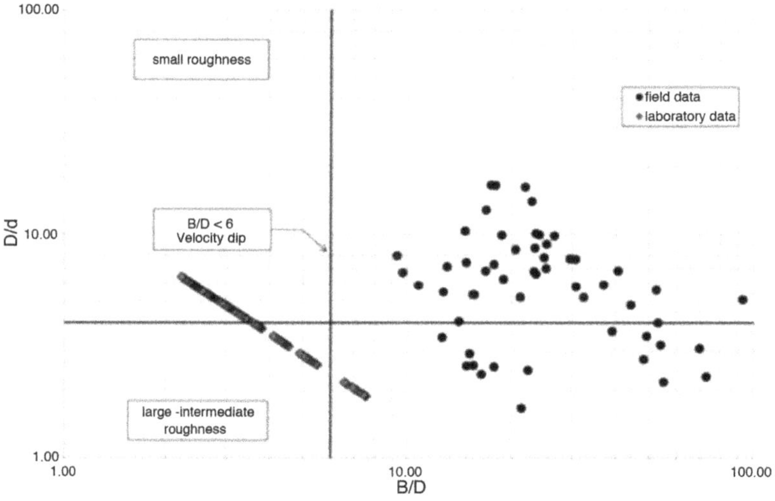

Fig. 7.4. **Relative submergence versus aspect ratio observed in field and laboratory data.**

In such a sense, while the laboratory data are enough representative of possible natural flow conditions in the case of large-intermediate roughness, this is not properly compelling in the case of small roughness flow due to the invariance of the flow width.

Once operated such filtering, keeping out the laboratory data with *D*/d>4, the theoretical approach discussed in the previous paragraph suggests a robust result in the analysis of real data, allowing us to assume the ratio between mean and maximum velocities as variable once shallow flow condition occurs.

The Eq. (7.11), indeed, can be apply to the present dataset, taking into account the constrain introduced by the two main regimes of large-intermediate roughness ($D/d<4$) and small roughness ($D/d>4$) obtaining that:

$$\begin{cases} A_{\Phi} = 0.16 & \\ B_{\Phi} = 0.44 & \Rightarrow \dfrac{D}{d} < 4 \\ & \text{and} \\ A_{\Phi} = 0 & \\ B_{\Phi} = 0.66 & \Rightarrow \dfrac{D}{d} > 4 \end{cases} \qquad (7.12)$$

The Fig. 7.5 plots on the whole dataset, undifferentiated between laboratory and field data, the Eq. (7.12). The result allows us to give more emphasis to employ entropy velocity law during operative activities in field. In fact, once defined the roughness condition, velocities measurements can be collected in the cross section at all stage levels, low, medium, high and flood, obtaining the maximum one (assumed as the maximum values between the whole measures) to be used in the Eq. (7.2) together with the value of $\Phi(\mathrm{M})$ corresponding to the observed relative submergence. Of course, for $D/d > 4$, $\Phi(\mathrm{M})$ is constantly equal to 0.66.

The clear view of the suitability of the proposed model arises from Fig. 7.6 which reports the observed mean velocity, U_m, for the whole investigated dataset and the corresponding values U_{mcomp}, obtained through the Eq. (7.2) using $\Phi(\mathrm{M})$ derived by the Eq. (7.1).

The comparison is strongly enlightening and reassures about an immediate use of the proposed model at all stages even if the field data present a wider spreading around the bisector than the laboratory data

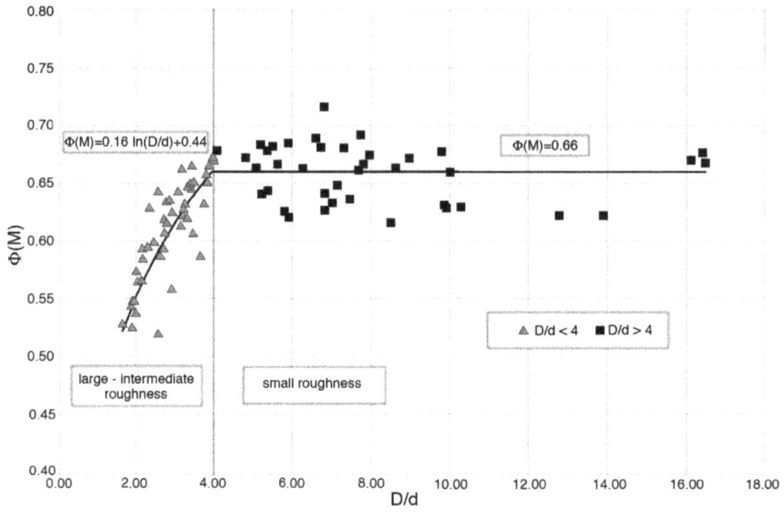

Fig. 7.5. **Generalized relationship between velocities ratio and relative submergence.**

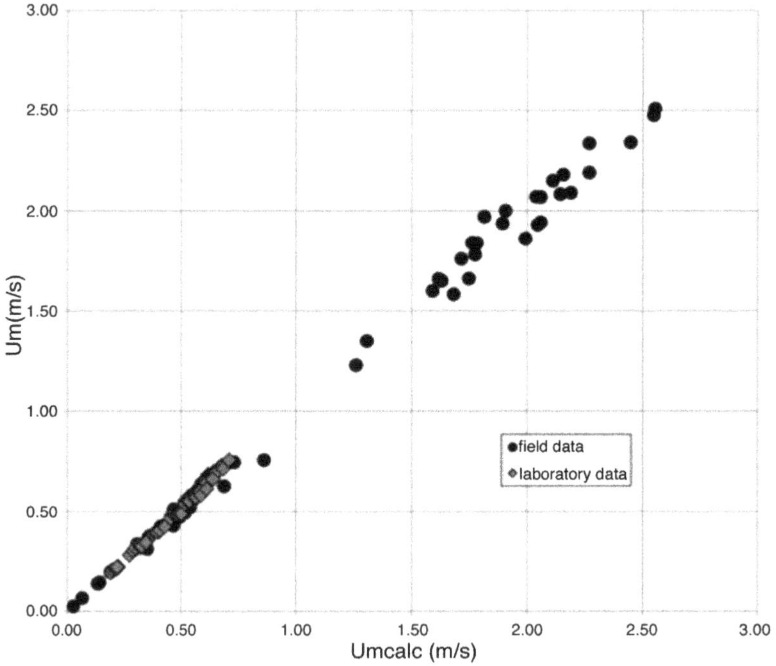

Fig. 7.6. Observed mean velocity versus computed mean velocity.

ones. In such a case, the better performance might be justified through the objective different level of data accuracy during the two measurement settings. Laboratory data collection can be assumed much more precise than the field data. Nevertheless the final response still remain notably good.

However, the robustness of operative response of the proposed model, obtained coupling Eq. (7.1) and Eq. (7.2), can be clearly outlined through the assessment of the observed water discharges. Fig. 7.7 shows the percentage errors between the observed water discharge and calculated one by the use of Eq. (7.1) and differentiated among field and laboratory data as well. The plot shows a very good response with relatively low values of uncertainty in any case confined in +/- 20%, even if laboratory data still present smaller spreading than field ones may be due to the better lab-measurement facilities.

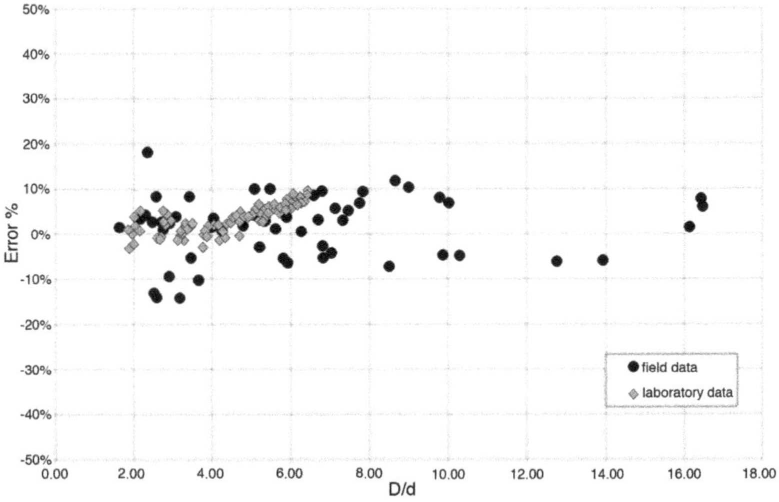

Fig.7.7. Observed percentage error in water discharge assessment.

Thus, Eq. (7.1) might be operatively employed during monitoring and forecasting procedures for water discharge assessment in natural and artificial open channel at all different stages low, medium and high flow or, rather, at different roughness scale small, intermediate and large.

The results can be easily extended to operative river monitoring activities.

7.4 CONCLUSION

A simple but effective theoretic-analytical formulation, based on entropy velocity profile law and classical relationships for uniform flow and friction factor, is developed to propose a general logarithmic relationship existing between the parameter $\Phi(M)$, defined as mean cross section velocity over maximum velocity, and the ratio water depth/bed roughness (D/d).

The [$\Phi(M)$-(D/d)] relationship has been applied to a set of experimental velocity data collected both in laboratory and in field, showing a good response of the theoretical model and selecting different behaviour depending on the roughness scale. The ratio $\Phi(M)$ results strongly depending on the ratio depth/roughness for values of D/d less than 4, when large and intermediate roughness scale occurs, while it might be assumed constant to 0.66 for small roughness scale (D/d>4) and flooding stages according to literature. The result allow us to give more emphasis to employ entropy velocity law during operative activities in field.

A comparison between the measured and the evaluated discharges indicates good performance of the proposed model improving the suitability of the entropy velocity distribution not only at high stage but in shallow water flows as well, enforcing the model as a strongly operational tool.

7.5 REFERENCE

Ardiclioglu, M., Atabay, S., Omran, M. (2007) Estimation of discharge and shear stress distribution in natural channels based on entropy concept, *32nd IAHR Congress*, Venice, Italy.

Bathurst, J.C., Li, R.M. & Simons, D.B. (1981) Resistance equation for large-scale roughness, *J. Hydraulics Division*, 107(HY12), pp. 1593-1613.

Bathurst, J.C. (1985) Flow resistance estimation in mountain rivers, *J. Hydr. Eng.*, 111(4), 625-643.

Chiu C.-L. (1987) Entropy and probability concepts in hydraulics *J. Hydr. Eng.*. ASCE, 113 (5), 583-600.

Chiu C.-L. (1989) Velocity distribution in open channel flow *J. Hydr. Eng.*. ASCE, 115 (5), 576-594.

Chiu C.-L. (1991) Application of entropy concept in open channel flow *J. Hydr. Eng.*. ASCE, 117 (5), 615-628.

Chiu, C.-L., and Hsu, S. M. (2006), Probabilistic approach to modelling of velocity distributions in fluid flows, *J. Hydrol. Eng.*, 316(1-4), 28-42.

Chiu C.-L., and Said C.A., (1995) Maximum and mean velocity and entropy in open channel flow *J. Hydr. Eng.*. ASCE, 121 (1), 26-35.

Cui H. and Singh V. P., (2013) Two-Dimensional Velocity Distribution in Open Channels Using the Tsallis Entropy. *J. Hydrol. Eng.*, 18:3, 331-339

Chiu, C. L., and Tung, N. C. (2002), Maximum velocity and regularities in open-channel flow, *J. Hydr. Eng.*, 128(8), 803-803.

De Araújo J.C., Chaudhry F.H. (1998) Experimental evaluation of 2-D entropy model for open channel flow, *J. Hydr. Eng.*. ASCE, 124 (10), 1064-1067.

Ferro, V. (2003) ADV measurements of velocity distributions in a gravel-bed flume, Earth Surface Processes and Landforms, 28, 707-722.

Greco M. (1999): Entropy velocity distribution in a river, *Proceedings of I.A.H.R. Symposium on River Coastal and Estuarine Morphodynamics*, Vol. II, 121-130, 6-10 September, Genoa;

Greco M. (2000): Entropy velocity profile: application in natural river, Environmental Engineering & Health Sciences: Proceedings of the International *Symposium on Environmental Engineering and Health Sciences: A Joint Effort for the XXI Century*, Water Resources Publications, LLC., ISBN: 1-887201-17-3, 397-405, OCT 26-30, 1998, Cholula, Mexico;

Greco M. and Mirauda D., (2004) Expeditive methodology for river discharge evaluation, *Second International Conference on Fluvial Hydraulics - RIVER FLOW 2004*, Napoli, Italia, (eds.) Greco, Carravetta & Della Morte, Taylor & Francis Group, London, ISBN 90 5809 658 0, Vol. 1, pp. 487-493;

Greco M., (2013) Effect of bed roughness on 1-D entropy velocity distribution in open channel flow, *Hydrology Research*, ISSN 1998-9563 doi:10.2166/nh.2013.122

Greco M. and Mirauda D., (2014): Entropy parameter estimation in large-scale roughness open channel, *J. Hydrol. Eng.*, ISSN: 1084-0699 eISSN: 1943-5584 doi: 10.1061/(ASCE)HE.1943-5584.0001009 (Apr. 10, 2014);

Greco M., Mirauda D., Volpe Plantamura A., (2014): Manning's roughness through the entropy parameter for steady open channel flows in low submergence, *Procedia Engineering*, ISSN 1877-7058, Published by Elsevier Ltd, doi: 10.1016/j.proeng.2014.02.084, Vol. 70, pp.773-780;

Luo, H. and Singh, V. (2011). Entropy Theory for Two-Dimensional Velocity Distribution. *J. Hydrol. Eng.*, 16(4), 303–315.

Marini, G., De Martino, G., Fontana, N., Fiorentino, M., and Singh, V. P. (2011). Entropy approach for 2D velocity distribution in open channel flow. *J. Hydraulic Res.* 49(6), 784–790.

Mirauda, D., Greco, M., and Volpe Plantamura, A. (2011). Influence of the entropic parameter on the flow geometry and morphology. Proc. World Academy of Science, Engineering and Technology, *ICWEE*, 1357–1362, Phuket, Thailand.

Moramarco, T. and Singh, V.P., (2010) Formulation of the entropy parameter based on hydraulic and geometric characteristics of river cross sections, *J. Hydr. Eng.*, 15, No. 10, pp. 852–858.

Moramarco, T., Saltalippi, C. & Singh V.P. (2004) Estimation of mean velocity in natural channels based on Chiu's velocity distribution equation, *J. Hydrol. Eng.*, 9(1), pp. 42-50.

Nezu, I. and Nakagawa, H. (1993), Turbulence in open-channel flows, *Balkema*, Rotterdam, Netherlands.

Rouse, H. (1950) Engineering hydraulics, *Proceedings of the fourth hydraulics conference*, Ed. Wiley.

Shannon, C.E. (1948). A mathematical theory of communication. *The Bell System Technical J.* 27, 623–656.

Streeter V.L. (1961), *Handbook of fluid dynamics*, McGraw-Hill, 1° edition

Tsallis, C. (1988), Possible generalization of Boltzmann-Gibbs statistics, *J. Stat. Phys.*, 52(1-2), 479-487.

Xia R. (1997) Relation between mean and Maximum velocities in a natural river, *J. Hydrol. Eng.*. ASCE, 123 (8), 720-723.

Yang, S.Q., Tan, S.K., Lim, S.Y., (2004) Velocity distribution and dip-phenomenon in smooth uniform open channel flows, *J. Hydr. Eng.*, 130, No. 12, pp.1179-1186.

CHAPTER 8

ASSESSING SOIL MOISTURE DYNAMICS TO CHARACTERIZE HYDROLOGICAL PATTERNS IN AN OLIVE TREE PLANTED CATCHMENT

Antonio J. Espejo-Pérez[1*], Juan V. Giráldez[1,2], Karl Vanderlinden[3], Luca Brocca[4], Aura Pedrera[3], and Gonzalo Martínez[3]

[1] Dept. of Agronomy, University of Córdoba. da Vinci Bldg, Cra Madrid km 396, 14071, Córdoba, Spain. *g82espea@uco.es

[1\2] Institute of Sustainable Agriculture, CSIC, Córdoba, Spain.

[3] IFAPA, Centro Las Torres-Tomejil, Seville, Spain.

[4] Research Institute for Geo-Hydrological Protection, National Research Council. Perugia, Italy.

ABSTRACT

The soil moisture dynamics of a small mid mountain experimental catchment in southern Spain is described. The catchment's annual average rainfall is 861 mm yr^{-1} due to the influence of the atmospheric exchange in the neighborhood of the Gibraltar Strait. The soil formed on the decomposing shallow rock has a great sand content, exhibiting a great variation enhanced by the presence of tree roots and by agricultural operations. Soil under olive tree canopies presented a lower bulk density and moisture content, and a higher saturated hydraulic conductivity than the soil in the inter row areas, mainly as a result of compaction produced at the harvest time. The rainfall-runoff response has been explored with two simple models that closely describe the evolution of soil moisture. The performance of the models evaluated with the Nash-Sutcliffe coefficient was above 0.90 for soil water and 0.60 for runoff volume.

8.1 INTRODUCTION

The frequent dry periods in semiarid areas, interspersed with irregular rain bursts, make effective soil and water control difficult in agricultural and urban settings, and complicate landscape planning. Advances in technology for soil and water monitoring allow to instrument and monitor entire catchments to improve the interpretation of their behaviour (*e.g.* Spence, 2010), which is necessary for the purpose of advising on better soil and water management practices in agriculture.

The rainfall-runoff response of a catchment depends on the soil water dynamics (Hrachowitz et al. 2013). Despite its importance, historical series of soil water measurements are not generally available, in contrast to rainfall or other meteorological data (Palecki and Bell, 2013). In some cases (Camici et al. 2011) a simple probabilistic model can be adopted to generate soil water content data. In other cases soil water can be estimated through remote sensing (Massari et al. 2013). Recently the emergence of low-cost sensors provides another means of continuously monitor soil moisture with promising results for agro-environmental applications (Vereecken et al. 2008).

Nowadays a plethora of rainfall-runoff models is available, but these are often difficult to use due to the excessive number of parameters to be calibrated. Therefore the use of simple models can be more adequate for the description of hydrological processes with a few parameters (*e.g.* Brocca et al. 2011), but their internal validation is still required to assess their reliability and robustness.

Many Mediterranean catchments are cropped with olive trees, which are usually not considered in the interpretation of soil water dynamics and runoff generation. The purpose of this chapter is to describe of the most relevant results obtained in a small experimental olive tree planted catchment in southern Spain during the last four years, identifying hydrological patterns and the influence of the trees.

8.2 CATCHMENT DESCRIPTION

The experimental catchment is located in Setenil de las Bodegas (province of Cádiz), 36° 52.2' N, 5° 7.8' W, Fig. 8.1.

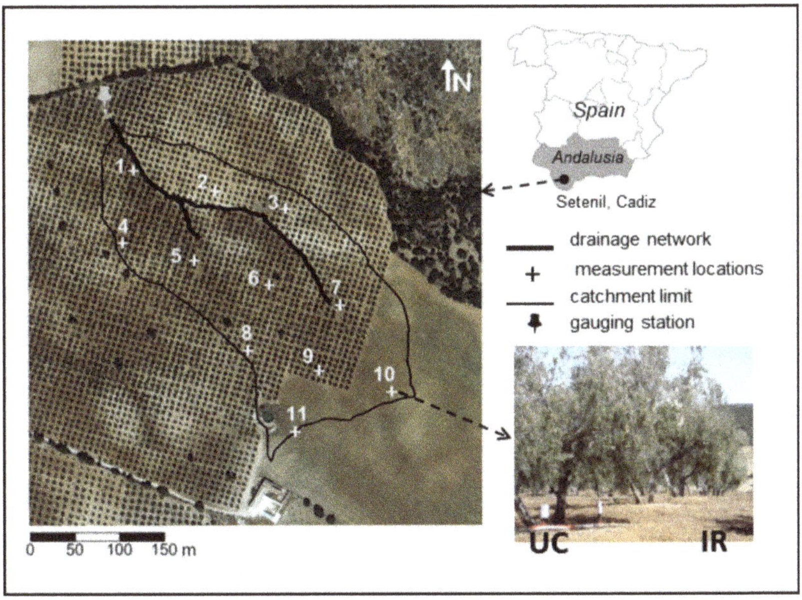

Fig. 8.1. Map location, aerial image of the experimental catchment with the 11 measurement locations where measurements are made at the inter-row areas, IR, and under the olive canopies, UC, and a detail of some of the trees in it.

The climate is sub-humid Mediterranean with an Atlantic influence. A representative daily evolution of rainfall, Penman-Monteith evaporation rate, ET, and average air temperature is plotted in Fig. 8.2. A wavy pattern is evident for air temperature and ET, reaching the greatest values at Summer season, with rates around 35ºC and 9 mm d^{-1}, respectively. Most of the rainfall occurred from October to April, leaving a prolonged summer rainless period. As a result the annual potential evaporation rate doubles the rainfall.

The soil subgroup is an intergrade between Lithic and Typic Rhodoxeralf (Soil Survey Staff, 1999, pp. 269-270; García del Barrio et al. 1971). The catchment is cropped with 18 year old olive trees, except for a small area in the south-eastern part of the catchment where trees are younger. In both zones, trees were planted on a 6 × 6-m grid. Until 2012 the soil was frequently tilled to remove weeds. Recently the farmer has reduced the tillage intensity shifting to minimum tillage, except for the area with young trees where the soil is still frequently tilled. The soil is

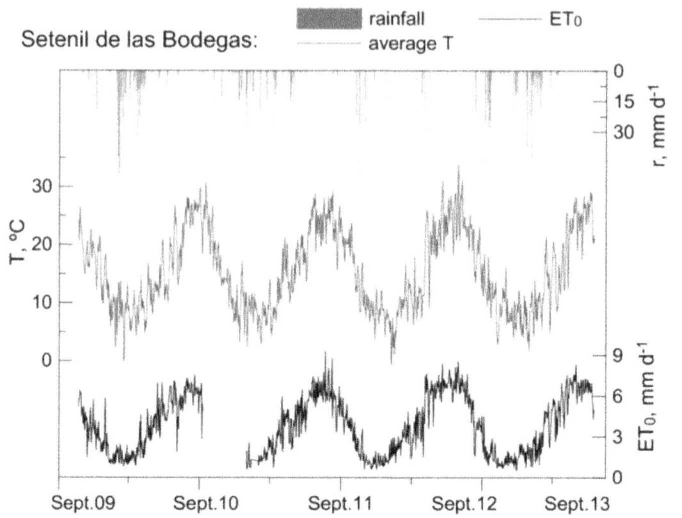

Fig. 8.2. Climate characteristics of Setenil de las Bodegas.

formed on a shallow bedrock consisting of calcarenites which limits its depth from 0.05 to 1.2 m. Fig. 8.3 shows the digital elevation model (DEM), with a mean elevation and slope of 740 m and 10.5 % respectively. The catchment is drained by a gully running from the southeast to the northwest, with a depth of 2 m near the outlet.

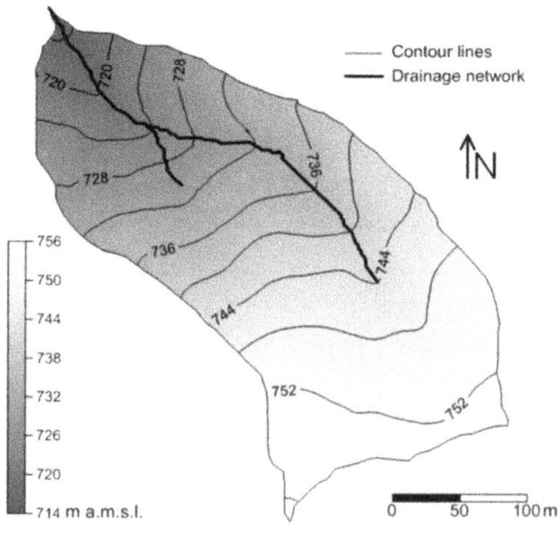

Fig. 8.3. Digital elevation model of Setenil catchment.

The catchment is equipped with a gauging station to measure runoff, (Taguas et al. 2009), and a rain gauge for rainfall, both recorded at one-minute intervals. The remaining meteorological data used in this study were collected at a nearby automated weather stations at hourly temporal resolution (CAP, 2013), located at a distance of 6.6 km from the catchment. Occasionally rainfall data gaps were filled using data collected by this station. A sensor network consisted of 98 10HS sensors devices (Decagon Devices, Pullman, WA) was used to measure soil water content at 5 minutes resolution, previously described by Espejo et al. (2014). The sensors were placed at 11 locations, and at each measurement location they were installed horizontally and oriented against the slope direction at depths of 0.05, 0.15, 0.25, 0.35 and 0.45 m if soil depth was sufficient, at the IR and UC areas. UC location was placed at a distance of 0.75 m from the tree trunk towards the south-east and IR in the north-east direction. Sensors were calibrated by using gravimetric soil moisture samples at both locations, IR and UC. No differences in the sensors performance were observed between IR and UC, and therefore a common exponential fit similar to the one proposed by Mittelbach et al. (2011) was adopted for them (R^2=0.73).

8.3 RAINFALL REGIME

A period of 4 years between September 1, 2009 and August 31, 2013 was analyzed. The rainfall events were determined excluding the events with a depth lower than 5 mm. Inter-event periods lasted at least 6 hours (Huff, 1967) and produced less than 2 mm rainfall. A two-parameter Gamma probability density function was fitted to rainfall depth, Fig. 8.4, (*e.g.* Bury 1999, Chapter 13):

$$f(x;\lambda,\sigma)=\frac{1}{\sigma\cdot\Gamma(\lambda)}\cdot\left(\frac{x}{\sigma}\right)^{\lambda-1}e^{-x/\sigma}\quad x\geq 0\quad \sigma,\lambda>0 \tag{8.1}$$

The parameters λ and σ are, respectively the shape, and scale factors. The values of the parameters fitted with the maximum likelihood method were λ=2.21 and σ=8.96. According to the χ^2 test the hypothesis of the gamma fit to these data could not be rejected at the 0.05 probability level.

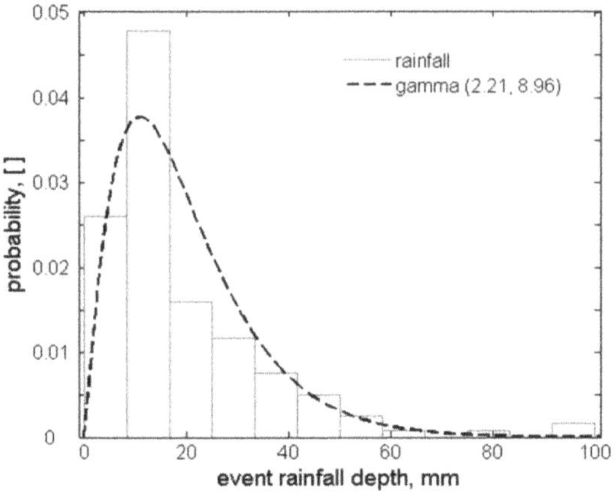

Fig. 8.4. Measured and fitted probability density function of event-rainfall depth.

The annual precipitation ranged from 345 mm in 2010-2011 to 1166 mm in 2009-2010. The 60-minute rainfall intensity, I_{60}, was 2.3 mm h^{-1}, and the 25th and 75th percentiles were 1.0 and 3.0 mm h^{-1}, respectively. Most of the rainfall occurred from October to April, Fig. 8.5. The largest amount of rainfall was accumulated from December to February, with a median value of near 110 mm. The largest dispersion in monthly rainfall was observed for December, with values ranging from 13 mm in 2011 to 416 mm in 2010. From June to August no rainfall was observed. The year 2011-2012 has an unusual dry winter period, receiving from December to March only 26 mm.

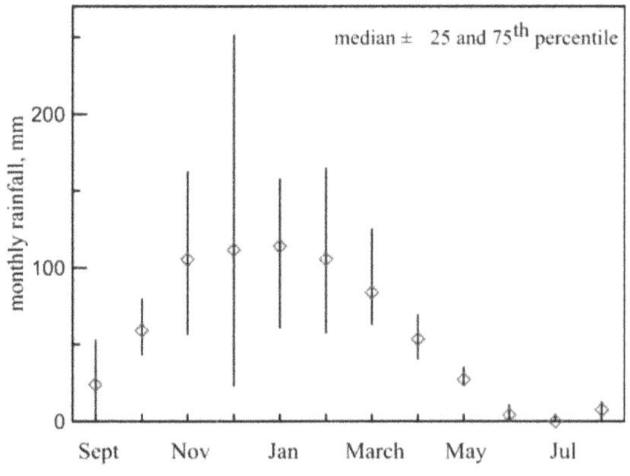

Fig. 8.5. Monthly distribution of rainfall.

8.4 EVOLUTION OF SOIL MOISTURE

Soil moisture showed a seasonal pattern in response to weather conditions, with a marked drying period during the summer, as indicated by Fig. 8.6 for the UC area at location 4. A similar pattern was observed at the other locations. Due to the relevance of the suction with respect to the gravitational gradient of soil water potential using the Boltzmann coordinate for the combined depth and time coordinates, Espejo et al. (2014) evaluated the hydraulic conductivity function of the soil for IR and UC at each location. Changes in topsoil moisture were greater than in deeper layers, highlighting the role of surface moisture in the energy exchange between soil surface and atmosphere (*e.g.* Betts, 2004).

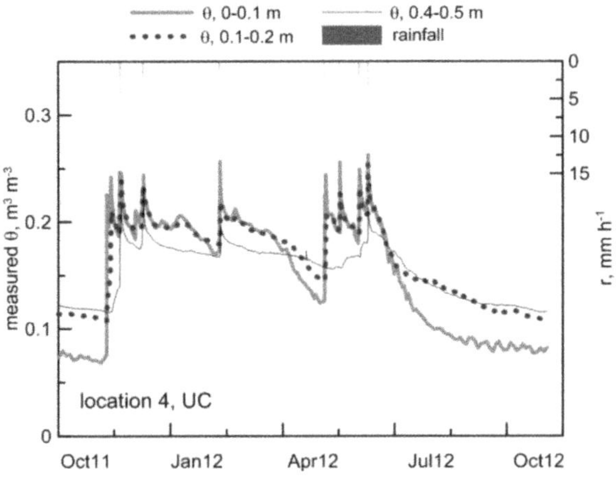

Fig. 8.6. Evolution of soil water content under the canopy, UC.

The performance of the soil water balance model (SWBM) developed by Brocca et al. (2008) with a single parameter, the saturated hydraulic conductivity, K_S, was investigated using observations from 2010 to 2013. The model was successful in reproducing hourly soil water dynamics in the top 0.2-m layer, with Nash-Sutcliffe and coefficients of determination of above 0.90 for UC and IR locations. Both the spatial mean soil water content, Fig. 8.7, and data collected at the individual monitoring points were used. Moreover, for UC sites, a more realistic representation of soil water dynamic was obtained when a simple formulation to describe the tree interception was introduced into the initial model scheme.

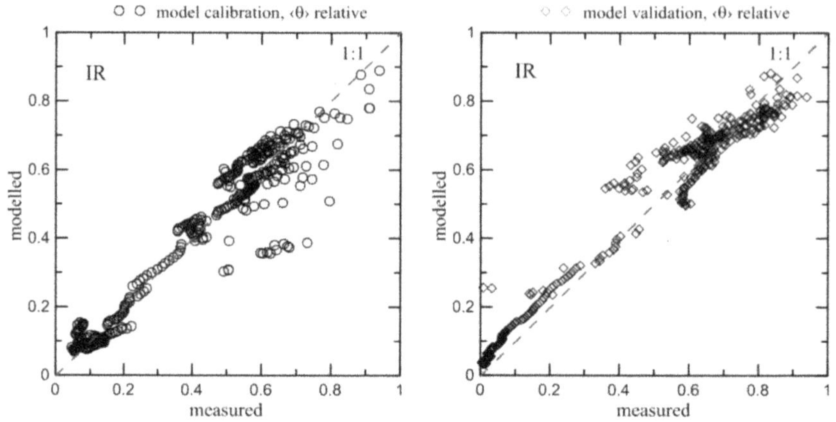

Fig. 8.7. Relationship between hourly modelled and measured spatial mean soil moisture, ‹θ›, in the top 0.2-m layer at the inter-row area, IR, for model calibration (2011-2012) and validation (2011-2013).

8.5 ANALYSIS OF THE OLIVE-CANOPY AREA INFLUENCE ON SOIL PROPERTIES AND SOIL WATER DYNAMICS

The average soil profile physical properties are summarised in Table 8.1, separately for IR and UC. The textural class of the soil is sandy loam, with profile-averaged clay and sand contents of 19 and 69%, respectively, at both IR and UC. The spatial mean profile bulk density was 1.60 Mg m^{-3} for both locations. However, the spatial mean 0.3-m topsoil bulk density was higher at IR as compared to UC, with average values of 1.74 and 1.59 Mg m^{-3} at IR and UC, respectively.

Topsoil K_S was measured in the laboratory with a constant head permeameter (EijkelkampAgrisearch Equipment, Giesbeek, The Netherlands). The average values were 2.9 and 16.7 cm h^{-1} at IR and UC, respectively. Soil depth ranged from 0.30 m, in the upslope areas north of the gully, to 0.88 m in the downslope area, with average values near 0.45 m at IR and UC.

Differences in soil water retention were found between IR and UC. The saturated water content ranged from 0.35 to 0.45 m^3 m^{-3} at IR and UC, respectively.

Table 8.1. Summary of the profile-averaged values of soil properties for inter row, IR, and under canopy, UC, areas.

	IR					
	z_{max}	clay	sand	ρ_b	k_s†	θ_{sat}†
location	m	%		Mg m^{-3}	cm h^{-1}	m^3 m^{-3}
1	0.80	19.3	69.8	1.67	41.2	0.33
2	0.30	12.4	68.2	1.67	13.2	0.32
3	0.50	20.4	64.4	1.68	7.5	0.35
4	0.30	20.1	69.9	-	8.8	0.35
5	0.76	17.7	68.2	1.59	1.4	0.39
6	0.60	19.5	68.5	1.69	0.5	0.35
7	0.70	19.2	70.9	1.70	1.1	0.31
8	0.20	18.9	72.9	1.53	0.4	0.43
9	0.50	18.4	65.9	1.56	0.5	0.28
10	0.30	20.5	70.5	1.51	1.2	0.35
11	0.30	23.6	65.8	1.57	20.1	0.37
average	0.48	19.1	68.6	1.62	2.9‡	0.35

	UC					
	z_{max}	clay	sand	ρ_b	k_s†	θ_{sat}†
location	m	%		Mg m^{-3}	cm h^{-1}	m^3 m^{-3}
1	0.70	18.0	70.3	1.71	43.7	0.39
2	0.30	13.0	65.4	1.33	87.9	0.44
3	0.30	20.9	63.3	1.71	79.6	0.41
4	0.49	19.8	70.3	1.65	10.1	0.42
5	0.70	16.8	70.3	1.60	30.2	0.49
6	0.60	18.1	70.9	1.56	2.9	0.57
7	0.88	19.5	70.8	1.73	9.7	0.37
8	0.50	19.7	71.1	1.52	5.2	0.49
9	0.30	20.6	70.9	1.66	14.7	0.54
10	0.50	22.8	67.2	1.61	10.9	0.42
11	0.40	24.4	64.8	1.55	13.2	0.46
average	0.52	19.4	68.7	1.60	16.7‡	0.45

†, average of the topsoil 0.20 m; ‡, in this case was used the geometric mean, and for the rest case the arithmetic mean.

At the different monitoring depths the measured soil water content was higher at IR than at UC for most of the two-year monitoring period. A comparison of the soil water profiles between IR and UC is presented during a recharge and discharge period in Fig. 8.8 for location 2, representative for all the monitoring locations. According to Espejo et al. (2014) the faster drying of the soil at UC as compared to IR can be explained by the lower soil water retention and higher k_s at UC. Similar results were found by Joffre and Rambal (1993), under Mediterranean evergreen oak trees located not far from this catchment. They attributed these differences to the higher suction and evaporative losses beneath canopies. Also Moran et al. (2010) observed that soil water content was lower at UC as compared to IR in scrubland, with increasing differences increased with depth. However, under similar conditions for the same plant species Bhark and Small (2003) reported that under canopies of grassland and shrubs frequently a higher soil water content was found as result of higher infiltration. Potts et al. (2010) found larger amounts of soil water beneath canopies of a mesquite canopy than at adjacent areas, and the differences increased with soil depth.

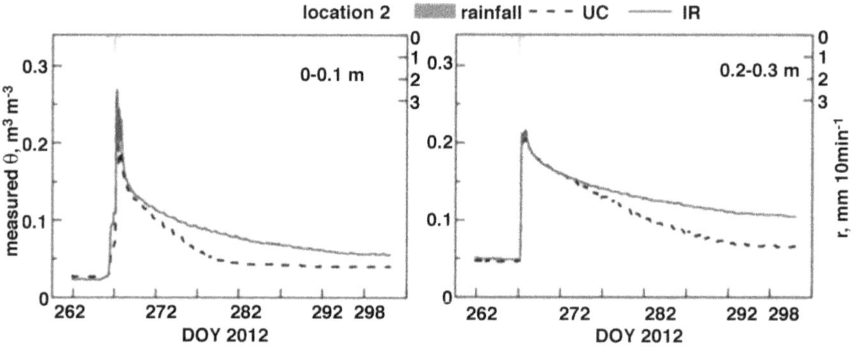

Fig. 8.8. Comparison of observed moisture profiles between the inter-row area, IR, and under the olive tree canopy, UC.

8.6 RUNOFF YIELD

The effect of antecedent soil water content on catchment discharge is displayed in Fig. 8.9. An apparent exponential relationship between accumulated runoff and antecedent soil moisture content was observed with a minimum threshold of around 0.60. These results indicate that the highest runoff losses are found in situations where the entire soil profile has been uniformly wetted.

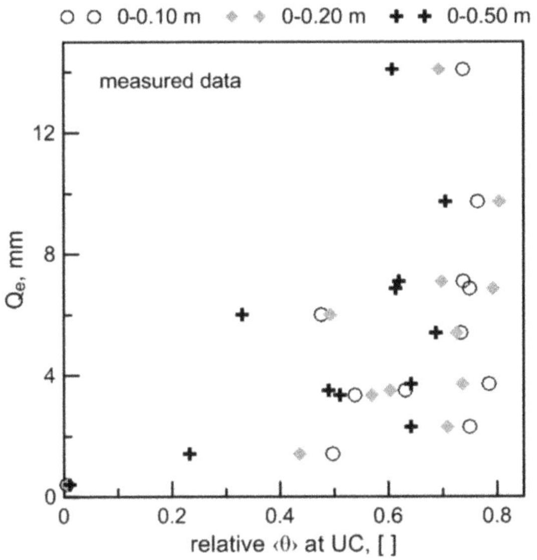

Fig. 8.9. Relationship between antecedent spatial mean measured, ‹θ›, soil moisture and runoff generation.

The lumped version of the MISDc model's developed by Brocca et al. (2011) has been applied to analyze the rainfall runoff response of this catchment. The model reliability and robustness were evaluated in terms of runoff generation. The model computes the soil water balance using the SWBM model. Also for runoff modelling the landscape was divided into IR and UC areas. For the latter an interception scheme was added to the conventional model. Calibration-validation results showed that the model performance was consistent in reproducing the ten-minute runoff hydrograph, with a Nash-Sutcliffe efficiency index near 0.65 at IR and at UC, respectively. No differences were found in modelled runoff between IR and UC. The annual average runoff ratio was 0.21, similar to that measured at the outlet. The model performance in reproducing accumulated runoff and peak flow is represented in Fig. 8.10, obtaining better results in the former.

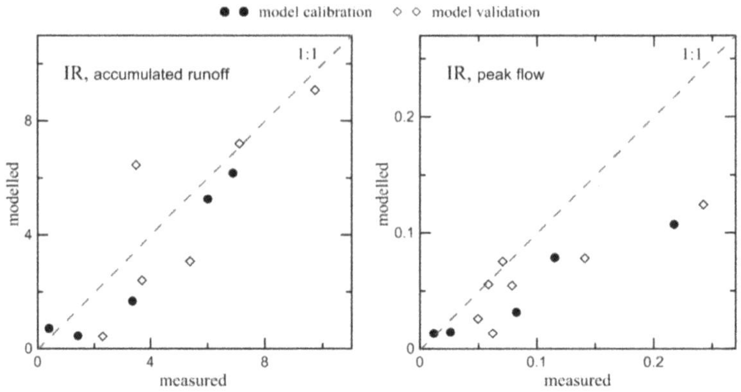

Fig. 8.10. Relationship between modelled and measured runoff for model calibration and validation at inter-row, IR, areas.

8.7 CONCLUDING REMARKS

The rainfall depth was described by a Gamma probability distribution function with an average annual amount of 861 mm. The runoff coefficient was 0.21. The trees influenced the soil water dynamics, while runoff modelling did not reveal any effect on the catchment discharge. The Nash-Sutcliffe coefficient for simulating soil water and runoff was 0.90 and 0.62, respectively. The influence of trees on soil water dynamics could be explained by the differences between UC and IR for relevant soil properties. The largest differences between UC and IR were observed for 0.3-m topsoil bulk density, saturated hydraulic conductivity and water retention. Intensive field monitoring efforts, complemented with simple modelling approaches allowed us to improve our knowledge of hydrological processes in this catchment, and to advise on improvements in soil and water management strategies.

8.8 ACKNOWLEDGEMENTS

This work was funded by the Spanish Ministry of Economy and Competitiveness and FEDER, through Projects AGL2009-12936-C03-03 and AGL2012-40128-C03-03, and by the Junta de Andalucía through grant AGR-4782 and the predoctoral fellowship P-09. We thank the staff of the Soil and Water Group at IFAPA Las Torres-Tomejil for their assistance with installing and operating field equipment, field sampling and soil analysis, especially M. Morón, the Dept. of Agricultural Engineering of the University of Cordoba and the farm owner, Alonso Zamudio, who allowed us to keep the instruments in the catchment. The first author acknowledges the kind hospitality of CNR-IRPI during a research stage there.

8.9 REFERENCES

Betts, A. K. 2004. Understanding hydrometeorology using global models. *Bulletin of the American Meteorological Society*, 85, 1673-1688.

Bhark, E. W. and E. E. Small. 2003. Association between plant canopies and the spatial patterns of infiltration in shrubland and grassland of the Chihuahuan desert, New Mexico. *Ecosystems*, 6, 185-196.

Brocca, L., F. Melone and T. Moramarco. 2008. On the estimation of antecedent wetness condition in rainfall-runoff modelling. *Hydrological Processes,* 22, 629–642.

Brocca, L., F. Melone and T. Moramarco. 2011. Distributed rainfall-runoff modelling for flood frequency estimation and flood forecasting. *Hydrological Processes*, 25, 2801-2813.

Bury, K. A. 1999. Statistical distributions in Engineering. Cambridge Univ. Press. Cambridge. U.K.

Camici, S., A. Tarpanelli, L. Brocca, F. Melone and T. Moramarco. 2011. Design soil moisture estimation by comparing continuous and storm-based rainfall-runoff modelling. *Water Resources Research*, 47, W05527, doi:10.1029/2010WR009298.

CAP, Consejería de Agricultura y Pesca de la Junta de Andalucía. 2013. Red de Alerta e Informacion Fitosanitaria (RAIF). Available at: http://www.juntadeandalucia.es/agriculturaypesca/portal/servicios/estadisticas/servicio-de-informacion-agroclimatica/red-de-alerta-e-informacion/datos-de-las-estaciones-agroclimaticas.html, accessed on 06/05/2013.

Espejo, A. J., J. V. Giráldez, K. Vanderlinden, E. V. Taguas and A. Pedrera. 2014. A method for estimating soil water diffusivity from moisture profiles and its application across an experimental catchment. *Journal of Hydrology*, 516, 161-168.

García del Barrio, I., L. Malvárez and J. I. González. 1971. Mapas provinciales de suelos. Cádiz. Ministerio de Agricultura. Madrid. Spain.

Hrachowitz, M., H. H. G. Savenije, G. Blöschl, J. J. McDonnell, M. Sivapalan, J. W. Pomeroy, B. Arheimer, T. Blume, M. P. Clark, U. Ehret, F. Fenicia, J. E. Freer, A. Gelfan, H. V. Gupta, D. A. Hughes, R. W. Hut, A. Montanari, S. Pande, D. Tetzlaff, P. A. Troch, S. Uhlenbrook, T. Wagener, H. C. Winsemius, R. A. Woods, E. Zehe and C. Cudennec. 2013. A decade of predictions in ungauged Basins (PUB)-a review. *Hydrological Sciences Journal*, 58, 1198-1255.

Huff, F. A. 1967. Time distribution of rainfall in heavy storms. *Water Resources Research*, 4, 7-54.

Joffre, R. and S. Rambal. 1993. How tree cover influences the water balance of Mediterranean rangelands. *Ecology*, 74, 570-582.

Massari, C., L. Brocca, S. Barbetta, C. Papathanasiou, M. Mimikou and T. Moramarco. 2013. Using globally available soil moisture indicators for flood modelling in Mediterranean catchments. *Hydrology and Earth System Sciences*, 10, 10997–11033.

Mittelbach, H., F. Casini, I. Lehner, A. J. Teuling and S. I. Seneviratne. 2011. Soil moisture monitoring for climate research: Evaluation of a low cost sensor in the framework of the Swiss Soil Moisture Experiment (SwissSMEX) campaign. *Journal of Geophysical Research*, 116, D05111, doi:10.1029/2010JD014907.

Moran, M. S., E. P. Hamerlynck, R. L. Scott, J. J. Stone, C. D. Holifield Collins, T. O. Keefer, R. Bryant, L. DeYoung, G. S. Nearing, Z. Sugg and D. C. Hymer. 2010. Hydrologic response to precipitation pulses under and between shrubs in the Chihuahuan Desert, Arizona. *Water Resources Research*, 46, W10509, doi:10.1029/2009WR008842.

Palecki, M. A. and J. E. Bell. 2013. U.S. Climate reference network soil moisture observations with triple redundancy: measurement variability. *Vadose Zone Journal*, 13, doi:10.2136/vzj2012.0158.

Potts, D. L., R. L. Scott, S. Bayram and J. Carbonara. 2010. Woody plants modulate the temporal dynamics of soil moisture in a semi-arid mesquite savanna. *Ecohydrology*, 3, 20-27.

Soil Survey Staff. 1999. Soil Taxonomy, 2nd ed. USDA Agr. Hbk. 436, Washington, D.C.

Spence, C. 2010. A paradigm shift in hydrology: storage thresholds across scales influence catchment runoff generation. *Geography Compass*, 4, 819-833.

Taguas E. V., J. L. Ayuso, A. Peña, Y. Yuan and R. Pérez. 2009. Evaluating and modelling the hydrological and erosive behavior of an olive orchard micro-catchment under no-tillage with bare soil in Spain. *Earth Surface Processes and Landforms*, 34, 738-751.

Vereecken, H., J. A. Huisman, H. Bogena, J. Vanderborght, J. A. Vrugt and J. W. Hopmans. 2008. On the value of soil moisture measurements in vadose zone hydrology: A review. *Water Resources Research*, 44, W00D06, doi:10.1029/2008WR006829.

CHAPTER 9

EFFECTS OF CATCHMENT SUBDIVISION ON THE DIFFUSIVE WAVE TRANSFER FUNCTION FOR SEMI-DISTRIBUTED HYDROLOGICAL MODELLING

Roger Moussa[1]

[1]INRA, UMR LISAH, 2 Place Pierre Viala, 34060 Montpellier, France
Tel: +33 4 99 61 24 56, moussa@supagro.inra.fr

ABSTRACT

Semi-distributed hydrological models are often used because they require few spatial units, are more easy-to-parameterize than distributed models, and are well adapted to poorly gauged basins. However they require a spatial subdivision of the catchment adapted to the main processes. This spatial subdivision will impact the properties of the transfer function used to route free-surface flow on subcatchments and through the channel network. This paper aims to study the effects of catchment subdivision on the transfer function for various levels of basin subdivision in order to guide the modeller in the choice of the adequate spatial subdivision for semi-distributed modelling. The method is based on the comparison of GIUH transfer functions for different levels of catchment subdivision. The GIUH is calculated by routing an impulse response function from each subcatchment through the channel network using the diffusive wave model. The relationship between the spatial subdivision and the transfer function was studied for seven French basins. Results show that this methodology enables the modeller to compare the GIUH transfer function of various basins and to select the optimal parsimonious spatial subdivision of the basin into subcatchments that minimizes the number of subcatchments and gives acceptable results.

9.1 INTRODUCTION

In-between lumped and spatially distributed hydrological models (e.g. Abbott et al. 1986), semi-distributed models (e.g. Diskin and Simpson, 1978; Fortin et al. 2001; Moussa et al. 2007; Khakbaz et al. 2012) are often used for water resources management and flood protection because they are more easy-to-parameterize than distributed models, they are well adapted to poorly gauged basins and scarce hydrological data, and because they represent well the main hydrological processes which are generally simplified in lumped models. In all lumped, semi-distributed and distributed models, the transfer function plays an important role when modelling free-surface flows and flood events. Compared to distributed models where a fine spatial resolution is generally used, semi-distributed models require a subdivision of the catchment into few hydrological units (or subcatchments) adapted to the main processes and to the objective of the study. This spatial subdivision will impact the properties of the free-surface flow transfer function due to scale effects (Blöschl and Sivapalan, 1995; Sposito, 1998), and a question arises on what will be the effects of the spatial subdivision of the catchment on the transfer function? (Wooldridge et al. 2001; Ajami et al. 2004; Khakbaz et al. 2012)

The free-surface flow equations are based on the Saint-Venant equations or their approximations such as the diffusive wave equation or the kinematic wave equation. These equations depend on the basin morphology, the channel topology and the geometric and hydraulic properties of reaches. A first approximation of the transfer function can be calculated on the basis of the diffusive wave equation description of the dynamics incorporated in a geomorphologic instantaneous unit hydrograph (GIUH; see a synthesis in Rodriguez-Iturbe and Rinaldo, 1997; Snell and Sivapalan, 1994; Moussa, 1997).

This paper aims to study the effects of catchment subdivision on the transfer function for various levels of basin subdivision into subcatchments in order to guide the modeller in the choice of the adequate spatial subdivision for semi-distributed modelling. The methodology is based on a comparison of the GIUH transfer function calculated for various levels of catchment subdivision. Seven French basins with different channel network shapes were used in the applications. For each basin, various levels of basin subdivision were defined. Then an impulse rainfall was routed using the diffusive wave model in order to calculate the resulting hydrograph at the basin outlet, which represents the GIUH transfer function. The analytical solution of the diffusive wave equation with lateral flow uniformly distributed was used because it is well adapted to represent lateral flows on reaches and is unconditionally stable (Moussa, 1996; Moramarco et al. 1999; Fan and Li, 2006; Cimorelli et al. 2014). Finally, for each of the seven basins studied, the resulting GIUH transfer functions were compared for various levels of basin subdivision and analysed function of the level of subdivision and of the channel network shape.

9.2 THE DIFFUSIVE WAVE MODEL

9.2.1 General Equation

The diffusive wave equation describes both convection and diffusion and can be written as

$$\frac{\partial Q}{\partial t}+C\left(\frac{\partial Q}{\partial x}-q\right)-D\left(\frac{\partial^2 Q}{\partial x^2}-\frac{\partial q}{\partial x}\right)=0 \tag{9.1}$$

where x [L] is downstream distance, t [T] time, C [LT^{-1}] and D [L^2T^{-1}] are non-linear functions of the discharge Q(x,t) [L^3T^{-1}] and are generally known as celerity and diffusivity, respectively. The term q(x,t) [L^2T^{-1}] represents the lateral flow distribution. Let $Q_a(t)$ [L^3T^{-1}] be the total lateral flow hydrograph

$$Q_a(t)=\int_0^L q(x,t)\,dx \tag{9.2}$$

L [L] being the length of the river reach. For this problem the boundary conditions are Q(x,0) [L^3T^{-1}] and the upstream inflow Q(0,t) [L^3T^{-1}] with

$$Q(x,0)=Q(0,0)+\int_0^x q(\alpha,0)\,d\alpha \tag{9.3}$$

and for x=L

$$Q(L,0)=Q(0,0)+Q_a(0) \tag{9.4}$$

The diffusive wave equation depends on the two parameters, celerity and diffusivity. Numerical methods based on the finite-difference or finite element techniques are generally used to resolve the diffusive wave equation (Moussa and Bocquillon, 1996).

9.2.2 Solution of the Diffusive Wave Equation with Lateral Flow

In the particular case where C and D are constant, and without lateral inflow q(x,t)=0, Hayami (1951) by means of the perturbation has derived an analytical solution of Eq. (9.1). Moussa (1996) extended this solution to the case of lateral inflow (or outflow) uniformly distributed to a channel reach. In the particular case of a semi-infinite channel, no physical downstream boundary condition exists, and the diffusive wave equation has an analytical solution

$$Q(x,t)=\left[Q(0,0)+\frac{Q_a(0)}{L}x+\frac{C}{L}\int_0^t\left(Q_a(\lambda)-Q_a(0)\right)d\lambda\right]+$$
$$\frac{x}{2(\pi D)^{1/2}}exp^{Cx/2D}\times$$
$$\int_0^t\left[Q(0,t-\tau)-Q(0,0)-\frac{C}{L}\int_0^{t-\tau}\left(Q_a(\lambda)-Q_a(0)\right)d\lambda\right]\times$$
$$\frac{exp^{-Cx/4D\left(x/C\tau+C\tau/x\right)}}{\tau^{3/2}}d\tau \tag{9.5}$$

Let I(t) and O(t) be respectively the upstream inflow minus the base flow and the downstream outflow minus the base flow, we have

$$I(t)=Q(0,t)-Q(0,0) \tag{9.6}$$

$$O(t)=Q(L,t)-Q(L,0) \tag{9.7}$$

and let K(t) $[T^{-1}]$ be the Hayami kernel function defined as

$$K(t)=\frac{L}{2(\pi D)^{1/2}}\frac{exp^{CL/4D(2-(L/Ct)-(Ct/L))}}{t^{3/2}} \quad \text{with } \int_0^\infty K(t)dt=1 \tag{9.8}$$

We define a function $\Phi(t)$ that characterizes the lateral flow as

$$\Phi(t) = \frac{C}{L} \int_0^t \left(Q_a(\lambda) - Q_a(0) \right) d\lambda \tag{9.9}$$

With the notations Eqs. (9.6), (9.7), (9.8) and (9.9), the general Eq. (9.5) can be expressed as

$$O(t) = \Phi(t) + \left(I(t) - \Phi(t) \right) * \mathrm{K}(t) \tag{9.10}$$

The symbol (*) represents the convolution relation. In the particular case of flood routing without lateral flow ($Q_a(t) = 0 \Rightarrow \Phi(t) = 0$), the Eq. (9.10) is reduced to a simple convolution integral:

$$O(t) = I(t) * K(t) = \int_0^t I(\tau) K(t - \tau) d\tau \tag{9.11}$$

This analytical solution is a linear model unconditionally stable (Eqs. 9.10 and 9.11), requires only an upstream boundary condition, which makes it very convenient to implement for networks of channel reaches (Moramarco et al. 1999; Fan and Li, 2006; Cimorelli et al. 2014).

9.3 CATCHMENT SUBDIVISION INTO SUB-CATCHMENTS

The transfer function proposed herein is based on a subdivision of the catchment into subcatchments (defined as surface zones) connected to reaches (defined as linear elements). An impulse function over each subcatchment represents the input hydrograph overland flow. Hydrographs are routed on subcatchments and through the channel network using the analytical resolution of the diffusive wave Eq. (9.10). The total volume of input hydrographs is chosen equal to one, and consequently the transfer function, defined as the calculated hydrograph at the basin outlet, is a unit hydrograph.

9.3.1 Topological Structure

Figure 9.1.a shows an example of a basin subdivided into subcatchments with: i) six nodes: three source nodes (N_1, N_2 and N_3), two internal nodes

(N_4 and N_5) and the outlet (N_6); ii) five reaches: three source reaches (R_1, R_2 and R_3) and two internal reaches (R_4 and R_5); iii) thirteen subcatchments: three source subcatchments (So_1, So_2 and So_3), five subcatchments corresponding to right-banks of reaches (RB_1, RB_2, RB_3, RB_4 and RB_5) and five subcatchments corresponding to left-banks of reaches (LB_1, LB_2, LB_3, LB_4 and LB_5).

All reaches (internal and source) are linked to a right-bank and a left-bank subcatchment; only source reaches are linked to a source subcatchment. Let N_r be the total number of reaches, N_{rs} the total

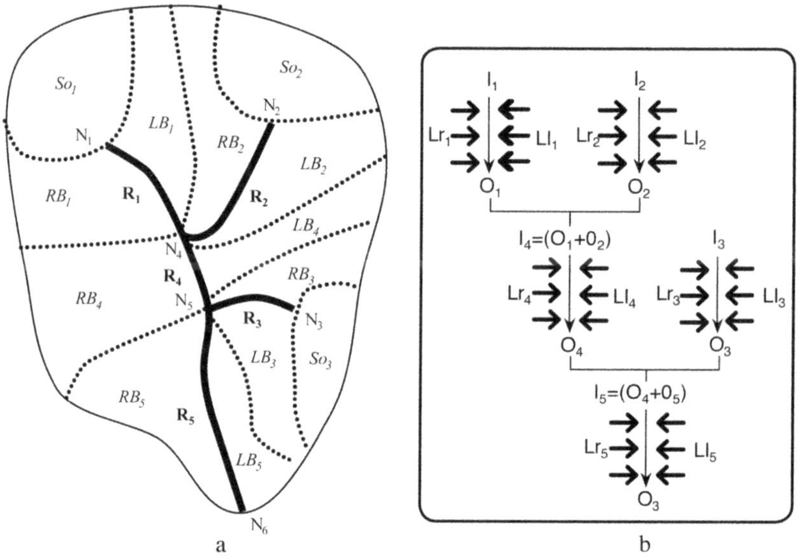

Fig. 9.1. a) Example of a channel network encoding: nodes (N_i), reaches (R_i in bold) and subcatchments (right-bank *RB_i*, left-bank *LB_i* and source basin *So_i* in italic). b) Topological structure of the transfer function scheme corresponding to sketch (a): inflow (I), outflow (O) and lateral flow (right-bank Lr and left-bank Ll) uniformly distributed on reaches.

number of source reaches and N_{ri} the total number of internal reaches. We have $N_r = N_{rs} + N_{ri}$ (in the example $N_r = 5$, $N_{rs} = 3$ and $N_{ri} = 2$). An internal node can have more than two-entry reaches but only one exit-reach. In the particular case where each internal node has only two entry-reaches, we have $N_{ri} = N_{rs} - 1$, and consequently the total number of source, right-bank and left-bank subcatchments is equal to $3\ N_{rs} + 2\ N_{ri} = 5\ N_{rs} - 2$ (in the example shown in Fig. 9.1a). Each reach 'i' ($1 \leq i \leq N_r$) is characterized by:

Nu_i : number of upstream node;

Nd_i : number of downstream node;

L_i : length [L];

S_i : mean slope [-];

As_i : area of the source subcatchment So_i [L^2];

Ar_i : area of the right-bank subcatchment RB_i [L^2];

Al_i : area of the left-bank subcatchment LB_i [L^2];

Ds_i : mean distance from the source subcatchment So_i to the reach R_i [L];

Dr_i : mean distance from the right-bank subcatchment RB_i to the reach R_i [L];

Dl_i : mean distance from the left-bank subcatchment LB_i to the reach R_i [L];

Ss_i : mean slope of the source subcatchment So_i [-];

Sr_i : mean slope of the right-bank subcatchment RB_i [-];

Sl_i : mean slope of the left-bank subcatchment LB_i [-].

Let A_o [L^2] be the total area of the basin

$$A_o = \sum_{i=1}^{N_r} \left(Ar_i + Al_i + As_i \right) \tag{9.12}$$

Note that for internal reaches, the terms As_i, Ds_i and Ss_i are not defined. The subdivision of the catchment into subcatchments and the calculation of their geometrical characteristics are made automatically from Digital Elevation Models (DEMs).

9.3.2 Basin Subdivision from DEMs

DEMs are first used to extract the drainage network. The D8 method used herein compares the elevation of each DEM pixel with the elevations of its eight adjacent pixels (Band, 1986; Gurnell and Montgomery, 1999). A flow direction is then assigned to point towards one of the eight adjacent pixels to which the steepest down-slope path exists. This phase of analysis generates two data sets. The first gives the flow direction for each pixel, or the drainage network and the second represents the flow accumulation data set. Then, in order to subdivide the basin into subcatchments for semi-distributed modelling, the method often used consists on defining the main channel network as pixels draining an area higher than a threshold area 'A_t' (Moussa, 2008). When A_t decreases the number of subcatchments and of reaches increase; the particular case $A_t = A_0$ corresponds to only one unit as for lumped modelling. It is recognized that different threshold areas A_t will result in substantially different spatial subdivision of the basin into subcatchments.

9.3.3 Subcatchments Characteristics Calculated from DEMs

We consider the channel network as a set of linear reaches and the subcatchments as a set of surface zones. The delimitation of subcatchments supposes that each pixel (belonging to the catchment) must belong to one of four categories: a source subcatchment (So_i), a right-bank subcatchment (RB_i), a left-bank subcatchment (LB_i) or a channel zone (R_i) of a reach number 'i'. Geometrical properties are then calculated from DEMs. In order to avoid a surface zone represented by the channel zone R_i, this latest is equally divided into right-bank and left-bank subcatchments (RB_i and LB_i). Once each pixel of the catchment is affected to one of the four categories, the geometrical properties of subcatchments are calculated. Let:

- a : grid size [L];
- d_k : hydrological distance through the flow path between a pixel P_k and the reach R_i [L];
- s_k : local slope of the pixel [-];

n_{Soi} : number of pixels belonging to the source subcatchment So_i [-] ;

n_{RBi} : number of pixels belonging to the right-bank subcatchment RB_i [-];

n_{LBi} : number of pixels belonging to the left-bank subcatchment LB_i [-];

n_{Ri} : number of pixels belonging to the channel zone R_i [-];

As'_i : area of the source subcatchment So_i [L^2];

Ar'_i : area of the right-bank subcatchment RB_i [L^2];

Al'_i : area of the left-bank subcatchment LB_i [L^2];

Ac'_i : area of the channel zone R_i [L^2];

Ds'_i : mean distance from the source subcatchment So_i to the reach R_i [L];

Dr'_i : mean distance from the right-bank subcatchment RB_i to the reach R_i [L];

Dl'_i : mean distance from the left-bank subcatchment LB_i to the reach R_i [L];

Dc'_i : mean distance from the channel zone R_i to the reach R_i [L];

Ss'_i : mean slope of the source subcatchment So_i [-];

Sr'_i : mean slope of the right-bank subcatchment RB_i [-];

Sl'_i : mean slope of the left-bank subcatchment LB_i [-];

Sc'_i : mean slope of the channel zone R_i [-].

The terms (As'_i, Ds'_i and Sc'_i), corresponding to source subcatchments, are not defined for an internal reach (e.g. R_4 and R_5 on Fig. 9.1.a). We have:

$$
\begin{aligned}
As'_i &= n_{Soi}\, a^2;\\
Ar'_i &= n_{RBi}\, a^2;\\
Al'_i &= n_{LBi}\, a^2;\\
Ac'_i &= n_{Ri}\, a^2;
\end{aligned}
\tag{9.13}
$$

$$
\begin{aligned}
Ds'_i &= \frac{1}{n_{Soi}} \sum_{P_k \in So_i} d_k;\\
Dr'_i &= \frac{1}{n_{RBi}} \sum_{P_k \in RB_i} d_k;\\
Dl'_i &= \frac{1}{n_{LBi}} \sum_{P_k \in LB_i} d_k;\\
Dc'_i &= 0
\end{aligned}
\tag{9.14}
$$

$$
\begin{aligned}
Ss'_i &= \frac{1}{n_{Soi}} \sum_{P_k \in So_i} s_k;\\
Sr'_i &= \frac{1}{n_{RBi}} \sum_{P_k \in RB_i} s_k;\\
Sl'_i &= \frac{1}{n_{LBi}} \sum_{P_k \in LB_i} s_k;\\
Sc'_i &= \frac{1}{n_{Ri}} \sum_{P_k \in R_i} s_k
\end{aligned}
\tag{9.15}
$$

The mean slope S_i of a each R_i is given by

$$
S_i = \frac{Z_{Nui} - Z_{Ndi}}{L_i}
\tag{9.16}
$$

where Z_{NUi} [L] is the altitude of the upstream node Nu_i, Z_{NDi} [L] the altitude of the downstream node Nd_i, and L_i [L] the distance between Nu_i and Nd_i. In order to simplify, we consider that the reaches are vectors with an area equal to zero. The area of a channel zone Ac'_i of a reach R_i, is equally divided between the right-bank and the left-bank. Using notations of Section 9.3.1, we obtain:

$$As_i = As'_i \; ; \; Ar_i = Ar'_i + \frac{Ac'_i}{2} \; ; \; Al_i = Al'_i + \frac{Ac'_i}{2} \tag{9.17}$$

We suppose that half-number of channel-pixels will belongs to the right-bank and the other half to the left-bank. Each half pixel is located at a mean distance equal to a/4 from the pixel centre. This leads:

$$\begin{aligned} Ds_i &= Ds'_i \; ; \\ Dr_i &= \frac{1}{n_{RBi} + \frac{n_{Ri}}{2}} \left(\sum_{P_k \in RB_i} d_k + \frac{1}{2} \sum_{P_k \in R_i} \frac{a}{4} \right) \; ; \\ Dl_i &= \frac{1}{n_{LBi} + \frac{n_{Ri}}{2}} \left(\sum_{P_k \in LB_i} d_k + \frac{1}{2} \sum_{P_k \in R_i} \frac{a}{4} \right) \end{aligned} \tag{9.18}$$

$$\begin{aligned} Ss_i &= Ss'_i \; ; \\ Sr_i &= \frac{1}{n_{RBi} + \frac{n_{Ri}}{2}} \left(\sum_{P_k \in RB_i} s_k + \frac{1}{2} \sum_{P_k \in R_i} s_k \right) \; ; \\ Sl_i &= \frac{1}{n_{LBi} + \frac{n_{Ri}}{2}} \left(\sum_{P_k \in LB_i} s_k + \frac{1}{2} \sum_{P_k \in R_i} s_k \right) \end{aligned} \tag{9.19}$$

For each value of the threshold A_t, this procedure enables to automatically extract from DEMs the topological structure used for semi-distributed modelling and for the calculation of the GIUH transfer function: the channel network (topology, length, slope) and the subdivision of the basin into subcatchments (topology, area, mean distance to a reach, slope).

9.4 THE GIUH DIFFUSIVE WAVE TRANSFER FUNCTION

A three-step conceptual approach to calculate the GIUH transfer function is proposed: i) definition of the impulse function on each subcatchment; ii) routing the impulse function from the subcatchment to the reach using the diffusive wave model; iii) routing hydrographs through the channel network using the diffusive wave model with lateral flow uniformly distributed (Fig. 9.1b): the GIUH transfer function is defined as the calculated hydrograph at the basin outlet.

9.4.1 Definition of the Impulse Function

A schematic input hydrograph is defined proportional to a Hayami kernel function H(t)

$$H(t)=\left(\frac{\theta z}{\pi}\right)^{1/2}\frac{exp^{z(2-(t/\theta)-(\theta/t))}}{\left(\frac{t}{\theta}\right)^{3/2}} \quad \text{with} \quad \int_0^{\infty} H(t)\,dt=1 \qquad (9.20)$$

where θ [T] is a time parameter that represents the center of gravity of the hydrograph, z [-] a form parameter, and $\pi = 3.1416$. Let $Qs_i(t)$, $Qr_i(t)$ and $Ql_i(t)$ $[L^3T^{-1}]$ be respectively the hydrographs on the source, the right-bank and the left-bank subcatchment of a reach ‘i’ ($Qs_i(t)$ is not defined for internal reaches); these hydrographs are defined proportional to the correspondent subcatchment area:

$$Qs_i(t)=\frac{As_i}{A_0}H(t) \quad Qr_i(t)=\frac{Ar_i}{A_0}H(t) \quad Ql_i(t)=\frac{Al_i}{A_0}H(t) \qquad (9.21)$$

The two parameters θ and z are considered identical for all subcatchments of all reaches 'i'. With this definition of schematic hydrographs, there is no base flow because $Qs_i(0) = Qr_i(0) = Ql_i(0) = 0$. Combining Eqs. (9.12), (9.20) and (9.21), we show that the total volume (V $[L^3]$) of all input hydrographs is equal to 1:

$$V = \sum_{i=1}^{N_r} \left(\int_0^\infty Qs_i(t)\,dt + \int_0^\infty Qr_i(t)\,dt + \int_0^\infty Ql_i(t)\,dt \right) = 1 \tag{9.22}$$

9.4.2 Routing the Impulse Function from the Subcatchment to the Reach

The second step consists on routing hydrographs $Qs_i(t)$, $Qr_i(t)$ and $Ql_i(t)$ $[L^3T^{-1}]$ from the source, right-bank and left-bank subcatchments to the reach 'i'. Let $I_i(t)$, $Lr_i(t)$ and $Ll_i(t)$ $[L^3T^{-1}]$ be the routed hydrographs respectively. We suppose that the mean distance between the source, right-bank and left-bank subcatchments and the reaches is Ds_i, Dr_i and Dl_i respectively. Using the diffusive wave model without lateral inflow Eq. (9.11), we obtain:

$$\begin{aligned} I_i(t) &= Qs_i(t) * Ks_i(t); \\ Lr_i(t) &= Qr_i(t) * Kr_i(t); \\ Ll_i(t) &= Ql_i(t) * Kl_i(t) \end{aligned} \tag{9.23}$$

where $Ks_i(t)$, $Kr_i(t)$ and $Kl_i(t)$ are the Hayami kernel functions obtained by substituting in Eq. (9.8) the term L by Ds_i, Dr_i, and Dl_i respectively. This model needs the two parameters C and D of the diffusive (see Section 9.4.4).

9.4.3 Routing Hydrographs Through the Channel Network

Figure 9.1b shows the scheme used for routing hydrographs through the channel network of Fig. 9.1a. For a reach 'i' of the channel network, $I_i(t)$ is the input hydrograph to the channel, $Lr_i(t)$ and $Ll_i(t)$ are the input of lateral inflow from right-bank and left-bank subcatchments respectively, and $O_i(t)$ [L^3T^{-1}] the output hydrograph. Hydrographs are routed through the channel network from upstream to downstream and added at each internal node of the channel network. Using the diffusive wave model with lateral flow Eq. (9.10), $O_i(t)$ is given using Eqs. (9.10) and (9.9):

$$O_i(t) = \Phi_i(t) + \left(I_i(t) - \Phi_i(t)\right) * \mathrm{K}_i(t) \tag{9.24}$$

with

$$\Phi_i(t) = \frac{C}{L_i}\int_0^t \left(Q_{ai}(\lambda) - Q_{ai}(0)\right) d\lambda \tag{9.25}$$

$Q_{ai}(t)$ [L^3T^{-1}] being the total lateral flow hydrograph on reach R_i defined as the sum of output hydrographs from right-bank and left-bank subcatchments, $Q_{ai}(0) = 0$ because there is no base flow and

$$Q_{ai}(t) = Lr_i(t) + Ll_i(t) \tag{9.26}$$

The function $K_i(t)$ is the Hayami kernel function obtained by substituting in Eq. (9.8) the term L by the term L_i. This model needs also the two parameters C and D of the diffusive equation (see Section 9.4.4).

9.4.4 Properties of the Transfer Function

The GIUH transfer function needs four parameters: the two parameters θ and z that describes the form of the input hydrograph and the two parameters C and D of the diffusive wave model. The two parameters θ and z can be related to rainfall intensities and duration while the other two parameters C and D can either be considered constant over the whole subcatchments and reaches of the basin, or related to geometrical and hydraulic properties. In the applications, two cases are studied:

i) The two parameters C and D are considered constant and identical for all source, right-bank and left-bank subcatchments and for all reaches. Let $\overline{C}$ and $\overline{D}$ be the respectively the constant values, we have

$$C = \overline{C} \quad \text{and} \quad D = \overline{D} \tag{9.27}$$

ii) In order to take into account the effect of slope on the celerity C, and using Manning's law hypothesis, C is considered function of the slope. D is considered constant for all subcatchments and reaches.

$$C = \overline{C}\sqrt{\frac{Slope_i}{\overline{S}}} \quad \text{and} \quad D = \overline{D} \tag{9.28}$$

where $Slope_i$ is equal to S_i when routing on reaches, and $Slope_i$ equal to Ss_i, Sr_i and Ss_i when dealing with source, right-bank and left-bank subcatchments respectively. $\overline{S}$ [-] is a constant representing a mean slope value (e.g. mean slope of all pixels). In this case, the model needs only three parameters $\overline{C}$, $\overline{D}$, and $\overline{S}$.

The routing model through the channel network enables the calculation of the hydrograph on each node of the channel network, and consequently the identification of the GIUH transfer function of each subcatchment.

9.5 ANALYSIS OF THE EFFECTS OF THE SUB-DIVISION OF THE CATCHMENT ON THE TRANSFER FUNCTION

This section is divided in three parts. First, the methodology adopted to analyse the relationships between the level of basin subdivision corresponding to a value of A_t and the transfer function is presented. Second, a detailed application case on the Hérault basin shows the different steps of the methodology, the main properties and limitations. Finally, the methodology was applied on seven basins in order to compare the GIUH transfer functions and to analyse the effects of various basins topology on the relationship between the level of basin subdivision refinement and the resulting GIUH transfer function.

9.5.1 Methodology

The DEM data used in the applications are in the form of the French National Geographic Institute (IGN) block with 50 m grid resolution. For each value of A_t corresponds a spatial subdivision of the catchment into subcatchments. A reference transfer function $TF_{ref}(t)$ was calculated for the lowest value of A_t noted A_{ref}. For values of $A_t > A_{ref}$, the transfer function $TF_{At}(t)$ was calculated and compared to the reference one using three error criteria characterizing the whole shape of the hydrograph:

- Nash and Sutcliffe (1970) criteria error noted NS (with $NS \leq 1$).
- Error on discharge $eQ = (Qmax_{At} / Qmax_{ref})$: this gives the ratio of the maximum amplitude of $TF_{At}(t)$ noted $Qmax_{At}$ $[L^3T^{-1}]$ over the maximum amplitude of $TF_{ref}(t)$ noted $Qmax_{ref}$ $[L^3T^{-1}]$.
- Error on time $eT = (Tmax_{At} / Tmax_{ref})$: this gives the ratio of the time of occurrence of the maximum amplitude of $TF_{At}(t)$ noted $Tmax_{At}$ [T] over the time of occurrence of the maximum amplitude of $TF_{ref}(t)$ noted $Tmax_{ref}$ [T].

The transfer function at the scale A_t (TF_{At}) is close to the reference (TF_{ref}) if the three error criteria are close to one.

The numerical applications aim to study the evolution of the error criteria for various values of A_t. The optimal value of A_t, and consequently the optimal spatial discretization of the basin into subcatchments, corresponds to the value that gives acceptable values of the error criteria and gives the most parsimonious basin subdivision by minimizing the number of subcatchments and reaches. The time of calculation is proportional to the total number of reaches, and source, right-bank and left-bank subcatchments; when A_t decreases, the time of calculation increases. The optimal threshold A_{opt} is defined as the maximum value of the threshold A_t verifying simultaneous conditions on error criteria. In the applications, we consider two sets of conditions as example:

$$\text{Conditions 1: } NS > 0.90 \text{ and } 0.8 < eQ < 1.2 \text{ and } 0.8 < eT < 1.2 \quad (9.29)$$

$$\text{Conditions 2: } NS > 0.95 \text{ and } 0.9 < eQ < 1.1 \text{ and } 0.1 < eT < 1.1 \quad (9.30)$$

Let A_{opt1} and A_{opt2} be the threshold areas verifying conditions 1 and 2 respectively. As conditions 1 are less constraining than conditions 2, we have $A_{opt1} \geq A_{opt2}$.

9.5.2 Application on the Hérault Basin

This application illustrates on the Hérault basin the steps of the calculation of the transfer function from DEMs for different values of A_t.

9.5.2.1 *Topological Structure of the Hérault Basin*

The Hérault basin (2617 km²) was subdivided into subcatchments for eleven different values of the threshold A_t (km²): 4, 8, 16, 32, 64, 128, 256, 512, 1024, 2048 and A_0 = 2617 km². The reference threshold is A_{ref} = 4 km². For three values of A_t = 4,32 and 256 km², Fig. 9.2 shows the effects of varying the threshold area on the topology of the channel network. Table 9.1 shows the evolution of the total number of reaches and source, left-bank, right-bank, and channel zone subcatchments for the various values of A_t. Three cases can be distinguished around a particular value of the threshold noted A_1: A_1 is the threshold area for which the total number of source nodes varies from two (for $A_t \leq A_1$) to one (for $A_t > A_1$); for the Hérault basin, we have A_1 = 429.25 km².

Table 9.1. Effects of varying the threshold area A_t on the topologic properties of reaches and subcatchments of the Hérault basin.

Al(km^J)	Reaches' properties			Subcatchments' properties				
	Total number of reaches N_r	Number of source reaches N_{rs}	Number of internal reaches N_{ri}	Total number of source, right-bank and left-bank subcatchments N_s	Total area of source subcatchments (km^2)	Total area of right-bank subcatchments (km^2)	Total area of left-bank subcatchments (km^2)	Total area channel zone subcatchments (km^2)
4	340	173	167	856	811	798	806	202
8	184	94	90	462	861	826	784	146
16	86	45	41	217	817	810	885	105
32	44	23	21	111	860	819	864	74
64	21	11	10	53	861	749	957	50
128	7	4	3	18	535	975	1069	38
256	5	3	2	13	978	848	763	28
512	1	1	0	3	797	1167	632	21
1024	1	1	0	3	1038	1079	482	18
2048	1	1	0	3	2049	446	115	7
2617	0	0	0	1	2617	0	0	0

First case: $A_t < A_1$. In this case, the total number of reaches N_r, and the total number of source, left-bank and right-bank subcatchments N_s decreases when A_t increases: from N_r = 340 and N_s = 856 for A_t = 4 km² to N_r = 5 and N_s = 13 for A_t = 256 km². The number of source and internal reaches decreases also when A_t increases. However, the total area of source subcatchments, the total area of right-bank subcatchments and the total area of left-bank subcatchments seem constant for various values of threshold A_t: the total area of source subcatchments ranges between 20.5 % and 37.4 % of the total basin area, the total area of right-bank subcatchments between 28.6 % and 37.3 %, and the total area of left-bank subcatchments between 29.1 % and 40.8 %. In the other side, the total area of pixels belonging to the channel zone decreases when A_t increases; it represents 7.7 % of the total basin area for $A_t = 4$ km² and 1.1 % for $A_t = 256$ km². When A_t decreases and tends to zero, the channel network fills the entire basin and consequently the zone representing the channel zone represents the total basin area.

Figure 9.2 (right) presents an example of the basin subdivision for $A_t = 256$ km²: the total number of reaches is 5 (3 source and 2 internal reaches) and the total number of source, right-bank and left-bank subcatchments is 13.

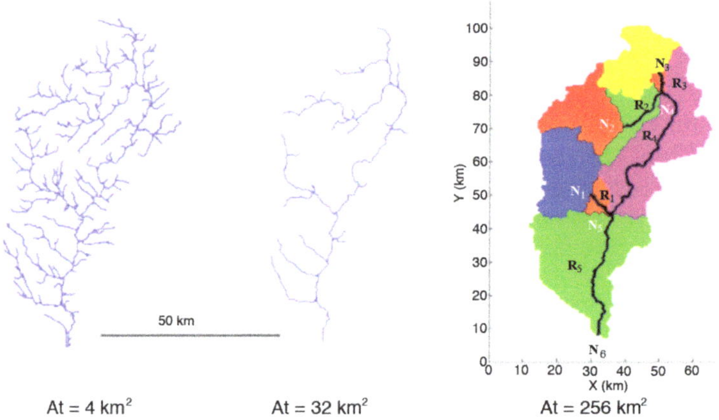

Fig. 9.2. Effects of varying the threshold area A_t on the channel network and the Hérault catchment subdivision for A_t = 4,32 and 256 km². The subdivision into subcatchments is shown for A_t = 256 km².

Second case: $A_1 < A_t < A_0$. In this case, the basin is represented with only one reach, one source, one right-bank and one left-bank subcatchment. When A_t increases from $A_t = 512$ km² to $A_t = 2048$ km², the area of the source subcatchment increases from 30.4 % of the total basin area to 78.3 %, the area of right-bank subcatchment decreases from 44.6 % to 17.0 %, the area of left-bank subcatchment decreases from 24.1 % to 4.4 % and the total area of pixels belonging to the channel zone decreases from 0.8 % to 0.3 %.

Third case: $A_t = A_0$. For this particular maximal value of the threshold, the whole basin is considered as one source subcatchment. There are no right-bank and left-bank subcatchments and no reaches.

9.5.2.2 Estimation of the parameters

In the applications, the impulse function parameters were chosen $\theta = 15$ minutes and $z = 5$. This corresponds to a rain event of about 30 minutes with the peak intensity at about 15 minutes. The mean values of the flood routing parameters $\bar{C}$ and $\bar{D}$ were calculated by analysing the main flood events on two stations: inflow hydrograph at Moulin-Bertrand, and outflow at Agde near the basin outlet; the distance between the two stations is about 90 km with $\bar{C} = 1.3\ m.s^{-1}$ and $\bar{D} = 5000\ m^2.s^{-1}$ (Moussa and Bocquillon, 2009).

9.5.2.3 Effects of the slope on the transfer function

This application compares the transfer functions for three values of A_t = 4,32 and 256 km² (Fig. 9.3) using two methods to calculate C: i) C is considered constant for all reaches and subcatchments (27) and ii) C is variable and defined function of the slope (28); in this case, the model needs three parameters ($\bar{C}$, $\bar{D}$, and $\bar{S}$). In both cases, D was considered constant for all reaches and subcatchments. For numerical applications we considered C and D equal to the values defined above ($C = \bar{C}$ and $D = \bar{D}$), and for the mean slope $\bar{S}$ a value equal to the mean slope of the principal channel reaches proportionally to the reaches' length (e.g. $\bar{S} = 0.00231$ for basin subdivision with A_t = 256 km^2).

Figure 9.3a shows TF_{ref}, TF_{32} and TF_{256} for the case when C is constant and Fig. 9.3b for the case when C is function of the slope. In Fig. 9.3a, the three transfer functions present the same shape; the differences between TF_{ref}, TF_{32} and TF_{256} are due to the number of subcatchments and reaches used in simulations. The same results are observed on Fig. 9.3b. However when comparing Figs. 9.3a and 9.3b, we observe that the transfer functions differ. Differences are essentially due to the relationship between C and the slope (e.g. Eq. (9.28)), to the value of the parameter $\bar{S}$, and to the accuracy of DEMs. The method using C function of the slope presents the advantage to take into account the spatial variability of C.

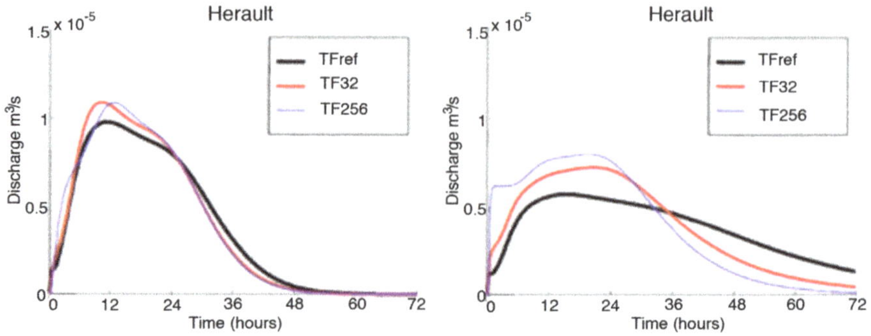

Fig. 9.3. Comparison of three transfer functions (noted TF_{ref}, TF_{32} and TF_{256}) for the Hérault river for respectively three thresholds A_t = 4, 32 and 256 km² (Fig. 9.2) in two cases: a) C constant for all reaches and subcatchemnts; b) C is function of the slope.

9.5.2.4 Relationships between basin subdivision and the error criteria

Transfer functions were calculated for the eleven values of A_t presented in Table 9.1. The parameters C and D were considered constant and identical for all reaches and subcatchments with $C=\bar{C}$ and $D=\bar{D}$. The reference transfer function TF_{ref} corresponds to A_t = 4 km². Table 9.2 shows the relationship between the level of basin subdivision corresponding to different values of A_t and the three error criteria for the Hérault basin: i) the Nash-Sutcliffe criterion is higher than 0.95 for $A_t \leq 1024\,\text{km}^2$; ii) the maximum amplitude $Qmax_{At}$ of the transfer function at the scale A_t don't exceed more than 12 % the maximum

Table 9.2. Error criteria (NS, eQ and eT) when comparing the transfer function calculated for a threshold A_t to the reference transfer function (herein for A_t = 4 km²) for the Hérault river.

A_t (km²)	NS	eQ	eT
4	1.000	1.000	1.000
8	0.990	1.080	0.905
16	0.981	1.110	0.905
32	0.980	1.115	0.905
64	0.976	1.072	0.869
128	0.981	1.096	1.095
256	0.968	1.113	1.117
512	0.961	1.015	1.876
1024	0.953	1.057	1.869
2048	0.469	1.700	1.606
2617	-0.539	2.329	1.365

amplitude $Qmax_{ref}$ of the reference transfer function $1.00 < eQ < 1.12$ for $A_t \leq 1024\,km^2$; iii) the time of occurrence $Tmax_{At}$ of the maximum amplitude of the transfer function at the scale A_t varies around +/- 13 % of the time of occurrence $Tmax_{ref}$ of the maximum amplitude of the reference transfer function $0.87 < eT < 1.12$ for $A_t \leq 256\,km^2$. In the particular case corresponding to $A_t = A_0 = 2617\ km^2$ (the whole basin represented by only one source subcatchment), we obtain high error criteria: the Nash-Sutcliffe criterion is –0.54, the error on discharge is 2.33 and the error on time 1.37. Conditions 1 (29) are verified for A_{opt1} = 256 km^2 (basin subdivided into 5 reaches and 13 subcatchments), and conditions 2 (30) are verified for A_{opt2} = 16 km^2 (basin subdivided into 86 reaches and 217 subcatchments).

Finally, Table 9.3 presents a sensitivity analysis of the optimal values A_{opt} such as NS > 0.9 for various values of the two parameters C (from 0.4 to 3 ms^{-1}) and D (from 500 to 20000 m^2s^{-1}) on the Hérault basin. Results show that A_{opt} depends on the values of C and D (A_{opt} increases when C decreases and when D increases), and consequently the choice of the adequate basin subdivision will depend on the parameters of the transfer function.

Table 9.3. Optimal threshold area A_{opt} (km^2) which minimizes the number of subcatchments and reaches, and verifies the criteria NS > 0.9 for various values of the two parameters celerity and diffusivity, for the Héarult river.

Diffusivity (m^2s^{-1})		Celerity (ms^{-1})				
		0.4	1	1.3	1.6	3
500		256	256	256	128	128
1000		512	512	512	256	256
5000		1024	1024	1024	512	512
10000		1024	1024	1024	1024	1024
20000		1024	1024	1024	1024	1024

9.5.3 Comparison of the Transfer Functions for Seven Basins

In order to analyse the effects of various basins shapes and channel networks topology on the effects of catchment subdivision on the transfer function, the methodology was applied on seven basins located southern France: Tech (738 km^2), Têt (1356 km^2), Agly (1101 km^2), Aude (5346 km^2), Orb (1582 km^2), Hérault (2617 km^2) and Vidourle (860 km^2).

Altitudes range between 0 (all outlets are at the sea level) and 2832 m. The main geometric characteristics of the seven basins for the reference threshold $A_{ref} = 4$ km² are: the total number of reaches varies between 106 (Vidourle) and 687 (Aude); the total number of subcatchments varies between 267 (Vidourle) and 1723 (Aude). For each of the seven basins, the total area of source subcatchments, the total area of right-bank subcatchments and the total area of left-bank subcatchments are of the same order. The total area of source subcatchments varies between 28.3 % of the total basin area (Têt) and 35.4 % (Tech); the total area of the right-bank subcatchments varies between 27.9 % (Tech) and 31.1 % (Têt); the total area of left-bank subcatchments varies between 28.6 % (Tech) and 33.0 % (Vidourle); the total area of channel zone pixels varies between 7.6 % (Agly) and 8.2 % (Aude). For $A_{ref} = 4$ km², we observe that for seven basins having different topological structures and different areas, the total area of channel zone pixels is approximately 8 % of the total basin area while the total area of source subcatchments, the total area of right-bank subcatchments, and the total area of left-bank subcatchments represents each approximately 31 % of the total basin area (Moussa et al. 2011). For all seven basins, the variation of these geometrical properties with respect to A_t presents similar results as those discussed in Section 9.5.2.1 for the Hérault basin.

Simulations were conducted using the same parameters as above: $\theta = 15$ minutes, z = 5, and C and D were considered constant and identical for all reaches and subcatchments with $C = \bar{C} = 1.3\,m.s^{-1}$ and $D = \bar{D} = 5000\,m^2.s^{-1}$. The values of the threshold A_t considered in the applications were (in km²): 4, 8, . . . , 2^j, . . ., A_0 (j is an integer number $j > 2$).

Figure 9.4 shows the evolution of the three criteria error (NS, eQ and eT) function of A_t/A_0 for the seven basins.

We observe the same general shape of the curves, and that the maximum error criteria is observed for the highest values of $A_t = A_0$ corresponding to a lumped modelling approach. However, the relationships differ for the different basins. When comparing the results for the seven basins in the cases corresponding to $A_t \leq 128\,km^2$, we observe: NS is always superior than 0.99 for the Agly basin, while it ranges between 0.80 and 0.95 for the Têt and the Orb; eQ ranges between 1.00 and 1.09 for the Orb while it ranges between 0.95 and 1.16 for the Têt; eT ranges between 0.98 and 1.02 for the Agly while it ranges between 0.81 and 1.00 for the Têt.

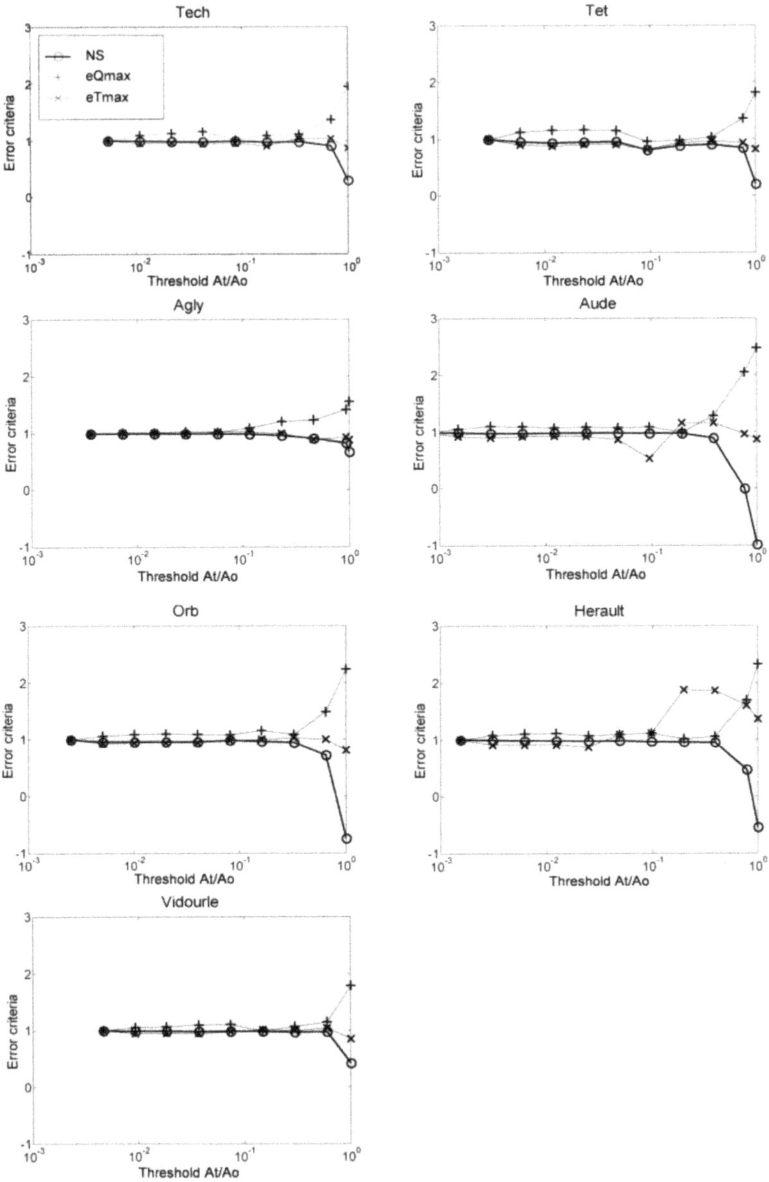

Fig. 9.4. Comparison for the seven basins (Tech, Têt, Agly, Aude, Orb, Hérault and Vidourle) of the relationship between the error criteria (NS, eQ and eT) and the level of basin subdivision (corresponding to different values of the threshold area A_t).

Considering conditions 1 (9.29) and conditions 2 (9.30), the corresponding optimal values of the threshold, A_{opt1} and A_{opt2} respectively, were calculated for the seven basins (Table 9.4). We observe that the optimal A_t varies function of the channel network shape. For conditions 1, A_{opt1} varies between 64 km² (Têt) and 512 km² (Vidourle); the corresponding total number of reaches varies between 1 (Tech, Orb and Vidourle) and 15 (Têt). For conditions 2, A_{opt2} varies between 4 km² (Têt) and 128 km² (Agly and Aude); the corresponding total number of reaches varies between 5 (Agly and Orb) and 192 (Aude). The total number of reaches (and consequently the total number of subcatchments) needed to obtain acceptable error criteria is not systematically proportional to the basin area.

Table 9.4. Optimal values of the threshold area A_{opt1} and A_{opt2} verifying respectively conditions 1 (29) and conditions 2 (30) for the seven basins.

Basin	Conditions 1			Conditions 2		
	A_{opt1} (km²)	A_{opt1}/A_0	Number of reaches (N_r)	A_{opt2} (km²)	A_{opt2}/A_0	Number of reaches (N_r)
Tech	256	0.346	1	8	0.011	59
Têt	64	0.047	15	4	0.003	167
Agly	128	0.116	5	128	0.116	5
Aude	256	0.048	9	16	0.003	192
Orb	512	0.324	1	128	0.081	5
Hérault	256	0.099	5	8	0.003	184
Vidourle	512	0.596	1	32	0.037	17

For example: i) with conditions 1, the Aude (5346 km²) needs only a subdivision with 5 reaches, while the Têt (1356 km²) needs a subdivision with 15 reaches; ii) with conditions 2, the Têt (1356 km²) and the Orb (1582 km²) which have the same area approximately needs very different levels of basin subdivision: the Têt needs a subdivision with 167 reaches while the Orb needs only a subdivision with 5 reaches.

Let $TF_{ref,}$ TF_{opt} and TF_{tot} be the transfer functions corresponding to three different values of $A_t = A_{ref}$, $A_t = A_{opt1}$, and $A_t = A_0$ respectively. Figure 9.5 compares the shape of the transfer function hyetographs for the seven basins. We observe that the transfer function corresponding to A_{opt1} (TF_{opt}) is very similar to the reference transfer function (TF_{ref}) while the transfer function obtained in the particular case when the whole basin is considered as only one source subcatchment (TF_{tot}) differs from TF_{ref}.

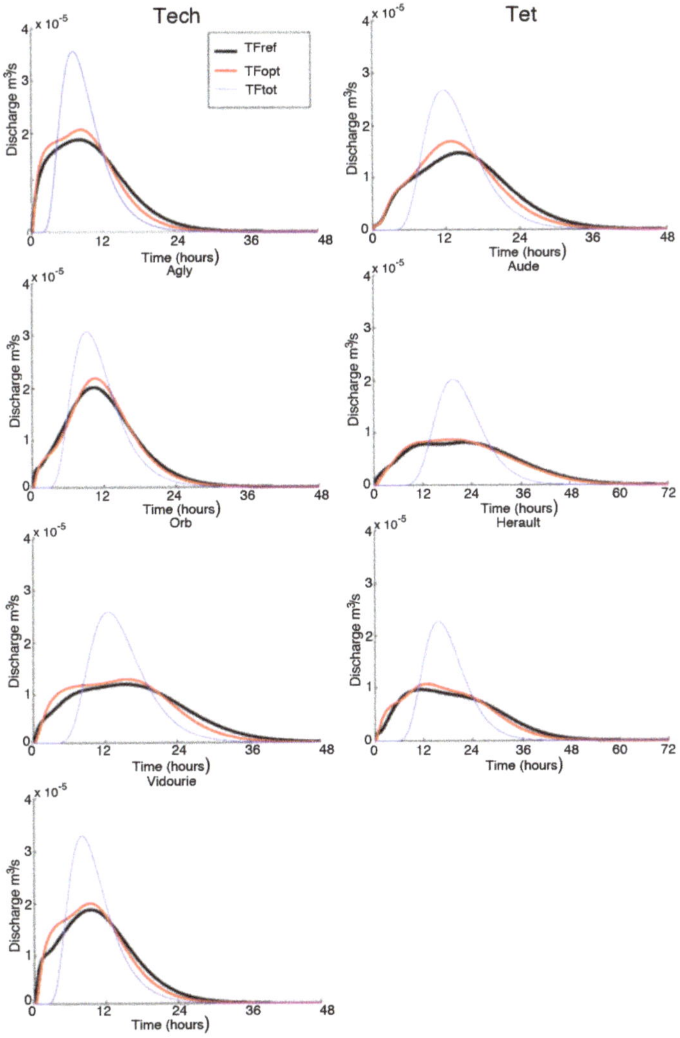

Fig. 9.5. Comparison of the transfer functions (TF) for the seven basins (Tech, Têt, Agly, Aude, Orb, Hérault and Vidourle): TF_{ref} for $A_t = A_{ref} = 4$ km²; TF_{opt} for $A_t = A_{opt1}$ (verifying conditions 1 in Table 9.4); TF_{tot} for $A_t = A_0$.

The calculation of the transfer function needs parameters related to the impulse function (θ and z) and to the diffusive wave model (C and D). Results are sensitive to these parameters, and conclusions obtained for a set of parameters may not be generalized for all parameter ranges. However, for a set of parameters, this methodology enables the modeller:

i) to automatically calculate the GIUH transfer function from DEMs taking into account the topology of the channel network and then to compare the transfer function of various basins; ii) to identify the optimal spatial subdivision of the basin into subcatchments that gives the most parsimonious basin subdivision with the lowest number of subcatchments and reaches.

9.6 DISCUSSION AND CONCLUSIONS

A methodology is proposed to calculate the GIUH transfer function, using DEMs for applications in semi-distributed hydrological modelling and poorly gauged basins in order to analyse the effects of catchment subdivision on the diffusive wave transfer function. The channel network is extracted from DEMs and the basin subdivided into subcatchments considering pixels draining a zone superior than a threshold area A_t. The GIUH transfer function is then calculated at the basin outlet by routing an impulse response function from each subcatchment through the channel network using the diffusive wave model with lateral flow uniformly distributed. For each value of the threshold A_t corresponds a spatial subdivision of the basin into subcatchments, and consequently the transfer function will depend on the spatial discretization of the catchment. Special attention was given to study the evolution of the shape of the transfer function of the catchment subdivision in order to guide the modeller in the choice of the appropriate subdivision, which minimizes the number of spatial units and reaches.

The transfer function is based on a Hayami linear model of the resolution of the diffusive wave equation with lateral flow. This method is unconditionally stable, is well adapted for routing hydrographs through a channel network, and ends as a simple computer program. This model is function of two parameters, the celerity C and the diffusivity D. These two parameters can either be considered constant for all reaches and subcatchments, or variable and related to hydraulic properties.

Mean values of C and D can be obtained by analysing inflow and outflows hydrographs on river reaches. The values of C and D may vary for various flood events, and consequently the transfer function shape varies function of C and D.

A methodology to help the modeller in the choice of the adequate parsimonious catchment subdivision for a given surface runoff modelling problem is proposed: i) the transfer function is calculated for various catchments' subdivision corresponding to various values of the threshold area A_t; ii) a reference transfer function is defined for the finest resolution corresponding to the lowest value of A_t; and iii) error criteria are calculated in order to compare the transfer function calculated for a given catchment subdivision to the reference one. This analysis help the user to reduce the number of subcatchments by selecting the highest value of A_t that gives acceptable error criteria.

The calculation of the transfer function of seven basins located southern France, showed that, for the same error criteria, the optimal threshold A_t varies for the different basins. A_t depends on the basin shape and the topology of the channel network. Results showed that the total number of reaches and subcatchments needed to obtain acceptable error criteria is not systematically proportional to the basin area. This methodology enables the modeller to automatically calculate the GIUH transfer function, to compare the transfer function of various basins, and to select a parsimonious spatial subdivision of the basin into subcatchments for semi-distributed hydrological modelling.

9.7 REFERENCES

Abbott, M.B., Bathurst, J.C., Cunge, J.A., O'Connell, P.E., and Rasmussen, J. 1986. An introduction to the European Hydrological System - Système Hydrologique Européen, "SHE", 2, Structure of a physically-based, distributed modelling system, *Journal of Hydrology*, 87: 61-77.

Ajami, N.K., Gupta, H., Wagener, T., and Sorooshian S. 2004. Calibration of a semi-distributed hydrologic model for streamflow estimation along a river system, *Journal of Hydrology*, 298: 112-135

Band, L.E. 1986. Topographic partition of watersheds with digital elevation models, *Water Resources Research*, 22(1): 15-24.

Blöschl, G., and Sivapalan, M. 1995. Scale issues in hydrological modelling: a review, *Hydological Processes*, 9: 251-290.

Cimorelli, L., Cozzolino, L., Della Morte, R., and Pianese, D. 2014. Analytical solutions of the linearized parabolic wave accounting for downstream boundary condition and uniform lateral inflows, *Advances in Water Resources*, 63:57–76.

Diskin, M.H., and Simpson, E.S. 1978. A quasi linear spatially distributed model for the surface runoff systems, *Water Resources Bulletin*, 14(4): 903-918.

Fan, P., and Li, J.C. 2006. Diffusive wave solutions for open channel flows with uniform and concentrated lateral inflow, *Advances in Water Resources*, 29(7):1000-1019.

Fortin, J.P., Turcotte, R., Massicotte, S., Moussa, R., Fitzback, J., and Villeneuve, J.P. 2001. Distributed watershed model compatible with remote sensing and GIS data. I : Description of the model, *Journal of Hydrologic Engineering*, 6(2): 91-99.

Gurnell, A.M., and Montgomery, A.R. 1999. *Hydrological applications of GIS*, John Wiley and Sons, Chichester, 176 pp.

Khakbaz, B., Imam, B., Hsu, K., and Sorooshian, S. 2012. From lumped to distributed via semi-distributed: Calibration strategies for semi-distributed hydrologic models, *Journal of Hydrology*, 418–419: 61-77.

Hayami, S. 1951. On the propagation of flood waves, *Disaster Prev. Res. Inst. Bull.*, 1: 1-16.

Moramarco, T., Fan, Y., and Bras, R.L. 1999. Analytical solution for channel routing with lateral inflow, *Journal of Hydraulic Engineering*, 125: 707-713.

Moussa, R. 1996. Analytical Hayami solution for the diffusive wave flood routing problem with lateral inflow, *Hydrological Processes*, 10: 1209-1227.

Moussa, R. 1997. Geomorphological transfer function calculated from digital elevation models for distributed hydrological modelling, *Hydrological Processes*, 11(5): 429-449.

Moussa, R. 2008. Effect of channel network topology, basin segmentation and rainfall spatial distribution on the GIUH transfer function, *Hydrological Processes*, 22: 395-419.

Moussa, R., and Bocquillon, C. 1996. Algorithms for solving the diffusive wave flood routing equations, *Hydrological Processes*, 10: 105-123.

Moussa, R., and Bocquillon, C. 2009. On the use of the diffusive wave for modelling extreme flood events with overbank flow in the floodplain, *Journal of Hydrology*, 374: 116-135.

Moussa, R., Chahinian, N., and Bocquillon, C. 2007. Distributed hydrological modelling of a Mediterranean mountainous catchment - model construction and multi-site validation, *Journal of Hydrology*, 337: 35-51.

Moussa, R., Colin, F., and Rabotin, M. 2011. Invariant morphometric properties of headwater subcatchments, *Water Resources Research*, 47: W08518

Nash, J.E., and Sutcliffe, J.V. 1970. River flow forecasting through conceptual model. Part I: a discussion of principles, *Journal of Hydrology*, 10: 282-290.

Rodriguez-Iturbe, I., and Rinaldo, A. 1997. *Fractal River Basins, Chance and Self-Organization*. Cambridge University Press.

Snell, J.D., and Sivapalan, M. 1994. On geomorphological dispersion in natural catchments and the geomorphological unit hydrograph, *Water Resources Research*, 4(5): 2311-2323.

Sposito, G. 1998. *Scale dependence and scale invariance in hydrology*, Cambridge University Press, Cambridge, 423 pp.

Wooldridge, S., Kalma, J., and Kuczera, G. 2001. Parameterisation of a simple semi-distributed model for assessing the impact of land-use on hydrologic response. *Journal of Hydrology*, 254: 16-32.

CHAPTER 10

INTERPRETATIONS OF THE MUSKINGUM FLOOD ROUTING METHOD: A HISTORICAL PERSPECTIVE

Muthiah Perumal[1], Bhabagrahi Sahoo[2], Tommaso Moramarco[3]

[1]Professor and Head, Department of Hydrology, Indian Institute of Technology Roorkee, Roorkee-247667, INDIA p_erumal@yahoo.com

[2]Assistant Professor, School of Water Resources, Indian Institute of Technology Kharagpur, Kharagpur– 721302, INDIA bsahoo2003@yahoo.com

[3]Researcher, Research Institute for Hydrogeological Protection, Via Madonna Alta, 126, Perugia, ITALY, t.moramarco@irpi.cnr.it

ABSTRACT

The hydrological literature is replete with a multitude of flood routing methods with varied interpretations of the Muskingum routing method, being used widely world-wide. Consequently, during the last several decades, the classical Muskingum method originally proposed by McCarthy in 1938 has been transformed from its semi-empirical characteristics into a physically based model accounting for the physical process of flood wave movement. However, some of these theories and interpretations existing in the literature to describe the Muskingum model are contradictory to each other. In light of the above discussion, the objective of this Chapter is to review different approaches used for the physical interpretations of the Muskingum parameters and to critically examine whether these interpretations are consistent with the physical behavior of the Muskingum method.

In this study, it has been demonstrated that there is no need for employing the matched diffusivity concept in interpreting the Muskingum method. Among the variants of this method developed till recently, the variable parameter McCarthy-Muskingum (VPMM) method advocated by Perumal and Price in 2013 is the only one which theoretically substantiates the heuristic assumptions built into the classical Muskingum model in expressing the channel reach storage in terms of prism and wedge storages. Hence, the interpretation used in the development of the VPMM method may be treated as the ultimate theory to describe the Muskingum routing method.

10.1 INTRODUCTION

Since a long time, the technique of flood routing has evolved from its empirical and conceptual frameworks to a more descriptive framework with simplicity in the physical interpretations to deal with the flood problems under varied data availability scenarios. To deal with the age-old global problem of flooding induced by global climate change and anthropogenic activities, the subject of flood routing has been of much interest world-wide to the hydraulic and hydrologic research community. Among the multitude of river flood routing methods available in the literature (see Fig.10.1), which starts from pure black box to physically distributed models, the Muskingum model is studied extensively due to its simplicity and wide applicability in many situations of routing floods in rivers (Singh, 1988). Generally, the physically-based simplified routing models are preferred over the full dynamic wave models, since these models that solve the full Saint Venant equations of mass and momentum conservation, are prone to numerical problems, require a large database at finer spatio-temporal resolutions, and need a downstream boundary condition in the solution scheme which may not be present physically in a natural river. The classical Muskingum method proposed by McCarthy (1938) is a semi-empirical method for routing floods in rivers and channels. Much of the research on this method has focused on linking the two parameters of the method, viz., the travel time parameter, K and the weighting coefficient, θ with the channel and flow characteristics. Consequently, over the past seven decades from 1938, a number of theories and interpretations have been advocated by various researches world-wide. However, some of the theories and interpretations are contradictory to each other. This has resulted in a curiosity in the minds of the budding researchers and hydrologists: 'which theory is the most appropriate?'

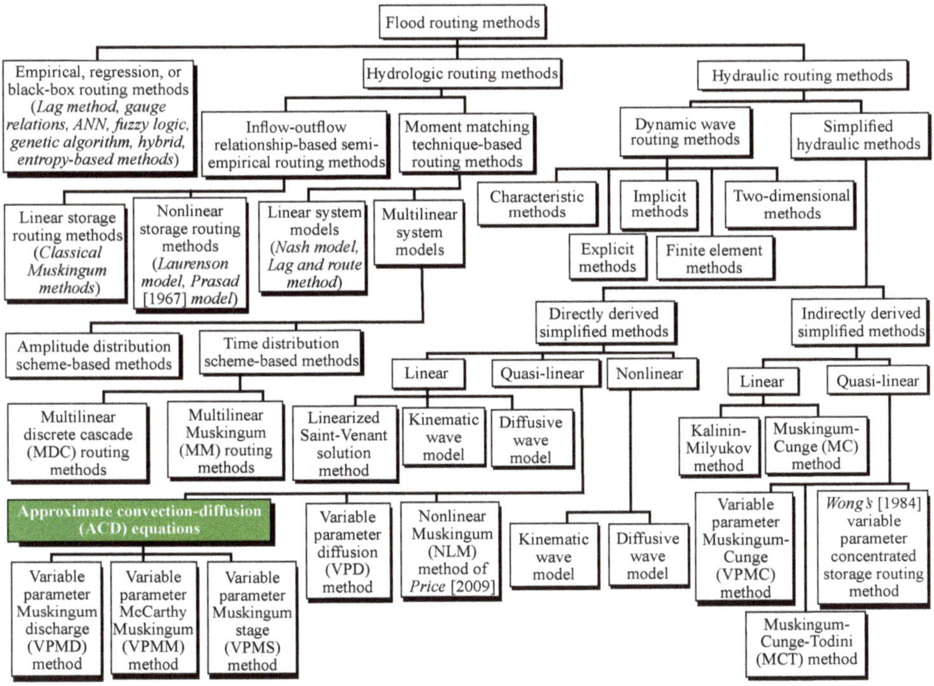

Fig. 10.1. Classification of flood routing methods available in the literature.

In light of the above discussion, the objective of this Chapter is to review different approaches used for the physical interpretations of the Muskingum parameters and to critically examine whether these interpretations are consistent with the physical behavior of the Muskingum method.

10.2 THE CLASSICAL MUSKINGUM FLOOD ROUTING METHOD

The classical Muskingum method (McCarthy, 1938), which is named after its first application to the Muskingum River, a tributary of the Ohio River in the USA, is a linear storage routing method. This method combines the lumped continuity equation with the linear storage equation expressed, respectively, as

$$\frac{dS}{dt} = I - O \qquad (10.1)$$

$$S = K\left[\theta I + (1-\theta)O\right] \tag{10.2}$$

to arrive at the difference equation which on simplification leads to the Muskingum routing equation as

$$O_{j+1} = C_1 I_{j+1} + C_2 I_j + C_3 O_j \tag{10.3}$$

where S is the storage volume, I is the inflow discharge, O is the outflow discharge, K is the travel time, θ is the weighting parameter, the suffix j denotes the ordinate at time $j\Delta t$, Δt is the routing time step, and the coefficients C_1, C_2, and C_3 are expressed as

$$C_1 = \frac{-K\theta + 0.5\Delta t}{K(1-\theta) + 0.5\Delta t} \tag{10.4a}$$

$$C_2 = \frac{K\theta + 0.5\Delta t}{K(1-\theta) + 0.5\Delta t} \tag{10.4b}$$

$$C_3 = \frac{K(1-\theta) - 0.5\Delta t}{K(1-\theta) + 0.5\Delta t} \tag{10.4c}$$

where $C_1+C_2+C_3 = 1.0$, which shows the mass conserving ability of the classical Muskingum method. The routing parameters K and θ of this method are computed by using the historical flood events by the trial and error approach. To calculate the value of θ, the storage S is plotted against the corresponding weighted discharge value $[\theta I+(1-\theta)O]$ in Eq. (10.2) for different trial values of θ resulting in various sizes of loops; and the value of θ which gives the narrowest loop of this plot is considered as the appropriate one for its use in the model. The solution of these Eqs. (10.1) and (10.2) is given by Nash (1959), Diskin (1967), and Dooge (1973).

Many of the textbooks (e.g., Chow et al. 1988) state that the value θ ranges between zero and 0.5. Perumal (2003) showed that by analysing the storage-weighted discharge relationship given by Eq. (10.2), more insight into the range of the assumed values θ could be brought out. Using Eq. (10.2), the outflow can be written as

$$O = -\frac{\theta I}{1-\theta} + \frac{S}{K(1-\theta)} \tag{10.5}$$

Differentiating Eq. (10.5) with reference to t gives

$$\frac{dO}{dt} = -\frac{\theta}{1-\theta}\frac{dI}{dt} + \frac{1}{K(1-\theta)}\frac{dS}{dt} \qquad (10.6)$$

When S is the maximum, $dS/dt = 0$, and Eq. (10.6) reduces to

$$\frac{dO}{dt} = -\frac{\theta}{1-\theta}\frac{dI}{dt} \qquad (10.7)$$

Equation (10.7) enables one to interpret about the positive and negative values of θ. For a typical flood routing case anyone of the situations as shown in Fig. 10.2 is realized. Applying Eq. (10.7) at the intersection of inflow and outflow hydrographs as in Fig. 10.2a, where the storage is the maximum, it is inferred that $\theta > 0$.

In a similar way, it can be recognized that Fig. 10.2b corresponds to the case when $\theta = 0$, i.e., when the storage as well as the outflow discharge is the maximum representing the storage as a function of outflow only (corresponding to the linear reservoir case). The last case, as shown in Fig. 10.2c, in which the peak outflow is less than the synchronous inflow, is realized when the channel reach length is less than that corresponding to the linear reservoir case. For such a situation both dO/dt and dI/dt are negative and thus, necessitating $\theta < 0$. Szilagyi (1992) confirmed this inference using some typical flood observations of the Danube River at Budapest.

Recognizing the inability of the constant parameter classical Muskingum method or its equivalents in physically based forms, such as those introduced by Apollov et al. (1964), Dooge and Harley (1967), Cunge (1969) and Dooge et al. (1982), for modeling the nonlinear behaviour of flood wave movement in channels, Gill (1978) sought to modify the classical Muskingum method by replacing the linear storage equation of the classical Muskingum method by the nonlinear storage equation of the form

$$S(t) = K\left[\theta I(t) + (1-\theta)O(t)\right]^m \qquad (10.8)$$

where m is the exponent of the weighted discharge $\left[\theta I + (1-\theta)O\right]$. In addition to the two parameters employed in the classical Muskingum

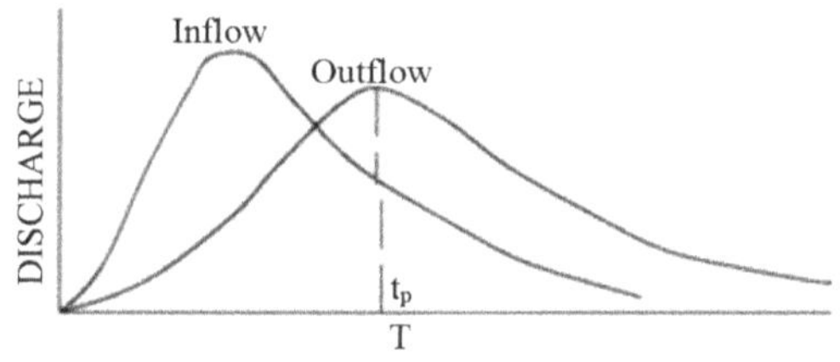

Fig. 10.2(a). Outflow hydrograph when θ>0.

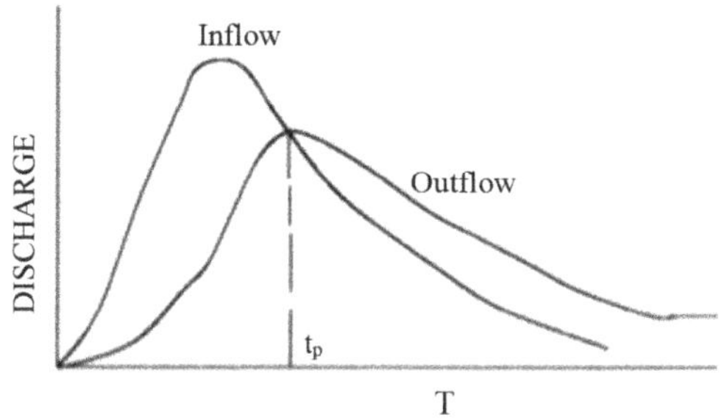

Fig. 10.2(b). Outflow hydrograph when θ=0.

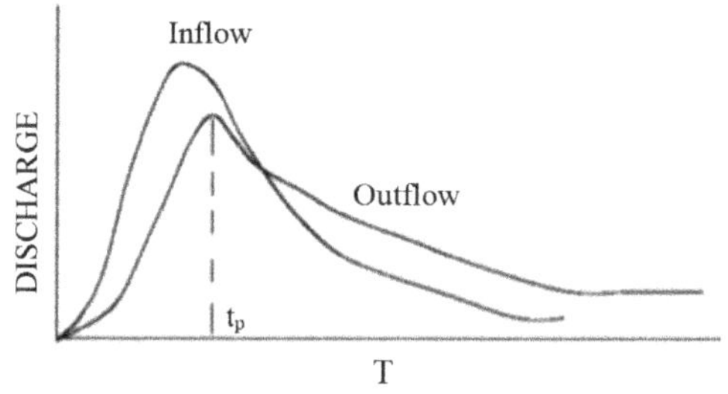

Fig. 10. 2(c). Outflow hydrograph when θ<0.

Fig. 10.2. Classical Muskingum solutions for varying θ ranges

method, Gill's storage equation employs a nonlinear exponent m. When m=1, the storage equation of the nonlinear Muskingum (NLM) method reduces to the storage equation of the classical Muskingum method.

It may be noted that an increase of one more parameter in the form of exponent 'm' associated with the modified form of the storage equation employed in the NLM method increases the flexibility of the method in simulating the observed outflow hydrograph of the calibration flood event. However, the close reproduction of the outflow hydrograph achieved in the calibration mode does not guarantee the set of estimated calibration parameter K, θ and m to be the universal or global optimum for the considered channel reach for simulating all the possible flood events that may be routed in that channel reach.

This statement implies that a set of calibrated parameters estimated from the recorded set of flood events in a routing reach is not only influenced by the channel characteristics, but also by the hydrograph characteristics characterized by the shape, peak and time of peak of the inflow hydrograph, and its rate of rise, besides the initial flow in the channel reach. If a different inflow flood hydrograph of a higher magnitude with differing inflow hydrograph characteristics and with a different associated initial flow in the reach other than that of the past calibrated event is to be routed in the same channel reach, then the parameters calibrated from the past event may not serve the intended purpose of successfully reproducing the observed hydrograph corresponding to this new inflow hydrograph.

Apart from the lack of investigation on the practical utility of the NLM method, none of the studies carried out so far on the parameter estimation of the NLM method by a number optimization techniques such as those proposed by Tung (1984), Yoon and Padmanabhan (1993), Mohan (1997) and Geem (2006, 2011) ever followed the conventional modeling protocol involving calibration and verification of the model. All these authors demonstrated only the superiority of one or other optimization technique employed in estimating the parameters of the NLM method based on the close reproduction of the observed outflow hydrograph available in the textbook by Wilson (1974), without ever discussing about the verification of the calibrated model using a new set of flood data observed for the same channel reach system.

The linear storage equation proposed by McCarthy (1938), which expresses the storage as a linear weighted function of the inflow and the outflow, forms the basis for the development of the classical Muskingum method. However, in this classical Muskingum equation, the routing parameters remain constant over the entire duration of the routing process. Since the flood waves are inherently nonlinear in nature as different discharges travel at varied magnitudes of celerity, the linear storage routing methods were perceived to be of limited use for flood routing studies.

Subsequently, several attempts have been made by various researchers (e.g., Dooge and Harley, 1967; Cunge, 1969; Dooge et al. 1982; Todini, 2007; Price, 2009; Perumal and Price, 2013) to link the routing parameters K and θ of the classical Muskingum method with the flow and channel characteristics using the hydrodynamics principles to transform it into a physically based method. Apart from the simplicity and wide applicability, the Muskingum method produces unrealistic initial routed outflow, namely, "negative flow" or "reduced flow" or "dip" which is well documented in the literature (Dooge, 1973; Weinmann and Laurenson, 1979; Perumal, 1992c, 1993b). This defect is also inherent with the different variants of the classical Muskingum method wherein the parameters are estimated based on the channel and flow characteristics. However, the classical Muskingum method is still popular in the present-day literature (e.g., Neitsch et al. 2005; Arnold and Fohrer, 2005; Gassman, 2007) for the simplicity in its model framework.

10.3 INTERPRETATIONS OF THE CLASSICAL MUSKINGUM METHOD

To gradually transform the classical Muskingum method into a physically-based model by linking its routing parameters with the measurable flow and channel characteristics, several researchers attempted through a number of interpretations of this method. All such interpretations were grouped by Kundzwicz (1986) under the following approaches: i) direct interpretation; ii) matching difference schemes, and iii) matching the impulse response of the Muskingum method with that of the linearized Saint-Venant equations using the method of moments approach (Dooge, 1973). However, with the recent developments in the Muskingum method, these interpretations can further be categorized as:

1. Classical interpretation of Muskingum storage equation (Chow, 1959);
2. Direct interpretation (Brakensiek, 1963; Strupczewicz and Kundzewicz, 1980);
3. Matching the impulse response of the Muskingum method with that of the linearized Saint Venant equations using the method of moments approach (Dooge, 1973);

4. Matching difference schemes (Harley, 1967; Cunge, 1969; Koussis, 1976; Dooge et al. 1982; Ponce and Yevjevich, 1978; Ponce and Chaganti, 1994; Todini, 2007);
5. Using the Kalinin-Milyukov routing method (Apollov et al. 1964);
6. Approximations of convection-diffusion equations (CDEs) (Price, 2009);
7. Using the heuristic interpretation of McCarthy (Perumal and Price, 2013).

A brief description of all these interpretations is discussed in the following sections.

10.3.1 Classical Interpretation of Muskingum Storage Equation

Chow (1959) interpreted the classical Muskingum storage equation by using the concepts that:

i) Both inflow (*I*) and outflow (*O*) discharges at the inlet and outlet sections of the prismatic river reach are uniquely related with the respective stages (*y*).

ii) The stages at the inlet and outlet sections of the Muskingum reach are uniquely related to the reach storages (*S*) established at the upstream reach of these respective sections.

Based on the above concepts, using the power function relationship, it can be written as

$$I = ay^n,\ O = ay^n,\ S_i = by^m,\ S_o = by^m \qquad (10.8a, b, c, d)$$

where S_i and S_i are the storages at the inlet and outlet sections, respectively; and *a*, *b*, *m* and *n* are the empirical values. The implicit assumption behind the set of Eq. (10.8) is that the channel or river reach is prismatic.

Using Eqs. (10.8a, b, c, d), one can obtain

$$S_i = b\left(\frac{I}{a}\right)^{m/n},\ S_o = b\left(\frac{O}{a}\right)^{m/n} \qquad (10.9a, b)$$

By weighing the reach storages at the inlet and outlet sections using the weighting factor θ:

$$S = \theta S_i + \left(1 - \theta\right) S_o \tag{10.10}$$

Now, by using Eqs. (10.9a, b) in (10.10), one can obtain after generalizing it as

$$S = K\left(\theta I^{m/n} + (1 - \theta) O^{m/n}\right) \tag{10.11}$$

where $K = ba^{-m/n}$, the storage coefficient.

If $m/n=1$, the above Eq. (10.11) yields the classical Muskingum storage equation.

There is a logical flaw in this interpretation, since the storage Eq. (10.11) was derived using one-to-one steady state relationship between stage and discharge which is not valid for unsteady flow conditions. Further, there is no physical basis in suddenly equating the exponent $m/n=1$. In view of these shortcomings, one may consider that the Chow's (1959) interpretation of the classical Muskingum method is not appropriate.

Further, it may be noted that the so called NLM method by Gill (1978) was introduced at the same time as that of the variable parameter Muskingum (VPM) method proposed by Ponce and Yevjevich (1978) which also aimed at for accounting the nonlinear characteristics of flood wave movement by varying the parameters of the method at every time step, but retaining the routing equation of the classical Muskingum method. As no investigation has been made to demonstrate the comparative evaluation of the performances of the NLM and VPM methods, both of which have been proposed for the same purpose of accounting nonlinearity in the channel routing method, Perumal et al. (2013) studied the performance of the NLM method advocated by Gill (1978), first by calibrating the parameters of the method using the genetic algorithm (GA) technique of Mohan (1997) for a number of given sets of hypothetical inflow-outflow hydrographs and verifying this calibrated model for routing different sets of inflow and the corresponding outflow hydrographs in the respective channel reaches. The solutions of the Saint-Venant equations obtained by routing a given inflow hydrograph in different hypothetical channels were used as the benchmark hydrographs for calibration purpose of this NLM method, and a similar set of benchmark solutions were obtained in the same set of channels for verification purpose, and compared the performances of the NLM method and the physically based variable parameter McCarthy-Muskingum (VPMM) method (Perumal and Price, 2013) in reproducing the same sets of benchmark hydrographs used for calibration as well as verification of these considered models. The study demonstrated that the use of the VPMM method, which is capable of accounting for nonlinear behavior of flood wave movement process without involving calibration process, is more reliable for field applications than the conceptually based NLM method which performs relatively poorly both at the calibration as well as the verification modes of the routing process. However, a detailed study involving similar calibration based NLM methods is required to arrive at a final conclusion regarding the appropriateness of the NLM based methods.

10.3.2 Direct Interpretation from the Saint-Venant Equations

Brakensiek (1963) was probably the first in the English literature who attempted to interpret the Muskingum method using the hydrodynamic principle based on the following concepts:

i) Under subcritical flow conditions, the unsteady flow in a river is more close to the kinematic wave (KW) movement;

ii) The Muskingum method is more suitable for modeling KWs in a river reach which follows a one-to-one relationship between stage and discharge; and

iii) The reach storage can be expressed as: Reach storage = Average of the flow areas of upstream and downstream boundaries × Length of channel reach, L.

If the outflow discharge with mean velocity V is give by the power function: $Q = aA^b$, then the Muskingum routing parameters can be derived using the hydrologic mass conservation equation assuming a uniform channel section as (Brakensiek, 1963)

$$Q = aA^b, \; K = \frac{L}{V}, \; \theta = \frac{1}{2b} \tag{10.12}$$

where A = flow area corresponding to flow Q; and L = reach length.

However, the interpretation by Brakensiek (1963) for the Muskingum parameters is not consistent with the parameter relationships established by the current theories. Moreover, this theory uses the one-to-one relationship between the stage and the discharge similar to Chow (1959), which is not valid for unsteady flows.

Subsequently, Strupczewski and Kundzewicz (1980) attempted the interpretation of the Muskingum method directly from the Saint-Venant equations, but they could not express the flood storage within the reach as a linear function of the weighted discharge due to their interpretation of a one-to-one relationship between the stage and the discharge for depicting the storage equation of the Muskingum method. However, this interpretation is not appropriate for describing flood waves when the one-to-one relationship between the stage and the discharge is not valid. Further, Strupczewski and Kundzewicz (1980) could not express the parameters K and θ explicitly in terms of channel and flow characteristics. Hence, their approach for interpreting the physical basis of the Muskingum method is not appropriate.

10.3.3 Interpretation based on the Moment Matching Approach

By matching the first and second moments of the instantaneous unit hydrograph (IUH) of the linearized Saint-Venant equation with the corresponding moments of the Muskingum IUH, and using the Chezy's friction law in wide rectangular channels, Dooge (1973) arrived at the following expressions for the routing parameters K and θ of the Muskingum method, respectively, as

$$K = \frac{\Delta x}{c_o} \tag{10.13}$$

$$\theta = \frac{1}{2} - \frac{Q_o \bar{y}_r}{2 S_o A_o c_o \Delta x}\left[1-(m-1)^2 F_o^2\right] \tag{10.14}$$

where $\bar{y}_r$ is the average hydraulic mean depth, A_o is the area of cross section corresponding to Q_o, and m = 3/2 or 5/3 for the Chezy's or Manning's friction law used, respectively.

10.3.4 Interpretation based on the Matched Diffusivity Schemes

Although, Dooge and Harley (1967) presented the Muskingum parameter relationships with the wide rectangular channel and flow characteristics through the moment matching technique (Nash, 1960), and Dooge et al. (1982) later arrived at the same for any shape of prismatic channel and for any type of friction law, it is the Cunge's (1969) matched diffusivity approach which has become more popular as the "Muskingum–Cunge (MC) method." It was Price (1973) (e.g., NERC, 1975) who first coined the term "Muskingum–Cunge method." By matching the numerical diffusivity of the approximate linear kinematic wave equation, derived from the classical Muskingum difference equation with the physical diffusivity of the linear convection–diffusion equation, Cunge (1969) arrived at the relationship for K as given in equation (10.13) and θ for wide rectangular channels as

$$\theta = \frac{1}{2} - \frac{Q_o}{2 S_o B_o c_o \Delta x} \qquad (10.15)$$

Using a more refined difference scheme to represent the Muskingum routing equation and based on the matched diffusivity concept, Koussis (1976, 1978, 1980, 1983) arrived at an alternate relationship for θ different from that given by Eq. (10.15). However, a comparison between the Cunge's and Koussis' approaches shows that both the approaches lead to the same coefficients of the Muskingum routing equation and the differences in the form of the weighting parameter of both the schemes are only apparent (Perumal, 1989).

On the basis of the state trajectory variation method applied to the Saint-Venant equations, Dooge et al. (1982) established the generalized relationships for K and θ for prismatic channels having any shape of cross sections, in which the relationship for K is same as Eq. (10.13) and the relationship for θ is expressed as given in Eq. (10.14).

Note that when the Froude number, $F_o = 0$, Dooge et al.'s (1982) expression (10.14) for θ reduces to Cunge's (1969) Eq. (10.15). Perumal (1992a) studied the differences in the routing results obtained using equations (10.14) and (10.15) for varying Froude number cases and showed that the use of latter equation makes a very insignificant improvement on the estimated numerical value of θ over that using the former equation. However, the use of Eq. (10.13) is advantageous as it retains the Vedernikov number (Jolly and Yevjevich, 1971; Ponce, 1991; e.g., Perumal, 1992a) which gives the amplification criterion of a flood wave while it moves downstream of a channel.

Todini (2007) was the first researcher to propose a mass conserving variable parameter Muskingum method, known as the Muskingum-Cunge-Todini (MCT) method based on Cunge's matched diffusivity approach (Cunge, 1969), but estimated the parameter, *K,* of the Muskingum method based on normal velocity (v_o) corresponding to the unsteady flow depth at the middle of the Muskingum reach, rather than using celerity as envisaged by Cunge (1969). The routing parameters of the MCT method is given as (Todini, 2007)

$$K = \frac{\Delta x}{v_o} \tag{10.16}$$

$$\theta = \frac{1}{2} - \left(\frac{Q_M v_{M,o}}{2 S_o B_M c_{M,o}^2 \Delta x} \right) \tag{10.17}$$

where Q_M is the discharge at the middle of the Muskingum reach, B_M is the water surface width corresponding to y_M, the midreach depth and $c_{M,0}$ is the normal wave celerity corresponding to Q_M.

10.3.5 Interpretation Based on the Extension of the Kalinin–Milyukov Method

The Kalinin–Milyukov routing method (Kalinin and Milyukov, 1958; Apollov et al. 1964) is a conceptual linear storage routing method derived from the hydrodynamics principles in which a given prismatic channel reach is subdivided into a number of subreaches (characteristic reaches) wherein the storage is a linear function of the outflow discharge. This method is based on the assumption of a one-to-one relationship between the stage and the discharge during unsteady flow condition in which the discharge at any instant of time at the outlet of the reach is related to the stage at the middle of the reach. Hence, the discharge at any instant of time is a function of the depth of flow and the slope of the water surface at that section. During unsteady flow condition, it is assumed that the slope of the water surface remains constant over the length of the characteristic reach. Further, during the transformation of a steady flow to unsteady flow, the discharge at the outflow section of the reach does not change.

The number of characteristic reaches, N_c, required for routing a discharge hydrograph using the Kalinin–Milyukov method in a given reach of length Δx is given by (Miller and Cunge, 1975)

$$N_c = \frac{S_o B_o c_o \Delta x}{Q_o} \tag{10.18}$$

and the reservoir coefficient, K, of each characteristic reach is given by

$$K = \frac{Q_o}{S_o B_o c_o^2} \tag{10.19}$$

where, Q_0 is the reference discharge, B_0 is the top width of the section and c_0 is the wave celerity, both corresponding to Q_0.

Using the moment matching technique, Dooge (1973) linked the first and second moments of the Nash model instantaneous unit hydrograph (IUH) with the corresponding moments of the linearized Saint-Venant equations, when the Froude number, $F = 0$; and arrived at the same relationships for the number of linear reservoirs in series and the reservoir coefficient as given by Eqs. (10.18) and (10.19), respectively. Note that when the Nash model parameter representing the number of linear reservoirs in a reach is an integer, the Nash model becomes a conceptual representation of the Kalinin–Milyukov method. Since the Nash model can operate on non-integer N_c values, the flood routing using the Nash model with the parameters estimated using Eqs. (10.18) and (10.19) is more flexible than the Kalinin–Milyukov method. This avoids the interpolation of the outflow hydrograph when the last characteristic reach of the Kalinin–Milyukov method does not coincide with the outflow section of the given routing reach (Koussis, 1980). Furthermore, Kundzewicz (1982) found that the routing solution of the Kalinin–Milyukov method, when compared with that of the Muskingum solution, was relatively insensitive to the variation in the reference discharge used for the estimation of the model parameters. When compared with other two parameter routing methods, Dooge (1973) showed that the Kalinin–Milyukov method closely reproduces the linearized Saint-Venant solution.

As translated by Apollov et al. (1964), the Kalinin–Milyukov flood routing method commented on the classical Muskingum method (McCarthy, 1938), which is paraphrased as:

> "*The basic shortcoming of the method of calculation of transformation used in the USA is the insufficient (scientific theoretical) substantiation for some of its initial assumptions. Consequently, the whole calculation technique largely depends on the correctness of the empirical selection of such parameters as k, the length of the river stretch and the calculating time interval*".
>
> (Pages 53-54 of Apollov et al. 1964)

> "*American hydrologists determine the parameter k very approximately by trial and error. It is more easily and more accurately determined theoretically from (55.III)*".
>
> (Pages 57 of Apollov et al.,1964)

On the basis of the extension of the Kalinin–Milyukov method described by Apollov et al. (1964) for the interpretation of the Muskingum method, Perumal (1992c) brought out that only this method enables one to establish the reason behind the formation of negative or reduced outflow at the beginning of the Muskingum method solution. The negative flow at the beginning of Muskingum solution is due to the form of storage equation employed. The parameter relationships for K and □ as established by this approach are the same as given by Eqs. (10.13) and (10.15). Wong (1984) also independently made the interpretation of the Muskingum storage equation on the lines of Apollov et al. (1964).

10.3.6 Interpretation based on the Approximations of Convection-Diffusion Equations (CDEs)

Price (2009) proposed a volume conservative routing method based on an approximation of the Saint-Venant equations such that the parameters are determined using the channel and flow characteristics. This resulted in the same parameter relationships as developed by Todini (2007). Although the approaches followed by Todini (2007) and Price (2009) in arriving at the routing equations are different, the parameter relationships of these methods are identical. Therefore, both methods should yield the same routing results (Price, 2009).

10.3.7 Using the Heuristic Interpretation of McCarthy

Although the VPMC method has been widely used in practice (DAMBRK by Fread, 1990; HEC-HMS by USACE, 2010, 2013), this method is saddled with perceptible loss of mass (Ponce and Chaganti, 1994; Perumal and Sahoo, 2008). Subsequently, the well-known mass conservation deficiency of the VPMC method has been tackled by Todini (2007), Price (2009), and Perumal and Price (2013). However, only the variable parameter McCarthy-Muskingum (VPMM) method advocated by Perumal and Price (2013) could explain the validity of the heuristic assumption considered by McCarthy in 1938 in the development of the classical Muskingum method as a storage routing method. According to McCarthy (1938), the reach storage consists of two component storages, viz., the prism storage and the wedge storage. It may be pointed out that the successful application of the classical Muskingum method in the past as well as in current practices implies the validity of the heuristic assumption of storage concept introduced by McCarthy (1938). However, questioning the nature of this concept and its replacement by Cunge (1969) with the matched diffusivity concept to explain the success of the Muskingum method leads to contradictions in the minds of field engineers and researchers dealing with flood routing problems.

The VPMM routing method is developed from an approximation of the momentum equation of the Saint-Venant equations. Subsequently, this approximation is applied directly to the one-dimensional continuity equation of the Saint-Venant equations leading to a fully mass conservative routing method. It turns out that the form of the routing equation of this method is the same as the classical Muskingum method proposed by McCarthy in 1938. The development of the method based on hydraulic principle enables the characterization of the considered channel reach storage into prism and wedge storages as envisaged by McCarthy (1938) in the classical Muskingum method. The form of the storage equation arrived at from the development of the VPMM method is identical to that of the classical Muskingum method with the travel time,

K, and weighting parameter, *θ*, respectively, linked to flow and channel characteristics as (Perumal and Price, 2013)

$$K = \frac{\Delta x}{v_{M,0}} \tag{10.20}$$

$$\theta = \frac{1}{2} - \frac{Q_{M,0}\left[1 - \frac{4}{9}F_0^2\left(\frac{P}{B}\frac{dR}{dy}\right)_M^2\right]}{2S_0 B_M c_{M,0}\Delta x} \tag{10.21}$$

where $Q_{M,0}$ and $c_{M,0}$ denote the normal discharge and normal celerity, respectively, corresponding to the flow depth y_M at the middle of the Muskingum reach; F_M denotes the Froude number, R = hydraulic radius (A/P); P = wetted perimeter; and B = dA/dy, the top width of the water surface corresponding to flow depth y.

As a consequence of the ability of the VPMM method to explain the classical Muskingum method, it may be inferred that the attenuation introduced in the classical Muskingum method can be attributed to the reach storage effect, without the need to explain the same based on the matched diffusivity theory as advocated by Cunge (1969) in interpreting the classical Muskingum method.

10.4 MULTILINEAR MUSKINGUM METHOD

To circumvent the volume conservation problem inherent in the VPMC method, Perumal (1992c) proposed the multilinear Muskingum discharge (MMD) routing method based on the time distribution scheme wherein the outflow hydrograph is computed using the convolution technique that ensures conservation of mass. In this method, each ordinate of the inflow discharge hydrograph, Q_u is routed through a Muskingum sub-model. The reference discharge, Q_o, needed for computing the Muskingum model parameters is estimated as

$$Q_o = Q_b + a_q(Q_u - Q_b) \tag{10.22}$$

where Q_b is the initial steady discharge in the channel reach before the arrival of the flood, Q_u is the current value of inflow, and a_q is a coefficient with limits $0 < a_q < 1$.

The ordinates of the discrete unit hydrograph needed for convolution with the inflow hydrograph ordinates can be arrived by successive application of the Muskingum routing equation. The overall outflow discharge at time $j\Delta t$ is computed by adding the component outflows. In a hypothetical study, Perumal (1992b) established the suitability of this method for routing floods in wide rectangular channels with no lateral flow. The study by Perumal (1992b) and Sahoo (2013) showed that the MMD method is able to reproduce the Saint-Venant solutions closely when the inflow hydrograph is characterized by a narrow loop rating curve. The formation of wide loop rating curve for a given inflow hydrograph in a given channel reach implies the development of significantly higher longitudinal water surface gradient, $(1/S_0)(\partial y/\partial x)$. Hence, the higher the magnitude of $(1/S_0)(\partial y/\partial x)$, the poorer would be the simulation of the Saint-Venant solutions by the MMD method. In a comparative study between the MMD and VPMC routing methods, Perumal (1993a) showed that the performance of the MMD method is slightly better than that of the VPMC method in reproducing the peak flow characteristics when an inflow hydrograph having a wider loop is routed downstream, as the application of the VPMC method for routing such floods is saddled with loss of mass, with the loss being directly proportional to the size of the loop and the number of channel subreaches used for routing. However, in the MMD method, the principle of conservation of mass is not violated as the outflow hydrograph is obtained by convoluting the inflow hydrograph with the time-variant unit hydrograph of the Muskingum channel reach. Since the MMD method was originally developed for routing in rectangular channels, it could be further extended for flood routing studies in trapezoidal channels with or without floodplains. Samani and Jebelifard (2003) used the MMD method for design of circular urban storm sewer systems.

Although the MMD method uses the physically based parameter relationships and has the characteristics of conserving mass fully, still it may be considered as a semi-empirical method due to the use of the empirical constant in estimating the reference discharge required for evaluating the routing parameters. Due to the same reason, it has the limitation such as the characterization of a unique relationship between the Muskingum parameters and the inflow discharge, not withstanding whether the discharge to be routed is on the rising or falling limbs of the hydrograph. Nevertheless, the MMD method can play useful role in operational purposes such as the real-time flood forecasting.

10.5 CONCLUDING REMARKS

With different interpretations of the Muskingum method available in the literature to describe the flow process of flood wave movement, which are developed based on contradictory concepts, a critical analysis has been carried out herein for their theoretical correctness and consistency with the physical behaviour of the classical Muskingum method. Among the variants of the Muskingum method developed till recently, the VPMM method is the only one which theoretically justifies the heuristic assumption made by McCarthy in 1938 in expressing the channel reach storage of the Muskingum method in terms of prism and wedge storages. Hence, there is no need for any alternative interpretations of the classical Muskingum method such as the one proposed by Cunge in 1969 based on the matched diffusivity concept. Using the proposed approach, the parameters of the Muskingum method are related to channel and flow characteristics, with the travel time parameter K of the method having the identically same relationship as established by Todini (2007) and Price (2009); however, the weighting parameter of the method, although having a similar form of relationship as established by Todini (2007) and Price (2009), does not have the same relationship. Further, the VPMM method is fully volume conservative like those methods introduced by Todini (2007) and Price (2009). Besides, the VPMM method enables simultaneous estimation of the stage hydrograph corresponding to the routed discharge hydrograph.

10.6 REFERENCES

Apollov, B.A., G.P. Kalinin, and V.D. Komarov. 1964. Hydrological forecasting, translated from Russian. Israel Program for Scientific Translations, Jerusalem.

Arnold, J.G., and N. Fohrer. 2005. SWAT2000: current capabilities and research opportunities in applied watershed modeling. *Hydrol. Process.*, 19: 563–572.

Brakensiek, D.L. 1963. Estimating coefficients for storage flood routing. *J. Geophy. Res.,* 68(24), 6471–6474.

Chow, V.T. 1959. *Open Channel Hydraulics.* McGraw Hill, New York, USA.

Chow, V.T., D.R. Maidment, and L.W. Mays. 1988. *Applied Hydrology*. McGraw-Hill Book Co., Singapore.

Cunge, J.A. 1969. On the subject of a flood propagation method (Muskingum Method). *J. Hydraul. Res., IAHR,* 7(2): 205–230.

Diskin, M.H. 1967. On the solution of the Muskingum flood routing equation. *J. Hydrol.,* 5: 286–289.

Dooge, J.C.I. 1973. Linear theory of hydrologic systems. USDA, *Agric. Res. Serv., Tech. Bull., No.1468.*

Dooge, J.C.I., and B.M. Harley. 1967. Linear routing in uniform open channel. in: *Proc. Int. Hydrol. Symp.* (6–8 September, 1967), Fort Collins, Colorado, USA, Vol. 1, 57–63.

Dooge, J.C.I., and B.M. Harley. 1967. Linear routing in uniform open channel. In: *Proc. Int. Hydrol. Symp.* (6–8 September, 1967), Fort Collins, Colorado, USA, Vol. 1, 57–63.

Dooge, J.C.I., W.G. Strupczewski, and J.J. Napiorkowski. 1982. Hydrodynamic derivation of storage parameters of the Muskingum model. *J. Hydrol.*, 54(4): 371–387.

Fread, D.L. 1990. *DAMBRK: The NWS Dam-Break Flood Forecasting model.* National Weather Service, Office of Hydrology, Silver Spring, Marryland, USA.

Gassman, P.W., M.R. Reyes, C.H. Green, and J.G. Arnold. 2007. The Soil and Water Assessment Tool (SWAT): Historical development, applications, and future research directions. *Working Paper 07-WP 443,* Center for Agricultural and Rural Development, Iowa State University, Ames, Iowa. <www.card.iastate.edu>

Geem, Z.W. 2006. Parameter estimation of the nonlinear Muskingum model using the BFGS technique. *J. Irrig. Drain. Eng., ASCE,* 132(5): 474–478.

Geem, Z.W. 2011. Parameter estimation of the nonlinear Muskingum model using Parameter-Setting-Free-Harmony search. *J. Hydrol. Eng., ASCE,* 16(8): 684–688.

Gill, M.A. 1978. Flood routing by the Muskingum method. *J. Hydrol.,* 36(3–4): 353–363.

Jolly, J.P., and V. Yevjevich. 1971. Amplification criterion of gradually varied, single peaked waves. *Hydrol. Pap. No. 51,* Colorado State University, Fort Collins, USA.

Kalinin, G.P., and P.I. Milyukov. 1958. Priblizhennyi raschet neutstanovivshegosya dvizheniya vodnykh mass (Approximate computation of unsteady flow of water masses). *Trudy TSIP, 66.*

Koussis, A.D. 1976. An approximate dynamic flood routing method. *Proc. Int. Symp. On Unsteady Flow in Open Channels, Paper L1,* Newcastle-Upon Tyne, U.K.

Koussis, A.D. 1978. Theoretical estimation of flood routing parameters. *J. Hydraul. Div., ASCE,* 104 (HY1): 109–115.

Koussis, A.D. 1980. Comparison of Muskingum method difference schemes. *J. Hydraul. Div., ASCE,* 106(HY5): 925–929.

Koussis, A.D. 1983. Unified theory for flood and pollution routing. *J. Hydraul. Eng., ASCE,* 109(HY12): 1652–1664.

Kundzewicz, Z.W. (1986), Physically based hydrological flood routing method, *Hydrol. Sci. J., IAHS, 31*(2): 237–261.

Kundzewicz, Z.W. 1982. Parametric analysis of flood routing models (in German). *Report No. 23,* Mitteilungen, Inst. Wasserbau, III, University of Karlsruhe, Germany.

McCarthy, G.T. 1938. The unit hydrograph and flood routing. *Conf. North Atlantic Division*, U.S. Army Corps of Engineers, New London, Conn.

Miller, W.A., and Cunge, J.A. 1975. Simplified equations of unsteady flow. in: *Unsteady Flow in Open Channels,* Vol. I, K. Mahmood and V. Yevjevich (Eds.), Water Resour. Publ., Highlands Ranch, Colo., 183–249.

Mohan, S. 1997. Parameter estimation of nonlinear Muskingum models using genetic algorithm. *J. Hydraul. Eng., ASCE,* 123(2): 137–142.

Nash, J.E. 1959. Systematic determination of unit hydrograph parameters. *J. Geophys. Res.,* 64(1): 111–115.

Nash, J.E. 1960. A unit hydrograph study with particular reference to British catchments. *Proc. Inst. Civ. Eng.,* 17: 249–282.

Natural Environment Research Council (NERC). 1975. Flood Routing Studies. In: *Flood Studies Report,* Vol. III, Institute of Hydrology, Wallingford, UK.

Neitsch, S.L., J.G. Arnold, J.R. Kiniry, R. Srinivasan, and J.R. Williams. 2005. *Soil and Water Assessment Tool Theoretical Documentation* (version 2005). Temple, TX: Grassland, Soil and Water Research Laboratory, Agricultural Research Service, USA. <http://www.brc.tamus.edu/swat/doc.html>

Nijssen, B., D.P. Lettenmair, X. Liang, S.U. Wetzel, and E.F. Wood. 1997. Streamflow simulation for continental-scale river basins. *Water Resour. Res.,* 33(4): 711–724.

Perumal, M. 1989. Unification of Muskingum difference schemes. *J. Hydraul. Eng., ASCE,* 115(4): 536–543.

Perumal, M. 1992a. Discussion on the kinematic wave controversy. *J. Hydraul. Eng., ASCE,* 118(9): 1335–1337.

Perumal, M. 1992b. Multilinear Muskingum flood routing method. *J. Hydrol.,* 133(3–4): 259–272.

Perumal, M. 1992c. The cause of negative initial outflow with the Muskingum method. *Hydrological Sci. J., 37*(4): 391–401.

Perumal, M. 1993a. Comparison of two variable parameter Muskingum methods. *Proc. Symp. Extreme Hydrological Events, Precipitation, Floods and Droughts,* Yokohama, Japan, *IAHS Publ. No. 213*:129–139.

Perumal, M. 1993b. The cause of negative initial outflow with the Muskingum method–Closure. *Hydrol. Sci. J.,* 38: 153–154.

Perumal, M. 2003. Removing some anomalies of the Muskingum method. In: *Proc. Int. Conf. on Water and Environment (WE-2003),* V.P. Singh and R.N. Yadava (Eds.), December 15–18, 2003, Bhopal, India, Allied Publisher Pvt. Ltd., ISBN: 81-7764-549-8.

Perumal, M., A. Naren, and C.M. Rao. 2013. Appraisal of two forms of nonlinear Muskingum flood routing methods. *6th International Perspective on Water Resources and Environment,* January 7-9, 2013, IPWE-000304, ASCE, Izmir,Turkey.

Perumal, M., and Price, R.K. 2013. A fully volume conservative variable parameter McCarthy–Muskingum method: Theory and verification. *J. Hydrol.*, 502: 89–102, 10.1016/j.jhydrol.2013.08.023.

Perumal, M., and Sahoo, B. 2008. Volume conservation controversy of the variable parameter Muskingum–Cunge method. *J. Hydraul. Eng., ASCE,* 134(4): 475–485.

Ponce, V.M. 1991. Kinematic wave controversy. *J. Hydraul. Eng., ASCE,* 117(4): 511–525.

Ponce, V.M., and P.V. Chaganti. 1994. Variable parameter Muskingum–Cunge method revisited. *J. Hydrol.*, 162(3–4): 433–439.

Ponce, V.M., and V. Yevjevich. 1978. Muskingum-Cunge method with variable parameters. *J. Hydraul. Div., ASCE,* 104(HY12): 1663–1667.

Prasad, R. 1967. A nonlinear hydrologic system response model. *J. Hydraul. Div., Proc. ASCE,* 93(4): 201–221.

Price, R.K. (2009), Volume conservative non-linear flood routing, *J. Hydraul. Eng., ASCE,* 135(10): 838–845.

Price, R.K. 1973. Flood routing methods for British rivers. *Proc. Inst. Civ. Engrs.*, 55(12): 913–930.

Sahoo, B. 2013. Field application of the multilinear Muskingum discharge routing method. Water Resour. Manag. 27(5): 1193-1205, doi: 10.1007/s11269-012-0228-5.

Samani, H.M.V., and S. Jebelifard. 2003. Design of circular urban storm sewer systems using multilinear Muskingum flow routing method, *J. Hydraul. Eng., ASCE,* 129(11): 832–838.

Singh, V.P. 1988. *Hydrologic Systems: Rainfall-Runoff Modeling.* Vol. I, Prentice Hall, Englewood Cliffs, New Jersy.

Strupczewski, W.G., and Z.W. kundzewicz. 1980. Muskingum method revisited. *J. Hydrol.,* 48: 327–342.

Szilagyi, J. 1992. Why can the weighting parameter of the Muskingum channel routing method be negative? *J. Hydrol.*, 138(1–2): 145–151.

Todini, E. 2007. A mass conservative and water storage consistent variable parameter Muskingum-Cunge approach. *Hydrol. Earth Syst. Sci.,* 11: 1645-1659.

Tung, Y. K. 1984. River flood routing by nonlinear Muskingum method. *J. Hydraul. Div., ASCE,* 111(12): 1447□1460.

U.S. Army Corps of Engineers (USACE). 2010. HEC-RAS river analysis system user's manual (version 4.1). *Rep. CPD-68*, Hydrol. Eng., Cent., Davis, Calif.

U.S. Army Corps of Engineers (USACE). 2013. Hydrologic modeling system HEC-HMS user's manual (version 4.0). *Rep. CPD-74D*, Hydrol. Eng., Cent., Davis, Calif.

Weinmann, P.E., and E.M. Laurenson. 1979. Approximate flood routing methods: a review, *J. Hydraul. Div., ASCE,* 105(12): 1521–1526.

Wilson, E.M. 1974. Engineering Hydrology, MacMillan Education Ltd., Hampshire, U.K.

Wong, T.H.F. 1984. Improved parameters and procedures for flood routing in rivers. Ph. D. Thesis, Monash University, Victoria, Australia.

Yoon, J., and G. Padmanabhan. 1993. Parameter estimation of linear and nonlinear Muskingum models. *J. Water Resour. Plann. Manag. Div., ASCE,* 119(5): 600□610.

CHAPTER 11

POPULATION BASED EVOLUTIONARY SEARCH OPTIMIZATION METHODS IN HYDROLOGY

Gokmen Tayfur

Department of Civil Engineering, Izmir Institute of Technology
Urla, Izmir 35430 TURKEY, gokmentayfur@iyte.edu.tr

ABSTRACT

This Chapter provides an overview of population based evolutionary search optimization methods used in hydrology. While giving the details of the genetic algorithm (GA), the Chapter summarizes other evolutionary methods, namely; the differential evolution (DE), the particle swarm optimization (PSO), the harmony search (HS) and the ant colony (AC). In this last decade, these methods have gained popularity in optimization of model parameters (or variables), employed in hydrology. The application ranged from river flow stage forecasting, discharge prediction, flood routing in natural channels to reservoir operation.

11.1 INTRODUCTION

Models in hydrology vary from simple black-box to sophisticated distributed physics-based. Black box models whose parameters are identified by a regression analysis relate input to output without concentrating on a detailed description of the physical processes. The use of time-series methods is complicated by non-stationarity and non-linearity in the data that require, in turn, experience and expertise from a modeler (Imrie et al. 2000). The physics-based models that account for physical processes can provide more accurate results at the expense of overwhelming data requirements, simplifications in the governing equations, and sophisticated numerical techniques. Although these models are very complicated with many parameters, frequently the information that they are supposed to provide is simple such as the mean annual groundwater recharge rate over part of the basin or a flood rate in a natural channel. Singh and Woolhiser (2002) provided an overview of different models, a historical perspective and challenges in hydrologic modeling.

One of the recent developments in hydrologic modeling is the employment the evolutionary optimization methods, such as GA, DE, PSO, HS, and AC. The developments of these methods start in late seventies and the first applications to hydrological problems happened to be after 1990s. These techniques are effective at handling dynamic, nonlinear, and noisy data, especially when the underlying relationships are not fully understood (See and Openshaw 2000). These models are more flexible, self-adaptive and less assumption-dependent. The reason for an increasing interest in these models resides in their intrinsic generality, flexibility, and global performance in most applications where other models either tend to fail or become cumbersome.

11.2 GENETIC ALGORITHM (GA)

Holland (1975) was the first to propose GA as search procedure in combinatorial optimization and machine learning problems. GA is a nonlinear search and optimization method. It is inspired from the Darwinian evolutionary view, which is biological processes of natural selection and survival of fittest, and the Mendelian genetics. They make relatively few assumptions and do not rely on any mathematical properties of functions, such as differentiability and continuity. These make them more generally applicable and robust (Liong et al. 1995). They can mostly outperform conventional optimization techniques on difficult, discontinuous, and multimodal functions (See and Openshaw 2000).

Large and complex multidimensional decision spaces can be efficiently searched by GA. It can handle non-convexities that cause difficulties for traditional optimization methods (Sen and Oztopal 2001). GA can consider many points at once during the search process thereby reducing the chance of getting trapped in a local minimum. One may visualize GA in a way where troops are sent to a field to find the lowest location of the space (see Fig. 11.1) or the peak point of the

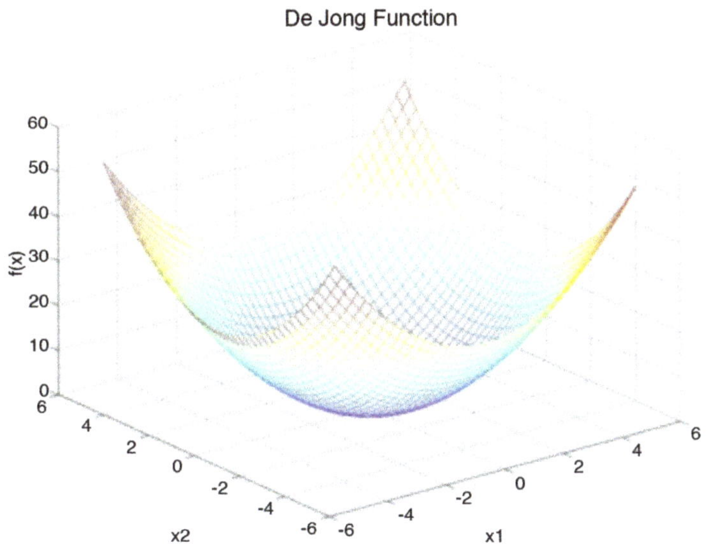

Fig. 11.1. De Jong's 1st Test Function [for two variables,

$$f(x)=\sum_{i=!}^{n} x_i^2, -5.12 \le x_i \le 5.12\,].$$

space, presented in Fig. 11.2. Certain number of troops, let us say 100, are initially dispersed over the whole field (space). They are then asked to find the lowest point (or the peak location) by searching the space. By the GA operations, eventually the troops are expected to reach the desired location. Note that the troops are the chromosomes in GA.

GA often uses Boolen algebra that is developed in late 19th century and based on '*2-based*' numbering system rather than classical *'10-based'*, developed by El-Kharezmi in late 8th century. In Boolen algebra numbers are expressed using digits of 0 and 1 only. For example 19 is expressed in Eq. (11.1) as 10011, which can be explicitly expounded as follows:

$$10011 = 1\bullet\mathbf{2}^4 + 0\bullet\mathbf{2}^3 + 0\bullet\mathbf{2}^2 + 1\bullet\mathbf{2}^1 + 1\bullet\mathbf{2}^0 \qquad (11.1)$$

Recall that, in a classical algebra, any number is expressed explicitly using '10-based' numbering system. For example, 2345 as shown by Eq. (11.2) can explicitly be expounded as:

$$2345 = 2\bullet\mathbf{10}^3 + 3\bullet 102 + 4\bullet 101 + 5\bullet 100 \qquad (11.2)$$

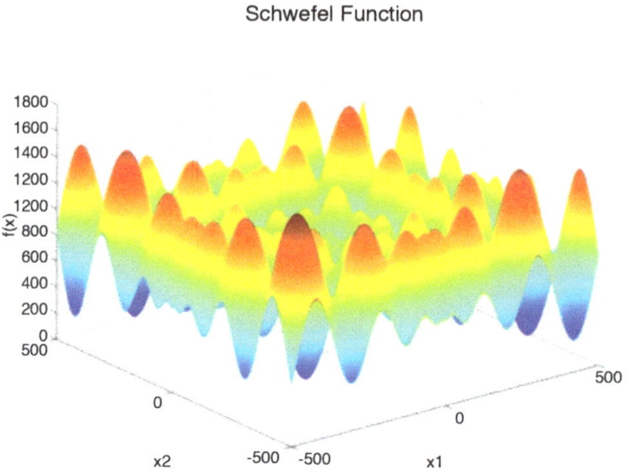

Fig. 11.2. Schwefel's 7th Test Function (for two variables).

$$f(x) = \sum_{i=!}^{n} x_i^2, \; -5.12 \le x_i \le 5.12 \,.$$

11.2.1 Basic Units of GA

The basic units of GA consist of *'bit'*, *'gene'*, *'chromosome'* and '*gene pool'*. Gene consists of bits, chromosome consists of genes and gene pool consists of chromosomes. Bits are zeroes (0) and ones (1). Like zeroes and ones in 10011 (on a 10_based number system it is 19). Gene stands for model parameter (or a decision variable). It corresponds to a characteristic of a person such as his/her hair color. For example; 10011 can be considered as a gene.

The combination of genes forms a chromosome. If there are three genes such as 101011, 110011, 111011, then the chromosome can be expressed as 101011*110011***111011**.

Gene pool consists of chromosomes. For example; the following shows a case when there are 7 chromosomes in the gene pool. Each chromosome has 15 bits and has the same length and consists of 3 genes of 5 bits length. Therefore, in total, there are 21 genes and 105 bits.

1110011100*10010*
1000111101*01111*
1100110011*00110*
1010101010*10101*
1001001001*11100*
1101010101*01001*
1111110001*01000*

11.2.2 Main Operations in GA

Main operations of GAs basically consist of: (1) Forming initial gene pool; (2) Evaluating fitness of each chromosome; (3) Selection; (4) Cross-over, and (5) mutation.

Initially chromosome population is randomly generated to form a gene pool. Each gene stands for a decision variable (model parameter) that needs to be optimized. Each chromosome is a possible solution for the variables. Depending upon the physical problem, optimal number of decision variables is first determined. The order of genes in a chromosome should be decided at the beginning although the order of chromosomes in the pool is not important. At least 15 chromosomes should be generated. Since genes are made up of bits of 0 and 1, these numbers for each gene can be generated by employing random number generators (Sen 2004).

Second operation is the evaluation of the fitness of each chromosome. This is carried out in two steps: (1) first, finding value of objective function for each chromosome; and (2) then, obtaining the fitness of each chromosome. The objective function is a function of the model parameters (decision variables) that need to be optimized. The value of the objective function for each chromosome can be directly found by substituting the chromosome into the function. For example; Tayfur and Moramarco (2008) predicted event_based hydrographs for Tiber River Basin in central Italy. They predicted event-based discharge hydrographs from level data employing GA, using the following RCM formulation:

$$Q_d(t) = \alpha \frac{h_d^{\beta}(t)}{h_u^{\gamma}(t - T_L)} + \eta \tag{11.3}$$

where Q_d is flow rate at downstream station, h_d is flow rate at downstream station, h_u is flow stage at upstream station, T_L is wave travel time, t is time, and α, β, γ, and η are the model parameters, whose optimal values are found by GA model. Here (in Eq. 11.3), each of these 4 parameters is a gene. The combination of these genes constitutes a chromosome. Tayfur and Moramarco (2008) expressed the objective function as follows:

$$MAE = \frac{1}{N}\sum_{i=1}^{N}\left|Q_m - Q_p\right| \tag{11.4}$$

where N is the number of observations, Q_m is measured flow discharge, and Q_p is predicted flow discharge, which is $Q_d(t)$ computed by Eq. (11.3).

The procedure is simply works as follows: initially random values are assigned for each gene (each parameter). In other words; 500 chromosomes are initially generated since they employed 500 chromosomes in their search procedure. Each of these chromosomes is substituted in Eq. (11.3) to obtain $Q_d\ (t)$ for each observation at time t, and then the computed $Q_d(t)$ value is substituted back into Eq. (11.4) to find the error caused by each chromosome for each observation. So, the total error produced by each chromosome for all the observations becomes objective function value for the corresponding chromosome. *Fitness function* is defined as follows (Sen 2004):

$$F\left(Ki\right)=\frac{f\left(Ki\right)}{\sum\limits_{i} f\left(Ki\right)} \tag{11.5}$$

Equation (11.5) gives the percentage of a chromosome in the pool. For the above example; the fitness of each chromosome is the ratio of total error produced by each chromosome to whole error, produced by all the chromosomes (500 of them). The fittest ones are the ones which produce minimum errors since the objective is to minimization of the error function, Eq. (11.4). Fitness determines the survival of individuals. Once the fitness of each individual in the population is evaluated, its fitness is used as the basis for selection.

Third operation is the *selection* implemented by eliminating low-fitness individuals from the populations and replacing them with the high-fitness individuals. This operation can be performed randomly based on the roulette wheel selection (Sen 2004). Figure 11.3 shows an example for 8 chromosome case, in which each slice is proportional to the fitness of each chromosome. The principle of selection based on turning the wheel and waiting until it randomly stops at one of the slices. The wheel is turned many times and the most occurrence ones are selected. Other selection process is the ranking where the chromosomes are ranked according to their fitness from the fittest to weakest. Then the fittest ones are copied on the weakest ones, resulting in the death of weak ones. Note that however after each operation in GA the number of chromosomes is the same.

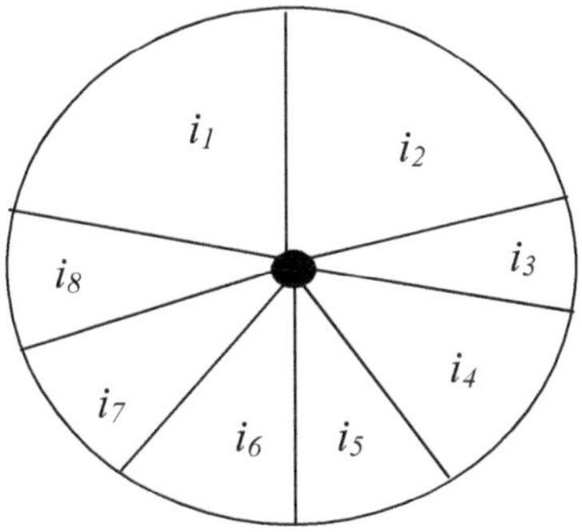

Fig. 11.3. Schematic representation of roulette wheel for 8-chromosome case.

The fourth operation is the *crossover* by which new chromosomes are generated, replacing the some of the existing ones. For this purpose, the chromosomes from the current generation are first selected for the recombination process based on their relative score (fitness). Generally 80% of the chromosomes are selected. The selected chromosomes are paired and they are called parent chromosomes. The pairing can be done randomly or preferable strong ones are paired with weak ones. Then, the genes of the parent chromosomes are interchanged. There are several ways of performing crossover operation (Goldberg 1989, Sen 2004, Tayfur 2012)

The final operation is the *mutation* operation by which GA seeks out new and novel solutions to minimize the chance of getting trapped in a local minimum. By this process, the next trend would faster converge to solution or may diverge from the solution, although it is expected that the algorithm would do better with the mutation operation. Generally up to 5% of all the bits are subjected the mutation operation by which particular bit is reversed (that is, 1 to 0 or 0 to 1).

Figure 11.4 presents crossover and mutation operations. According to Fig. 11.4, the parent chromosomes (190 and 50 at the top) are subjected to crossover by interchanging genes cut from third place from the left. As seen, new offsprings have values of 62 and 178 after the crossover operation. Then, subjecting the new offsprings to mutation by reversing the first 0 digits from left to 1, one has then newer offsprings whose values are 174 and 242 (last two chromosomes in Fig. 11. 4). That means, after these two operations, one continues to the search process, instead of the first two parent chromosomes (190, and 50), with new two offsprings of 174 and 242.

The flowchart presented in Fig. 11.5 shows GA operations in a compact form. A single iteration is performed in order to clearly show the GA operations. For this purpose; objective function of $f(x)=x^{0.25}$ within $0<x<60$ is minimized, working with 6 chromosomes, as presented in Table 11.1.

1	0	1	1	1	1	1	0	→ 190
0	0	1	1	0	0	1	0	→ 50

after cross-over

0	0	1	1	1	1	1	0	→ 62
1	0	1	1	0	0	1	0	→ 178

after mutation

1	0	1	1	1	1	1	0	→ 174
1	1	1	1	0	0	1	0	→ 242

Fig. 11.4. Schematic representation of crossover and mutation operations.

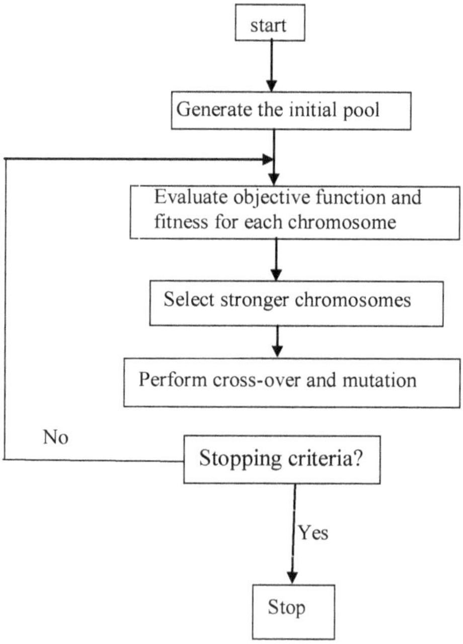

Fig.11.5. Schematic representation of GA algorithm.

Table 11.1. Computed values of objective function and fitness for each chromosome.

Chromosome (initial pool)	10_based	Objective function $f(x)= x^{0.25}$	Fitness (%)
1 1 1 0 1 1	59	2.77	79
1 1 0 0 0 0	48	2.63	80
1 0 0 1 0 0	36	2.45	82
0 1 0 0 0 1	17	2.03	85
0 0 0 1 1 1	7	1.63	88
0 0 1 1 0 1	13	1.90	86
		Total: 13.41	

Performing the selection operation by copying the 5th chromosome on the 1st chromosome and the 6th chromosome on the 2nd chromosome. This would result in the following chromosomes after the selection stage.

0 0 0 1 1 1 (7)
0 0 1 1 0 1 (13)
1 0 0 1 0 0 (36)
0 1 0 0 0 1 (17)
0 0 0 1 1 1 (7)
0 0 1 1 0 1 (13)

Now, for the crossover operation; the chromosomes are paired, i.e., parent chromosomes are formed. To do so; let us pair 1st chromosome with the 3rd chromosome (7-36), the 2nd chromosome with the 4th chromosome (13-7), and 5nd chromosome with the 6th chromosome (7-13):

0 0 0 1 1 1 (7)
1 0 0 1 0 0 (36)
0 0 1 1 0 1 (13)
0 1 0 0 0 1 (17)
0 0 0 1 1 1 (7)
0 0 1 1 0 1 (13)

Performing the crossover operation by cutting the strings of the parent chromosomes from the 3nd digit from the left side and interchanging the corresponding bits. This would result in the following new offsprings:

1	0	0	1	1	1	(39)
0	0	0	1	0	0	(4)
0	1	0	1	0	1	(21)
0	0	1	0	0	1	(9)
0	0	1	1	1	1	(15)
0	0	0	1	0	1	(5)

Now, we can apply mutation operation on the 1st and 3rd chromosomes. The first bit of 1st and the 2nd bit of 3rd chromosomes are replaced by 0 bit. This results in the following chromosomes at the end of first iteration:

0	0	0	1	1	1	(7)
0	0	0	1	0	0	(4)
0	0	0	1	0	1	(5)
0	0	1	0	0	1	(9)
0	0	1	1	1	1	(15)
0	0	0	1	0	1	(5)

As seen; at the end of first iteration, all the chromosomes have values less than 20 and 5 of them are less than 10. It should be kept in mind that this example is for the illustrative purpose and it is performed by bias since we decided the pairing, crossovering and mutation. This is of course is naturally done by the computer code by which it may take more iterations.

11.2.3 Remarks

The details of GA is given in elsewhere (Sen 2004 and Tayfur 2012). Several applications of GA in hydrology are presented in Tayfur (2012). GA has found wide application in water resources engineering, groundwater flow, sediment transport, and channel flows (Aral and Guan 1996; Guan and Aral 1998; Sen and Oztopal 2001; Jain et al. 2004; Wardlaw and Bhaktikul 2004; Fazal et al. 2005; Singh and Datta 2006; Kim et al. 2006, Tayfur 2009, Tayfur et al. 2009, Tayfur and Singh 2011, Tayfur and Karimi 2014).

GA is also employed together with ANN. It is used to obtain the optimal values of the connection weights of ANN. For example, Morshed and Kaluarachchi (1998) employed GA to predict connection weights of ANN in simulating flow and transport of solute in groundwater. GA was also used to train the radial basis functions (Harpham et al. 2004). Also, GAs can be used to better proportion the training and testing data sets for ANN. For example, Akkurt et al. (2003) applied GA to separate training and testing data sets from the complete original data set by using the GAs in modeling compressive strength of cement mortar.

Lastly, it needs to be pointed out that; unlike the ANN method, GA has a good extrapolation capability. For example; Tayfur and Moramarco (2008) calibrated the rating curve method's parameters in Equ (1) with low peak hydrograph data and successfully predicted high peak hydrographs. Similarly, they calibrated the same parameters with short wave travel time hydrograph and predicted long wave travel time hydrographs.

11.3 DIFFERENTIAL EVOLUTION (DE)

DE is a simple and powerful evolutionary optimization technique which is faster and robust (Storn and Price, 1997). It is a population based algorithm like genetic algorithm. It uses the similar operations like selection, crossover, and mutation. The main difference is that, while genetic algorithm relies on the crossover, DE relies on the mutation and the selection operations. While the mutation is used as a search mechanism and the selection directs the search toward the prospective regions in the search space. The DE algorithm also uses a non-uniform crossover by which it can take offspring vector parameters from one parent more often than it does from others. By using the components of the existing population members to construct trial vectors, the crossover operator efficiently shuffles the information about successful combinations. Thus, it enables the search for a better solution space.

An optimization task consisting of DE parameters can be represented by a D-dimensional vector. In DE, the population of NP (number of particle) solution vectors is randomly created at the beginning. This population is successfully improved by applying mutation, crossover and selection operators (Karaboga and Okdem, 2004). Normally, *NP* should be about 10 times the number of parameters in a vector (Storn and Price, 1997; Karahan, 2011).

For each target vector $X_{i,j,G}$, a mutant vector is produced by DE/rand/1/bin strategy (Price and Storn, 1997).

$$V_{i,j,G+1} = X_{R1,j,G} + F\left(X_{R2,j,G} - X_{R3,j,G}\right) \tag{11.6}$$

where, $\mathrm{j} \in \{1,2,...,\mathrm{D}\}$; $\mathrm{i, R1, R2, R3} \in \{1,2,...,\mathrm{NP}\}$ are randomly chosen and must be different from each other, *G* is the generation number. In Eq. (11.6), *F* is the mutation factor which has an effect on the difference vector $(\mathrm{X}_{\mathrm{R2,j,G}} - \mathrm{X}_{\mathrm{R3,j,G}})$.

The parent vector is mixed with the mutated vector to produce a trial vector as follows:

$$U_{i,j,G+1} = \begin{cases} V_{i,j,G+1}, if rand_j \leq CR \;\; or \;\; j \leq jrand \\ X_{i,j,G}, if rand_j > CR \;\; or \;\; j \neq jrand \end{cases} \tag{11.7}$$

where, $\mathrm{j} \in \{1,2,...,\mathrm{D}\}$; $\mathrm{rand_j} \in [0,1]$ is the random number; $\mathrm{CR} \in [0,1]$ is crossover rate and $\mathrm{jrand} \in \{1,2,...,\mathrm{D}\}$ is the randomly chosen index.

All solutions in the population have the same chance of being selected as parents without the dependence on their fitness value. The offspring produced after the mutation and crossover operations is evaluated. Then, the performance of the offspring vector and its parent is compared and the better one is selected. If the parent is still better, it is retained in the population as the following procedure:

$$X_{i,j,G+1} = \begin{cases} U_{i,j,G+1}, f\left(U_{i,j,G+1}\right) < f\left(X_{i,j,G+1}\right) \\ X_{i,j,G}, \;\; otherwise \end{cases} \tag{11.8}$$

The details of DE can be obtained from the literature (Price and Storn, 1997; Karaboga and Okdem, 2004; Karahan, 2011; Vasan and Simunovic, 2010; Gurarslan, 2011; among many).

11.4 PARTICLE SWARM OPTIMIZATION (PSO)

PSO is an algorithm tailored to optimize complicated numerical functions based on metaphor of human social interaction (Chau 2007). Although it is initially developed as a tool for modeling social behavior, the PSO algorithm has been recognized as a computational intelligence technique intimately related to evolutionary algorithms (Clerc and Kennedy, 2002). It is a populated search method for optimization of continuous nonlinear functions resembling the movement in a bird flock.

The basic principle of PSO algorithm is formed on the assumption that potential solutions are flown through hyperspace with acceleration towards more optimum solutions. Each particle (bird) adjusts its flying according to the flying experiences of both itself and its neighbours. During the process, the coordinates in hyperspace associated with its previous best fitness solution and the overall best value attained thus far by other particles within the group are kept and recorded in the memory (Chau 2007, Kumar and Reddy 2007).

The advantage of PSO algorithm is its relatively simple coding and hence low computational cost (Chau 2007). It is quite similar to a genetic algorithm with respect to the fitness concept and the random population initialization. However, the evolution of generations of a population in such a system is by cooperation and competition among the individuals themselves. The population responds to the quality factors of the previous best individual values and the previous best group values. The allocation of responses between the individual and group values ensures a diversity of response. The principle of stability is adhered to since the population changes its state if and only if the best group value changes. It is adaptive corresponding to the change of the best group value. The capability of PSO algorithm, in determining the global optimum with high probability and fast convergence rate, is demonstrated elsewhere (Chau 2007, Kumar and Reddy 2007).

PSO algorithm is mathematically presented in the following (Kayhan et al. 2010). Let

f: fitness function governing the problem

NP: number of particles in the swarm

D: dimension of the problem (e.g. number of decision variables)

$x_i = [x_{i1}, x_{i2}, ..., x_{iD}]^T$: vector containing current positions of the particles in each direction

$v_i = [v_{i1}, v_{i2}, ..., v_{iD}]^T$: vector containing the current velocities of the particles in each dimension

$\hat{x}_i = [\hat{x}_{i1}, \hat{x}_{i2}, ..., \hat{x}_{iD}]^T$: vector containing the current best position of each particle in each dimension and

$\hat{g} = [g_1, g_2, ..., g_D]^T$: vector that contains the global best position in each dimension ($\forall i = 1,2,...,NP$ and $\forall j = 1,2,...,D$), T is the transpose operator.

The new velocities of the particles are then calculated as follows:

$$V_i^{k+1} = \omega v_i^k + c_1 r_1 \left(\hat{X}_i - X_i^k\right) + c_2 r_2 \left(\hat{g} - X_i^k\right), \quad \forall i = 1,2,...,NP \qquad (11.9)$$

where k is the iteration index, ω is the inertial constant, c_1 and c_2 are the acceleration coefficients which are used to determine how much the particle's personal best and the global best influence its movement, and r_1 and r_2 are the uniform random numbers between 0 and 1. Note that the values of ω, c_1 and c_2 control the impact of previous historical values of particle velocities on its current one. A larger value of ω leads to global exploration whereas smaller values results with a fine search within the solution space. Therefore, suitable selection of ω, c_1 and c_2 provides a balance between the global and local search processes. Note that the terms $c_1 r_1 \left(\hat{x}_i - x_i^k\right)$ and $c_2 r_2 \left(\hat{g}_i - x_i^k\right)$ in Eq. (11.9) are called the *cognition* and *social* terms, respectively. The *cognition* term takes into account only the particle's own experience, whereas the *social* term signifies the interaction between the particles. Velocities of particles in a swarm are usually bounded with a maximum velocity $\mathrm{v}^{\max} = \left[\mathrm{v}_1^{\max}, \mathrm{v}_2^{\max}, \ldots, \mathrm{v}_D^{\max}\right]^T$ which is calculated as a fraction of the entire search space as follows (Shi and Eberhart, 1998):

$$v^{max} = \gamma\left(x^{max} - X^{min}\right) \tag{11.10}$$

where γ is a fraction $(0 \leq \gamma < 1)$, $\mathrm{v}^{\max} = \gamma\left(\mathrm{x}^{\max} - \mathrm{X}^{\min}\right)$ and $\mathrm{x}^{\min} = \left[\mathrm{x}_1^{\min}, \mathrm{x}_2^{\min}, \ldots, \mathrm{x}_D^{\min}\right]^T$ are the vectors that contain the upper and lower bounds of the search space for each dimension, respectively. After updating the velocity vector by Eqs (11.9) and (11.10), the new positions of the particles are then calculated as follows:

$$x_i^{k+1} = x_i^k + v_i^{k+1}, \quad \forall i = 1, 2, \ldots, NP \tag{11.11}$$

After obtaining the new positions by Eq. (11.11), the corresponding fitness values are calculated based on the new positions of the particles. Then, the values of $\hat{\mathbf{x}}_i$ and $\hat{\mathbf{g}}$ $(\forall i = 1, 2, \cdots, NP)$ are updated. This solution procedure is repeated until the given termination criterion is satisfied. The details of PSO can be obtained elsewhere (Shi and Eberhart, 1998; Clerc and Kennedy, 2002; Chau, 2007; Gurarslan and Karahan, 2011; Karahan, 2012; among many). In hydrology; Kumar and Reddy (2007) employed PSO for multi-purpose reservoir operation. Chu and Chang (2009) applied PSO to obtain optimal values of the Muskingum model parameters in flood routing. Karahan (2012) used PSO for contructing relations between rainfall intensity-duration-frequency. Chau (2007) applied PSO to train ANN when forecasting river stage data. Ayvaz (2013) applied PSO for predicting Manning's rougness coefficient in natural channels.

11.5 HARMONY SEARCH (HS)

HS is a heuristic optimization algorithm which is proposed by Geem et al. (2001). It is a stochastic random search method that is conceptualized based on the process of searching for perfect state of harmony. In HS algorithm, musical performance seeks to find a musically pleasing harmony as determined by some aesthetic standards. This relates to the optimization process which seeks to obtain a global optimal solution as determined by an objective function value. The harmonies in the musicians' memories are analogous to the optimization solution vector. In addition, the musicians' improvisations are analogous to the solution iterations in the optimization process (Geem et al. 2001).

A musically pleasing harmony can be found based on three musical rules: (i) by playing a note from harmony memory (HM); (ii) by playing a note which is closer to another note stored in HM; and (iii) by playing an arbitrary note from the entire note range. Combination of these rules allows finding a musically pleasing harmony (best state). Adaptation of these rules to an optimization problem is as follows:

(i) Generate a new solution vector from HM (memory consideration)

(ii) Replace a decision variable with a new one which is close to the current one (pitch adjusting)

(iii) Generate a solution vector from the possible random range (random selection).

Combined utilization of these rules allows the identification of the optimal or near optimal solutions. Although HS searches the optimal solution by considering multiple solution vectors as in GA, its reproduction process is different from GA. While GA generates a new offspring from two parents in the population, HS generates it from all of the existing vectors stored in HM. The algorithm has the following steps (Karahan et al. 2013):

Step 1: Generate random vectors $\left(x^1, x^2, ..., x^{HMS}\right)$ as many as HMS (harmony memory size), then store them in harmony memory (HM). The HM has the following structure:

$$HM = \left[\begin{array}{ccccc|c} x_1^1 & x_2^1 & \cdots & x_{N-1}^1 & x_N^1 & f(x^1) \\ x_1^2 & x_2^2 & \cdots & x_{N-1}^2 & x_N^2 & f(x^2) \\ \vdots & \vdots & \vdots & \vdots & \vdots & \vdots \\ x_1^{HMS-1} & x_2^{HMS-1} & \cdots & x_{N-1}^{HMS-1} & x_N^{HMS-1} & f(x^{HMS-1}) \\ x_1^{HMS} & x_2^{HMS} & \cdots & x_{N-1}^{HMS} & x_N^{HMS} & f(x^{HMS}) \end{array}\right] \qquad (11.12)$$

Step 2: Generate a new vector $\mathbf{X}'$. For each component x_i',

- with probability *HMCR* (harmony memory considering rate), pick the stored value from HM, $x_i' = x_i^{\text{int}(rand(0.1)^* HMS)+1}$
- with probability *(1-HMCR)*, pick a random value within the allowed range.

Step 3: Perform additional work if the value in Step 2 comes from HM

- with probability *PAR* (pitch adjusting rate), change x_i' by a small amount: $x_i' = x_i' + bw^* rand(0,1)$ for continuous variable.
- with probability *(1 – PAR)*, do nothing.

Step 4: If X' is better than the worst vector X^{Worst} in HM, replace X^{Worst} with X'.

Step 5: Repeat from Step 2 to Step 4 until termination criterion (e.g. maximum iterations) is satisfied.

The details of HS can be obtained from the literature (Geem et al. 2001; Ayvaz 2007; Karahan et al. 2013; among many).

The HS algorithm has been applied to various water resources engineering optimization problems including, river flood management (Kim et al. 2001), optimum design of water distribution network (Geem, 2006), aquifer parameter and zone structure identification (Ayvaz, 2007), groundwater management (Ayvaz 2009; Ayvaz and Elci 2013) and groundwater pollution source identification problems (Ayvaz 2010).

11.6 ANT COLONY (AC)

AC is ispired from the behavior of ant colony and it was first proposed by Dorigo (1992) who asked the question: 'how do ants, who are completly blind, manage to establish the shortest paths from thier nest to their feeding source and back ?' In fact, this puzzle had long been studied by the ethologists who found that 'ants communicate information by depositing pheromone (chemical) trails'. A moving ant leaves, in various quantities, some pheromone on the graound to mark its way. An isolated ant, moving essentially at random, can detect previously laid pheromone trail and follow it with high probability. Thus, reinforcing the track with its own pheromone. The more ants follow the track, the more attractive becomes the track. Thus, the probability with which an ant chooses a path increases with the number of ants that previously choose the same path.

11.6.1 AC Concept

Figure 11.4 shows AC conceptually where there are two paths from ant nest A location to food source B location. Between A and B locations, there is an obstacle. Short path (ACDEB) is 4 unit and long path (ACFEB) is 6 unit. That is; AC = CD= DE = EB = 1 unit and CF = FE = 2 unit. Let us assume that there are 16 ants in the nest and they are about to search for food.

Each ant moves one unit distance per unit time. Each ant deposits one unit pheromone (P) on the path after reaching the next node. At initial stage (T = 0), according to Fig. 11.6, 16 ants start moving in AC direction. At 1 unit time (T = 1), 16 ants arrive at C node, leaving 16 units pheromone on AC. At C, it is assumed that 8 ants choose CD and 8 ants choose CF paths. At the end of 4 unit time (T = 4), 8 ants on the short track arrive at the food source B, while other 8 is just half way of FE, as shwon in Fig. 11.7. At T = 5, when 8 ants on short route arrive back at node E, there is already 8 units of pheromone on ED and 8 units of pheromone on EF. Therefore, it is assumed that, with equal probability, 4 ants would choose ED and other 4 would chose EF.

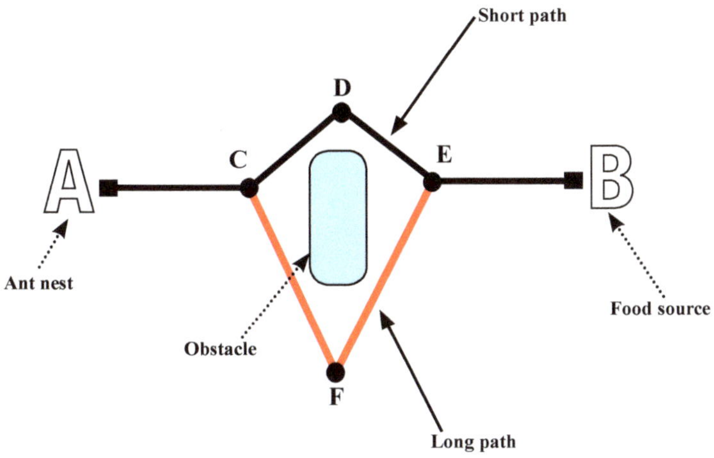

Fig. 11.6. Conceptual representation of ant colony behaviour.

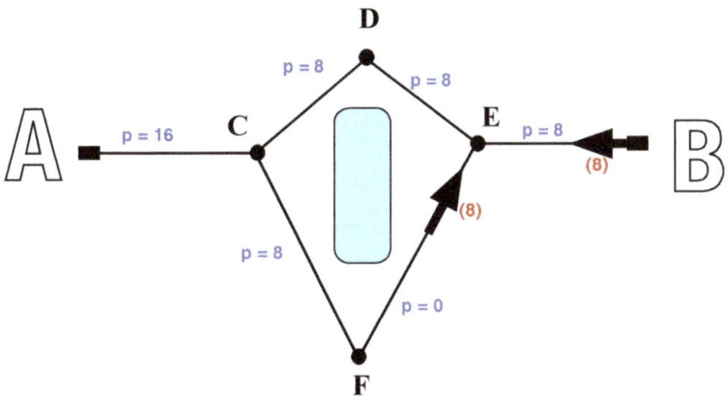

Fig. 11.7. Pheremone concentrations and ant locations at the end of 4 unit time.

As ants continue their journey back and fourth, as presented in Fig. 11.6, at T =13, 3 would be at node E from the food source. 5 on the short route would be at E for the food source. 1 on the long route would be at E for food source. 3 would be at F. 4 ants from the nest would be at node C. When 4 ants from the nest reach C, there is already 24 units of p on CD and 20 on CF. With high probability more ants would choose CD. Hence, it is assummed that 3 would chose CD and 1 CF. When 3 ants from the food source reach node E, there is already 24 p on DE and 17 on FE.

Therefore, it is assumed that 2 would choose ED and 1 EF. As seen in Fig. 11.8, the pheromone concentration on the short route is already higher and therefore as time progresses, more ants with high probability would choose the short path. Eventually all the ants would be on the short path.

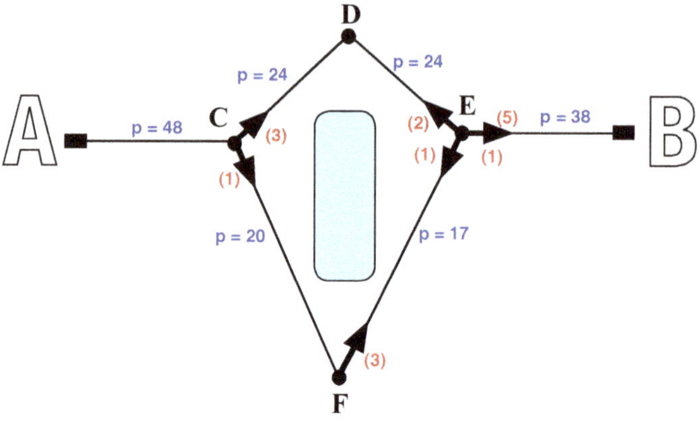

Fig. 11.8. Pheremone concentrations and ant locations at the end of 13 unit time.

11.6.2 AC Algorithm

The basic principle of positive reinforcement via the use of pheromone trails underlines the algorithm. It makes use of a colony of cooperating individuals and adopts a stochastic decision making policy using local information. It can incorporate features that are not found in natural ants. For example; artifical ants are generally given some memory and sight and live in an environment where time is discrete (Dorigo et al. 1996).
Pheromone updates may only occur once one or more ants have completed their tour. In order to implement AC, the combinatorial optimization problem under consideration is mapped onto a graph **G = (D,L,C)** where;

- D = (d1, d2,...dn) is a set of points (i = 1,2,...n) at which decisions have to be made,
- L = (l i(j)) is the set of options (j) available at each decision point (i) and
- C = (c i(j)) is set of costs associated with options L = (l i(j))

Figure 11.9 is shown as an illustrative purpose. A set of finite constraints may also be assigned over the elements of D and L. A feasable path over G is called a solution (f) and a minimum cost path is an optional solution (f*). The cost of solution is denoted by f(f) and the cost of the optional solution is represented by f(f*) (Caro and Dorigo 1998).

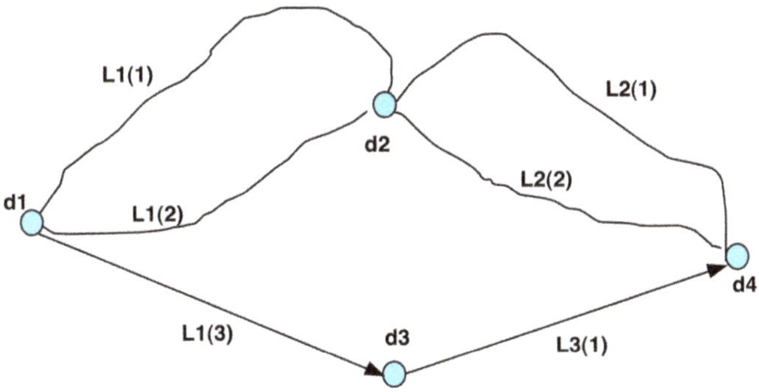

Fig. 11.9. Illustration of (D, L, C) mapping.

Figure 11.9 illustrates the concept where there are 4 decision points—d1, d2, d3, d4. A cost is associated with each of the available option. For example, if the various options represent distances between cities, the costs associated with the various options would be the lengths of the respective paths. Solutions would consist of the different paths connecting d1 to d4 and the costs of different solutions would be the total lengths of the paths.

Figure 11.10 shows the flowchart for AC algorithm. According to Fig. 11.10, the main steps of AC algorithm can be summarized as follows:

- Trial solutions are incrementally constructed as artiificial ants move from one decision point to the next until all decision points have been covered.
- The cost of the generated trial solution is calculated. The generation of a complete trial solution and calculation of the corresponding cost is referred to as a cycle (k)
- Pheromone is updated after the completion of one iteration (t), which consists of m cycles (i.e. the construction of m trial solutions) where m is the number of used ants. The generation of a complete trial solution and calculation of the corresponding cost is referred to as a cycle (k)

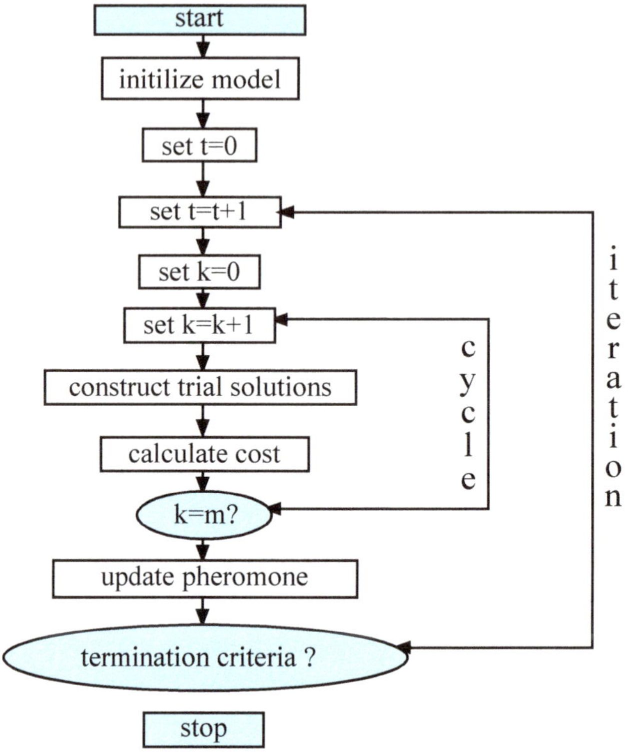

Fig. 11.10. Flowchart for AC algorithm.

There are cycles as many as ants. That is m cycles = m ants. One cycle is the completion of path from nest to source by one ant and the related cost. One iteration covers when all the ants cover the path.

At each decision point, the component to be added to the trial solution is exprssed as (Dorigo et al. 1996):

$$P_{i(j)}(k,t) = \frac{\left[p_{i(j)}(t)\right]^{\alpha}\left[\eta_{i(j)}(t)\right]^{\beta}}{\Sigma_{L_{i(j)}}\left[p_{i(j)}(t)\right]^{\alpha}\left[\eta_{i(j)}(t)\right]^{\beta}} \quad (11.13)$$

where; $P_{i(j)}$ (k,t) is the probability that option Li(j) is chosen at cycle k and iteration t; $p_{i(j)}$ (t) is the concentration of pheromone associated with option Li(j) at iteration t; η i(j) = L/c i(j) is the heuristic factor favoring options that have smaller 'local' costs; and α and β are the exponent

parameters that control the relative importance of pheromone and local heuristic factor, respectively.
For example; the following would give the probability that 10th ant at 20th iteration will choose path 3 when it is at node 1.

$$P_{1(3)}\left(10,\ 20\right)=\frac{p_{1(3)}\left(20\right)\ \eta_{1(3)}}{p_{1(1)}\left(20\right)\ \eta_{1(1)}p_{1(2)}\left(20\right)\ \eta_{1(2)}p_{1(3)}\left(20\right)\ \eta_{1(3)}} \tag{11.14}$$

The probability depends on the existing pheromone concentrations deposited on the paths and the vision of the ants which, in turn, depends on the cost of the paths. It should be noted that the addition of local heuristic factor $\eta_{i(j)}$ is analogous to providing real ants with sight, and it is sometimes called 'visibility'. Artificial ants can also be provided with memory to ensure that each decison point is only visited once.

Once an iteration has been completed, and m trial solutions have been constructed, the pheromone trails are updated in a way that reinforces good solutions. The general form of the pheromone update equation is given as follows (Dorigo et al. 1996):

$$p_{i(j)}\left(t+1\right)=\delta p_{i(j)}\left(t\right)+\Delta p_{i(j)} \tag{11.15}$$

where; $p_{i(j)}$ (t+1) : pheromone concentration associated with option L i(j) at iteration t+1 and ΔD p i(j) : change in pheromone concentration associated with option L i(j) as a function of the trial solutions found at iteration t. δ is less than 1 and it is pheromone persistance coefficient. It enables greater exploration of the search space and avoids premeture convergance to suboptimal solutions as it reduces the difference in pheromone concentration between options at each decision point. In addition, it reduces the likelihood that high cost solutions will be selected in future cycles. Also, it reduces the likelihood that high cost solutions would be selected in future cycles.

The details of AC algorithm can be obtained from Dorigo (1992), among many. There are some applications of AC algorithm in hydrology. For example; Abbaspour et al. (2001) employed AC for estimating unsaturated soil hydraulic parameters. Li et al. (2006) used AC for parameter estiamtion in groundwater. Ibanez et al. (2008) used AC for the optimal control of pumps in water distribution networks.

11.7 CONCLUDING REMARKS

Evolutionary optimization methods have found wide application in solving hydrological problems in recent years. The concept of GA is based on the biological processes of natural selection and the survival of the fittest. GA can sometimes become more complicated when there are many model parameters. Furthermore, crossover and mutation operations may prevent GA reaching an optimal solution.

DE is a population based algorithm like genetic algorithm. It uses the similar operations such as selection, crossover, and mutation. The main difference is that, while genetic algorithm relies on the crossover, DE relies on the mutation and the selection operations. The DE algorithm also uses a non-uniform crossover by which it can take offspring vector parameters from one parent more often than it does from others.

PSO is also similar to GA with respect to the fitness concept and the random population initialization. However, the evolution of generations of a population in such a system is by cooperation and competition among the individuals themselves. The basic principle of PSO algorithm is formed on the assumption that potential solutions are flown through hyperspace with acceleration towards more optimum solutions. Each particle adjusts its flying according to the flying experiences of both itself and its neighbours.

HS is a stochastic random search method that is conceptualized based on the process of searching for perfect state of harmony. In HS algorithm, the optimization process seeks to obtain a global optimal solution as determined by an objective function value.

AC is ispired from the behavior of ant colony. The basic principle of positive reinforcement via the use of pheromone trails underlines the algorithm. It makes use of a colony of cooperating individuals and adopts a stochastic decision making policy using local information. It can incorporate features, such as memory and sight that are not found in natural ants.

11.8 REFERENCES

Abbaspour, K. C., Schulin, R. and van Genuchten, M. T. 2001. Estimating unsaturated soil hydraulic parameters using ant colony optimization. *Advances in Water Resources,* 24(8), 827-841.

Abebe, A.J., Solomatine, D.P., and Venneker, R.G.W. 2000. Application of adaptive fuzzy rule-based models for reconstruction of missing precipitation events. *Hydrological Sciences Journal*, IAHS, 45(3): 425-436.

Akkurt, S., Ozdemir, S., Tayfur, G., and Akyol, B. 2003. The use of GA-ANNs in the modeling of compressive strength of cement mortar. *Cement and Concrete Research*, 33, 973-979.

Aral, M.M., and Guan, J. 1996. Genetic Algorithms in Search of Groundwater Pollution Sources. NATO Adv. Study Inst. on Ground Water Pollution Control and Remediation, Kluwer Acad. Publ., pp. 347-369.

Ayvaz , M.T. 2007. Simultaneous determination of aquifer parameters and zone structures with fuzzy c-means clustering and meta-heuristic harmony search algorithm. *Advances in Water Resources*, 30, 2326-2338.

Ayvaz, M.T. 2009. Application of harmony search algorithm to the solution of groundwater management models. *Advances in Water Resources*, 32, 916-924.

Ayvaz, M.T. 2010. A linked simulation-optimization model for solving the unknown groundwater pollution source identification problems. *J. Contaminant Hydrology*, 117(1-4), 46-59.

Ayvaz, M.T. 2013. A linked simulation-optimization model for simultaneously estimating the Manning's surface roughness values and their parameter structures in shallow water flows. *Journal of Hydrology*, 500, 183-199.

Ayvaz, M.T. and Elci, A. 2013. A groundwater management tool for solving the pumping cost minimization problem for the Tahatali watershed (Izmir-Turkey) using hybrid HS-Solver optimization algorithm. *Journal of Hydrology*, 478, 63-76.

Caro, G. and Dorigo, M. 1998. AntNet: Distributed Stigmergetic Control for Communications Networks. *Journal of Artificial Intelligence Research* (JAIR), 9, 317– 365.

Chau, K.W. 2007. A split-step particle swarm optimization algorithm in river stage forecasting. *Journal of Hydrology*, 34, 131-135.

Chu, H-J. and Chang, L-C. 2009. Applying particle swarm optimization to parameter estimation of the nonlinear Muskingum model. *Journal of Hydrologic Engineering*, 14(9), 1024-1027.

Clerc, M., Kennedy, J. 2002. The particle swarm-explosion, stability, and convergence in a multidimensional complex space. IEEE Transactions on Evolutionary Computation, 6(1), 58-73.

Dorigo, M. 1992. Optimization, Learning and Natural Algorithms. (in Italian). Ph.D. thesis, Dipartimento di Elettronica, Politecnico di Milano, Italy.

Dorigo, M. , Maniezzo, V. , and Colorni, A. 1996. Ant system: Optimization by a colony of cooperating agents. *IEEE Trans. Syst., Man, Cybern., Part B: Cybern.* , 26 (1), 29–41.

Fazal, M.A., Imaizumi, M., Ishida, S., Kawachi T. and Tsuchihara T. 2005. Estimating groundwater recharge using the SMAR conceptual model calibrated by genetic algorithm. *Journal of Hydrology*, 30(1-4), 56-78.

Geem, Z.W., Kim, J.-H., Loganathan, G.V. 2001. A new heuristic optimization algorithm: harmony search. *Simulation*, 76 (2), 60-68.

Geem, Z.W. 2006. Optimal cost design of water distribution networks using harmony search. *Engineering Optimization*, 38 (3), 259–280.

Goldberg, D. E. 1989. Genetic algorithms in search, optimization and machine learning, Addison-Wesley, Reading, Massachusetts, USA.

Goldberg, D.E. 1999. Genetic Algorithms, Addison-Wesley, USA.

Guan, J. and Aral, M. M. 1998. Progressive Genetic Algorithm for Solution of Optimization Problems with Nonlinear Equality and Inequality Constraints. *Journal of Applied Mathematical Modelling*, Vol. 23, pp. 329-343.

Guan, J. and Aral, M.M. 2005. Remediation System Design with Multiple Uncertain Parameters using Fuzzy Sets and Genetic Algorithm. *ASCE Journal of Hydrologic Engineering*, Vol. 10, No. 5, pp. 386-394.

Gurarslan, G. 2011. Identification of Groundwater Contaminant Source Locations and Release Histories by Using Differential Evolution Algorithm (In Turkish), PhD Dissertation, Pamukkale University, Denizli, Turkey.

Gurarslan, G., Karahan, H., 2011. A Parameter Estimation Technique for the Nonlinear Muskingum Flood Routing Model, 6th EWRA International Symposium-Water Engineering and Management in a Changing Environment, 2011, Catania, Italy.

Harpham, C., Dawson, C.D. and Brown, M.R. 2004. A review of genetic algorithms applied to training radial basis function networks. *Neural Computing and Applications*, 13(3), 193-201.

Holland, J.H. 1975. Adaptation in natural and artificial systems: an introductory analysis with applications to biology, control and artificial intelligence, University of Michigan Press, 183 pages.

Ibanez, K.L., Prasad, T.D. and Paechter, B. 2008. Ant colony optimization for optimal control of pump in water distribution networks. *J. Water Resources Planning and Management*, 134(4), 337-346.

Imrie, C.E., Durucan, S. and Korre, A. 2000. River flow prediction using artificial neural networks: generalization beyond the calibration range. *Journal of Hydrology*, 233, 138-153.

Jain, A., Bhattacharjya, R.K. and Sanaga, S. 2004. Optimal design of composite channels using genetic algorithm. *J. Irrigation and Drainage Engrg.*, ASCE, 130(4), 286-295.

Karaboga, D., Okdem, S., 2004. A Simple and Global Optimization Algorithm for Engineering Problems: Differential Evolution Algorithm. *Turkish J. Electrical Engineering*, 12 (1), 53-60.

Karahan, H., 2011. Obtaining regional rainfall-intensity-duration-frequency relationship curves by using differential evolution algorithm. *Scientific Research Project of TUBITAK* (108Y299), Denizli, Turkey (In Turkish).

Karahan, H., 2012. Determining rainfall-intensity-duration-frequency relationship using Particle Swarm Optimization. *KSCE Journal of Civil Engineering*, 16(4), 667-675.

Karahan, H., Gurarslan, G., Geem, Z.W. 2013. Parameter Estimation of the Nonlinear Muskingum Flood Routing Model Using a Hybrid Harmony Search Algorithm. *J. Hydrologic Engineering*, 18 (3), 352-360.

Kayhan, A.H., Ceylan, H., Ayvaz, M.T., Gurarslan, G. 2010. PSOLVER: A new hybrid particle swarm optimization algorithm for solving continuous optimization problems. *Expert Systems with Applications,* 37(10), 6798-6808.

Kim, T., Heo, J-H., Jeong, C-H. 2006. Multireservoir system optimization in the Han River basin using multi-objective genetic algorithms. *Hydrological Processes*, 20, 2057-2075.

Kim, J.H., Geem, Z.W., Kim, E.S. 2001. Parameter estimation of the nonlinear Muskingum model using harmony search. *J Am Water Resour Assoc*, 37 (5), 1131-1138.

Kumar, D.N., Reddy, M.J. 2007. Multipurpose reservoir operation using particle swarm optimization. *J. Water Resources Planning and Management*, 133 (3), 192–201.

Li S., Liu Y., and Yu H. 2006. Parameter estimation approach in groundwater hydrology using hybrid ant colony system. Computational Intelligence and Bioinformatics. *Lecture Notes in Computer Science*, 2006, 4115, 2006, 182-191.

Liong, S.Y., Chan, W.T. and ShreeRam, J. 1995. Peak flow forecasting with genetic algorithm and SWMM. *Journal of Hydraulic Engineering*, ASCE, 121(8), 613-617.

Mohan, S. 1997. Parameter Estimation of Nonlinear Muskingum models using Genetic Algorithm. *J. Hyraulic. Engg*, ASCE, 123(2), 137-142.

Morshed, J. & Kaluarachchi, J. J. 1998. Parameter estimation using artificial neural network and genetic algorithm for free-product and recovery. *Water Resources Research*, 34(5), 1101–1113.

See, L. and Openshaw, S. 2000. A hybrid multi-model approach to river level forecasting. *Hydrological Sciences Journal*, 45(4), 523-536.

Sen, Z. and Oztopal, A. 2001. Genetic algorithms for the classification and prediction of precipitation occurrence. *Hydrological Sciences Journal*, 46(2), 255-267.

Sen, Z. 2004. Genetic algorithms and optimization methods. Turkish Water Foundation, Istanbul, Turkey. (in Turkish).

Shi, Y., and Eberhart, R. 1998. A modified particle swarm optimizer. Proceedings of the IEEE international conference on evolutionary computation, Anchorage, Alaska, 69-73.

Singh, V.P. and Woolhiser, D.A. 2002. Mathematical modeling of watershed hydrology. *Journal of Hydrologic Engineering*, ASCE, 7(4), 270-292.

Singh, R. M. and Datta, B. 2006. Identification of Groundwater Pollution Sources Using GA-based Linked Simulation Optimization Model. *J. Hydrologic Engrg*., ASCE, 11 (2), 101-109.

Storn, R., Price, K.V., 1997. Differential evolution-a simple and efficient heuristic for global optimization over continuous spaces. *J. Global Optimization*, 11(4), 341-359.

Tayfur, G. and Moramarco, T. 2008. Predicting hourly-based flow discharge hydrographs from level data using genetic algorithms. *Journal of Hydrology*, 352 (1-2), 77-93.

Tayfur, G., Barbetta S., Moramarco T. 2009. Genetic Algorithm-Based Discharge Estimation at Sites Receiving Lateral Inflows. *Journal of Hydrologic Engineering*, 14(5), 463-474.

Tayfur, G. 2009. GA-optimized model predicts dispersion coefficient in natural channels. *Hydrology Research*, 40(1), 65-78.

Tayfur G, Singh VP, 2011. Predicting Mean and Bankfull Discharge from Channel Cross-Sectional Area by Expert and Regression Methods. *Water Resources Management*, 25(5), 1253-1267.

Tayfur, G. 2012. Soft Computing Methods in Water Resources Engineering. WIT Press, Southampton, England, 267 p.

Tayfur, G. and Karimi, Y. 2014. Use of principle component analysis in conjunction with soft computing methods for investigating total sediment load transferability from laboratory to field scale. *Hydrology Research*, (in press).

Wardlaw, R. and Bhaktikul, K. 2004. Comparison of genetic algorithm and linear programming approaches for lateral canal scheduling. *J. Irrigation and Drainage Engrg*., ASCE, 130(4), 311-317.

Vasan, A. and Simunovic, S. 2010. Optimization of water distribution network design using differential evolution. *J. Water Planning and Management*, 136(2), 279-287.

CHAPTER 12

COMBINING REGIONAL REGRESSION APPROACHES WITH GEOSTATISTICAL TECHNIQUES FOR PREDICTING FLOOD QUANTILES IN UNGAUGED BASINS

A. Pugliese[1], A. Castellarin[1], S. A. Archfield[2]

[1]DICAM, University of Bologna, Bologna, Italy
[2]US Geological Survey, Northborough, MA, USA
Email: A. Pugliese (alessio.pugliese3@unibo.it)

ABSTRACT

The study explores the application of two different spatial interpolation techniques, Generalized Least Squares (GLS) and Top-kriging, aiming to enhance the prediction performances by blending these two powerful methods. GLS is one of the most common approaches used in United States (US) for predicting flood quantiles in ungauged basins, Top-kriging predicts streamflow statistics along river networks (e.g. flood quantiles) taking both the catchment area and nested nature of catchments into account. We first applied GLS in cross-validation, through a comprehensive leave-one-out procedure, to predict flood quantiles over a wide region in the southeast of US by simulating the ungauged conditions at each and every site, then we used Top-kriging to model GLS residuals. The results reveal that combining GLS with Top-kriging increases the prediction capability of GLS. Blending the techniques increases the Nash-Sutcliffe efficiency from 0.67-0.76 to 0.75-0.91, depending on the frequency of the quantile; moreover blending GLS with Top-kriging reduces the prediction errors in the majority of the study catchments relative to using GLS alone.

12.1 INTRODUCTION

Prediction of streamflow indices in ungauged or poorly gauged catchments is a topical issue in hydrology. For instance, the International Association of Hydrological Sciences (IAHS) promoted the initiative named Predictions in Ungauged Basins (PUB), launched in 2003 and concluded in 2012, which was aimed at formulating and implementing international research efforts to engage and energise the scientific community, in a coordinated manner, towards a major advancement in the accuracy of predictions of streamflow indices (e.g. low-flow indices, annual flows, flood quantiles, etc.) in ungauged basins.

Concerning the prediction of flood quantiles in ungauged locations the scientific literature proposes several approaches. Among these, regional flood frequency analysis (RFFA) is one of the most used and applied in hydrology (see e.g. Dalrymple, 1960; Burn, 1990; Stedinger et al. 1993; Hosking and Wallis, 1997; Castellarin et al. 2001; Merz and Bloschl, 2005). According to the classical scheme that characterizes the index-flood procedure (Dalrymple, 1960), RFFA pools the hydrological information within hydrologically homogeneous groups of gauged basins and transfers it to the ungauged target site through a suitable standardization procedure. Hydrologically homogeneous pooling-groups of sites can be identified on a geographical or physiographical basis (see e.g., Burn 1990; Reed, 1999; Ouarda et al. 2001), but is generally associated with a certain degree of subjectivity (see e.g., Castiglioni et al. 2011). Another classical example of statistical regionalization approach is the Generalized Least Squares (GLS) linear regression. GLS linearly relates catchment descriptors with flood quantile magnitude, has been widely used as regional method for flood estimation (see e.g. Archefield et al. 2013, Griffis and Stedinger, 2007) and was showed to be highly reliable for continental US. As a result, US Geological Survey utilizes GLS as the standard method for the prediction of flood quantiles at ungauged sites in homogeneous hydro-climatic regions (Gotvald et al. 2009). GLS accuracy and robustness notwithstanding, a critical point of its application relies in the delineation of the homogeneous hydro-climatic regions, which may introduce some subjectivity into the analysis.

Concerning the issue illustrated above, the recent scientific literature reports on several applications of innovative regionalization techniques which are based on geostatistical methods and, as such, dispense with the delineation of hydrologically homogeneous regions (and the associated degree of subjectivity). Among these techniques, Top-kriging (TK), which is a block-kriging with variable support area, proved to be particularly successful in several studies dealing with the prediction of streamflow indices of different nature, e.g. flood quantiles, low-flows, water temperature, habitat suitability indices, flow-duration curves etc., along the stream network (see. Merz et al. 2008; Castiglioni et al. 2011; Vormoor et al. 2011; Archfield et al. 2013; Laaha et al. 2013, Pugliese et al. 2013, Ceola and Pugliese, 2014).

In this study we show the first application of TK to predict GLS residuals of flood quantiles over the stream network of a wide study area located in the southeast United States. The application is aimed at blending GLS and TK to improve the prediction accuracy of the regression-based technique, which, in a previous study on the same study area, was shown to be complementary with TK in terms of prediction accuracy (see Archfield et al. 2013). Blending GLS with TK could in principle enhance the accuracy and reliability of predicted flood quantiles and our study aims at assessing this hypothesis. We address this science question by applying GLS and TK separately, as well as blended together, in cross-validation to a study region consisting of 61 gauged basins. We focused on the prediction of flood quantiles associated with four arbitrarily selected recurrence intervals, i.e. 10, 50, 100 and 500 years.

12.2 REGIONAL METHODS FOR PREDICTING FLOOD QUANTILES

12.2.1 Generalized Least Squares Regression

Regression-based methods to estimate flood quantiles at ungauged locations relate catchment characteristics to the flood quantile of interest and then use this relation to estimate the flood quantile at an ungauged location. For a flood quantile of interest, the regression equation typically has the general form

$$Y = \sum_{i=1}^{M} a_i X_i + \varepsilon \tag{12.1}$$

where Y is a vector of the log-transformed values of the observed floods across the gauged locations, X_i's are the vectors of the log-transformed values of the observed catchment characteristics, a_i's are the coefficients estimated by the regression, M is the total number of catchment characteristics, and ε is the vector of the model residuals. When the true residuals of the regression model have the same variance and are independent, ordinary least squares (OLS) regression can be used to estimate the model coefficients. However, it is unlikely that rivers across a given spatial extent would experience flooding completely independent of one another, and therefore, the assumption of independence of the residuals is likely to be violated (Stedinger et al. 1993). Furthermore, because different record lengths are available to compute the empirical flood quantiles at each gauged location, the certainty for which the flood quantile is known differs across gauged locations, and the assumption of equal variance is also likely to be violated. To overcome the potential violations of these assumptions, generalized least squares regression is utilized. GLS accounts for the unequal variances and spatial correlation by weighting each flood quantile value by a function of its record length, estimated cross-correlation between floods, and an estimate of the variance determined from OLS regression (Stedinger and Tasker, 1985). Further technical details of GLS as applied to flood quantile estimation at ungauged locations can be found in Tasker and Stedinger (1989). Stedinger and Tasker (1985) found that when the cross-correlation between floods was greater than 0.6, GLS regression was able to estimate regression coefficients that result in more accurate estimates of the regression model coefficients than OLS regression. This is one reason GLS regression is commonly utilized as a method of regionalization, particularly for estimating flood quantiles.

12.2.2 Topological Kriging

Topological kriging or Top-kriging (TK in the reminder) is a powerful geostatistical procedure proposed by Skøien et al. (2006) which performs hydrological predictions at ungauged sites along stream-networks on the basis of the empirical information collected at neighbouring gauging stations. As kriging techniques, the spatial interpolation is obtained in Top-kriging by a linear combination of the empirical values; therefore, the unknown value of the streamflow index of interest at prediction location x_0, $Z(x_0)$ can be estimated as a weighted average of the variable measured in the neighborhood:

$$Z(x_0) = \sum_{i=1}^{n} \lambda_i Z_i(x_i) \tag{12.2}$$

where λ_i is the kriging weight for the empirical value $Z(x_i)$ at location x_i, and n is the number of neighbouring stations used for interpolation. Kriging weights λ_i can be found by solving the typical ordinary kriging linear system Eq. (12.3), with the constrain of unbiased estimation Eq. (12.3b):

$$\sum_{i=1}^{n} \gamma_{i,j} \lambda_i + \theta = \gamma_{0,i} \qquad i = 1,\ldots,n \tag{12.3a}$$

$$\sum_{j=1}^{n} \lambda_j = 1 \tag{12.3b}$$

where θ is the Lagrange parameter and $\gamma_{i,j}$ is the semi-variance between catchment i and j. The semi-variance is also referred to as variogram in geostatistics and represents the space variability of the regionalised variable Z.

A peculiar feature of Top-kriging is to consider the variable defined over a non-zero support S (i.e. the catchment drainage area) (Cressie, 1993; Skøien et al. 2006); this implies that the kriging system (2.3) remains the same, but the gamma values between the measurements need to be obtained by regularization, that is the smoothing effect of support area S on the point variogram, which is computed by applying an integral average of the variable Z over S. After this, the point variogram can be back-calculated by fitting aggregated variogram values to the sample variogram (details can be found in Skøien et al. 2006).

12.2.3 Blending GLS with TK for an Enhanced Prediction Capability

Regression based regionalization methods, such as GLS, proved to be very successful for flood quantile prediction in ungauged conditions due to their robustness and to the strength and soundness of their theoretical basis (Stedinger and Tasker, 1985). The main feature of GLS relative to other regression techniques is the capability of the approach to explicitly model the spatial correlation (i.e. intersite or cross-correlation) between flood sequences observed at different sites in the region, which is generally present within a given geographical region (see e.g., Stedinger and Tasker, 1985, Castellarin et al. 2005).

Regionalization approaches that make explicit use of geostatistics, such as Top-kriging (TK), differ substantially from GLS, even though some common elements can be acknowledged. One example is the fact that they both account for spatial correlation between series; TK does it through a variogram model that expresses variability between empirical values as a function of distance (and in some instances direction, see anisotropic variograms e.g. in Castellarin, 2014) between pairs of sampling locations (i.e. in TK catchments centroids). Another element are the linear functions that characterize both approaches: a linear function is used in GLS to relate catchments descriptors to flood quantiles (or their log-transformations), TK uses a linear weighting scheme for averaging the available observation and predicting the streamflow index of interest in the selected ungauged location.

Archfield et al. (2013) considered the prediction of flood quantiles in ungauged sites for the same study area considered herein, the authors investigated the effect in terms of prediction accuracy that can be obtained by blending together two different geostatistical prediction techniques, i.e. Canonical kriging (CK, see e.g. Chockmani and Ouarda, 2004, and Castellarin, 2014) and Top-kriging, and compared the results with the predictions obtained with the application of GLS technique. The main outcomes of the study are that TK performs better than GLS and that blending the TK and CK could in some cases enhance the performances obtained by using them separately.

What are the possible effects of blending GLS and Top-kriging together? Archfiled et al. (2013) did not specifically investigate this point, yet the science question is definitely an important one given the complementarities and affinities of GLS and TK when dealing with spatial cross-correlation, and the interoperability of the two techniques recalled above. We address the science question above by analysing the possibility to predict GLS residuals via TK, which enables one to blend the two procedures together, and by assessing whether or not blending GLS with TK increases the overall prediction performances. Figure 12.1 shows the work flow implemented in our study: (1) GLS model is applied for predicting flood quantiles associated with arbitrarily selected recurrence intervals (i.e., 10, 50, 100 and 500 years), simulating ungauged conditions through a leave-one-out cross-validation algorithm (LOOCV, see e.g. Archfield et al. 2013); (2) residuals between empirical flood quantiles and GLS predictions are computed; (3) Top-kriging is then applied in cross-validation (i.e. using an LOOCV procedure) to predict GLS residuals; (4) adjusted flood-quantile predictions are computed by summing the TK predicted residuals to GLS cross-validated predictions. We call this blending strategy GLS adjusted via TK or GLSTK.

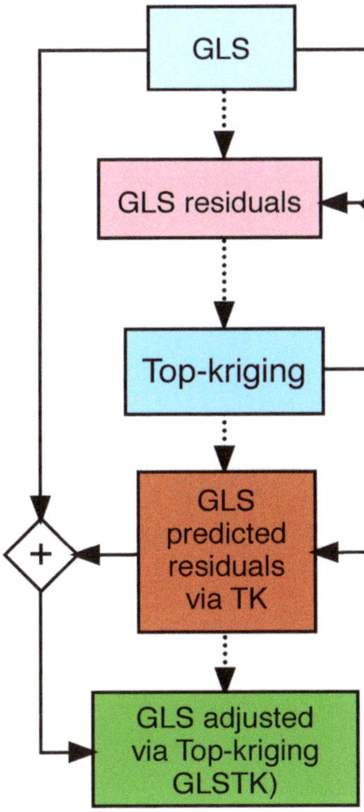

Fig. 12.1. Chart representing the work flow.

12.3 STUDY AREA AND DATA

The study area is located in the southeast United States (Fig. 12.2), and is characterized by a temperate climate and mean annual precipitation between 1000 and 1500 mm per year (Gotvald et al. 2009). The study

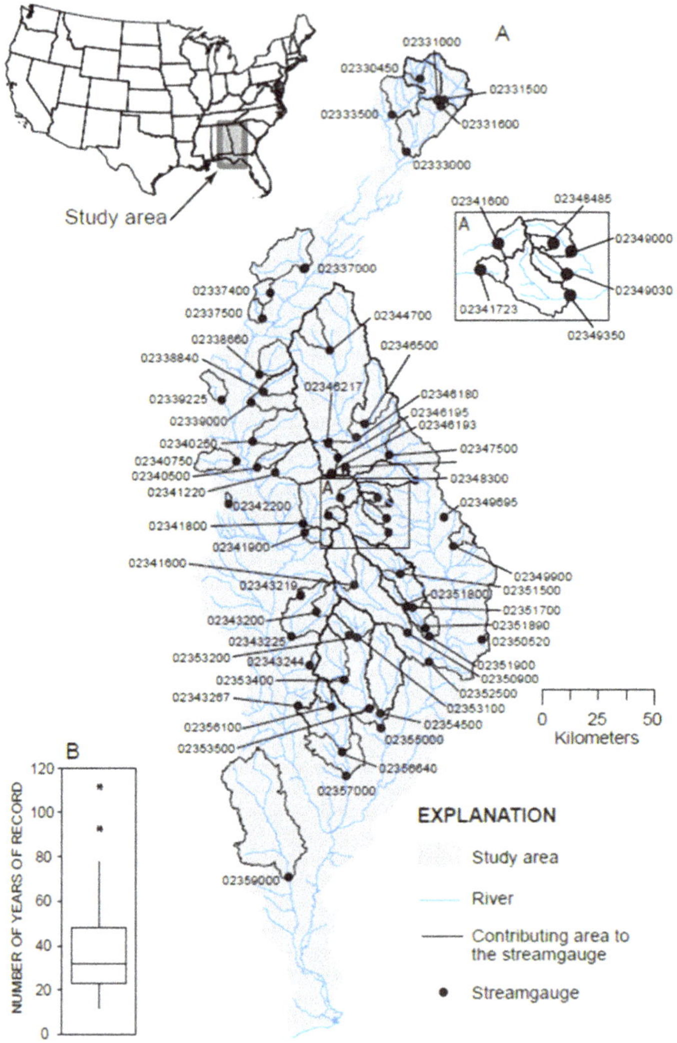

Fig. 12.2. Study Area (from Archfield et al. 2013, Fig. 1 on page 1577).

area was recently analysed in the regional study Gotvald et al. (2009) in which the GLS regression had been applied to set up a regionalization model that predicts flood quantiles associated with recurrence intervals between 10 and 500 years for rural ungauged catchments in the entire southeast United States. Study stream gauges are considered to have undeveloped upstream catchments and to have minimal regulation. Additionally, the study stream gauges have at least 10 years of annual peak-streamflow data (Fig. 12.2B) and were screened to ensure that no significant trends were present in the annual flood peak sequences (Gotvald et al. 2009).

A number of catchment characteristics describing the morphology, climate, land cover and soil properties of the contributing areas to the study stream gauges are available (see Table 12.1, Fig. 12.2, and Gotvald et al. 2009). Where applicable, the catchment characteristics listed in Table 12.1 were computed as an average over the drainage area. Empirical quantiles corresponding to the 10-, 50-, 100-, and 500-yr floods (see Table 12.2) were determined as described by the widely used guidelines outlined in Bulletin 17-B of the Hydrology Subcommittee of the Interagency Advisory Committee on Water Data (1982; Gotvald et al. 2009).

This interagency document describes how to obtain empirical estimates of flood quantiles by fitting a Pearson Type III distribution to the annual peak-streamflow time series at each of the study stream gauges and pooling at-site and regional information to estimate the parameters of the fitted distribution (Gotvald et al. 2009). Specific details of the procedure as applied to the study stream gauges can be found in Gotvald et al. (2009).

Table 12.1. Summary of catchment characteristics for the study area.

Catchment Characteristics	Unit of Measure	Minimum Value	Median Value	Maximum Value
Morphology				
A – Drainage Area	km^2	1.8	328.9	13752.1
DrD – Drainage density	km per km^2	0.75	1.28	5.76
ChL – Main channel length	km	2.8	43.1	410.5
ChS – Main channel slope	m per km	0.40	1.31	14.27
BPe – Catchment perimeter	km	5.7	113.7	1071.6
BSF – Catchment shape factor	-	2.48	5.30	12.25
MeE – Mean catchment elevation	M	47	174	758
MaE – Maximum catchment elevation	M	103	252	1351
MiE – Minimum catchment elevation	M	9	110	435
BMS – Mean catchment slope	%	0.98	5.32	33.04
Precipitation				
MAP – Mean annual precipitation	mm	1079	1300	1899
RQ1 – 24 h, 2 yr maximum	mm	101	101	126
RQ2 – 24 h, 10 yr maximum	mm	152	157	194
RQ3 – 24 h, 25 yr maximum	mm	177	181	226
RQ4 – 24 h, 50 yr maximum	mm	197	203	256
RQ5 – 24 h, 100 yr maximum	mm	203	221	282
Land cover and soil properties				
SDI – Soil drainage index	-	2.40	3.09	4.04
HSI – Hydrologic soil index	-	1.83	2.14	2.86
Imp – Percent of catchment that is impervious	%	0.14	0.56	7.95
For – Percent of catchment that is forested	%	3.54	59.00	94.63

Table 12.2. Summary of the empirical flood quantiles obtained from the Pearson Type III distribution.

Empircal Flood Quantiles	Unit of Measure	Minimum Value	Median Value	Maximum Value
Q10	m^3/s	3.5	168.8	1673.5
Q50	m^3/s	7.5	303.0	2489.1
Q100	m^3/s	9.9	368.1	2860.0
Q500	m^3/s	13.8	535.2	3822.8

12.4 IMPLEMENTATION OF THE REGIONAL METHODS TO THE STUDY AREA

12.4.1 Application of GLS in Cross-Validation

GLS was applied to the study area by closely following the methods described by Gotvald et al. (2009). Log-transformed values of the flood quantiles and basin attributes were used with the procedures described in Tasker and Stedinger (1989) and as implemented by the weighted multiple linear regression (WREG) program (Eng et al. 2009). Gotvald et al. (2009) initially tested all possible subsets of the 22 catchment characteristics shown (Table 12.1) for inclusion as predictor variables in the final regression equations. Based on an analysis of these regression subsets, Gotvald et al. (2009) determined that drainage area and percent of catchment in each of five hydrologic regions were able to best predict flood quantiles in the GLS regression equations. Hydrologic regions were determined by examining the spatial distribution of residuals obtained from an OLS regression of flood quantiles and catchment area and their correspondence with US Environmental Protection Agency Level III Ecoregions (Gotvald et al. 2009). For this study, we chose not to apply the hydrologic regions because they were defined based on the larger study area in Gotvald et al. (2009).

To compare the flood quantiles predicted by GLS with empirical values, a leave-one-out cross-validation approach was applied. In this validation procedure, each of the 61 study stream gauges was subsequently removed from the data set, and the GLS method, was then applied to estimate the flood quantiles at the removed stream gauge from the remaining 60 stream gauges. GLS residuals, i.e. the differences between empirical and predicted flood quantiles values, are site specific and can be interpolated with geostatistical techniques.

12.4.2 The role of the drainage area

In a preliminary analysis we tested if GLS residual obtained in cross-validation showed any empirical dependency on drainage area. Such a dependence is detrimental for the application of TK and should be removed before applying TK, and can in principle be removed by adopting a suitable scaling relation (for an extensive description of the GLS estimation procedure see Sec. 12.2.1). A power-law is generally a common choice for a scaling relationship that describes the variability of residuals as a function of drainage area, and it reads:

$$\gamma_{GLS} = aA^b + \varepsilon \qquad (12.4)$$

where γ_{GLS} are the GLS residual, A is the drainage area, a and b are regression coefficients and ε is the error term. Fitting Eq. (12.4) to GLS residuals obtained for the four considered flood quantiles (see Section 12.4.3) through an ordinary least squares approach, results in very similar estimates of the b exponent, which was therefore assumed to be constant and equal to 0.57.

12.4.3 Modelling GLS residuals through Top-kriging

Top-kriging can be applied to GLS residuals (standardized by A^b, see Section 12.4.2) as illustrated in the stepwise description by Skøien (2014). This task can be accomplished through the suite of R-functions included in the R-package "rtop", which can be accessed from the Comprehensive R Archive Network (CRAN, http://cran.r-project.org/).

The point sample variogram can be computed using the binned variogram technique, for which sample points are aggregated in distance classes or bins, under the hypothesis of isotropy (i.e. the variogram does not vary with direction). The sample variogram can then be described through a suitable mathematical function among the available theoretical models (e.g. exponential, Gaussian, spherical, fractal, etc.). Skøien et al. (2006) recommend the use of the exponential variogram.

Once the empirical variogram is modelled, the number *n* of neighbouring stations on which to base the spatial interpolation is set iteratively by the user on the basis of a first set of preliminary analyses, which aim at identifying the *n* value that produces the most accurate predictions in cross-validation. Our preliminary analyses show that choosing $n = 10$ enable us to obtain the best cross-validated results.

Figure 12.3 shows relative residuals (i.e. residuals divided by empirical flood quantiles) associated with GLS predictions (black circles) and predicted by applying TK (red dots) through a leave-one-out cross-validation procedure (see Sec. 12.4.1). The panels in the figure

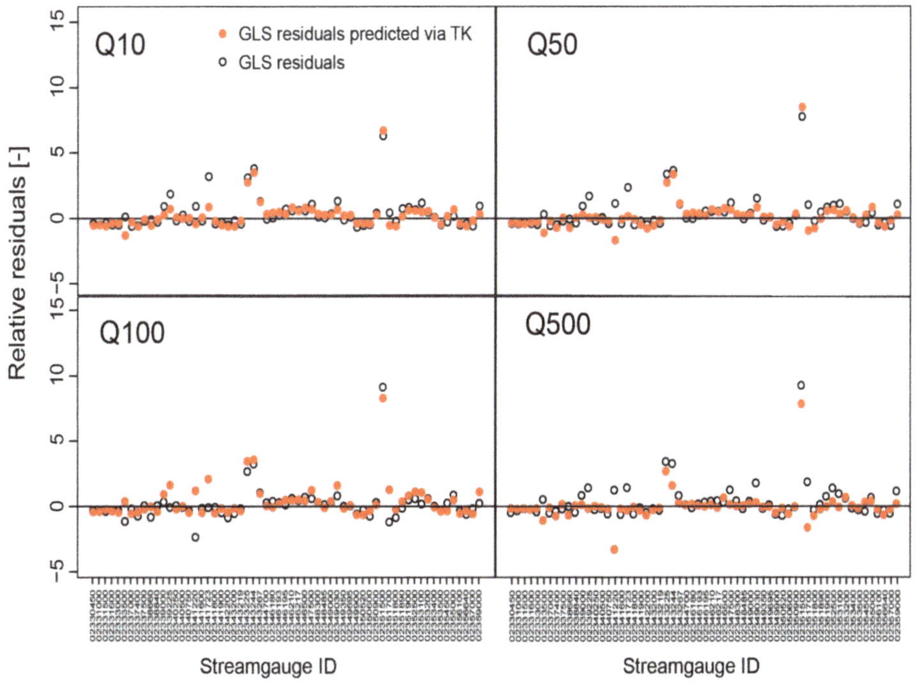

Fig. 12.3. Relative residuals (i.e., residuals divided by empirical flood quantiles) obtained by using GLS in cross-validation (black circles) and predicted by applying Top-kriging (red dots).

refers to flood quantiles associated with four different recurrence intervals: 10, 50, 100 and 500 years (i.e. Q10, Q50, Q100 and Q500, see Sec. 12.3 for details), and clearly show that in most cases TK provides the users with very accurate predictions of the residuals. It is worth noting that TK miss-predicted the residual sign for a significant percentage of basins over the total number of sites in the study area (i.e. 14.7%, 21.3%, 22.9% and 29.5% for Q10, Q50, Q100 and Q500, in this order).

12.5 FINAL RESULTS

Figure 12.4 reports the empirical flood quantiles against the predictions obtained in cross-validation (i.e. LOOCV algorithm) with the proposed blended approach, which we termed GLSTK (green dots), with GLS (black circles), or with TK (blue squares) for each one of the four recurrence intervals considered in the study. Table 12.3 explains the differences between methods in terms of Nash-Sutcliffe Efficiency, both for natural (NSE) and for Log-transformed (LNSE) flood quantiles.

As both Table 12.3 and Fig. 12.4 highlight, GLSTK performs better than GLS. Blending GLS with TK produces a significant improvement relative to GLS with an increase in terms of NSE ranging from 12% (Q500) to 18% (Q10), and in terms of LNSE from 11% to 24%.

Table 12.3. Nash-Sutcliffe efficiency for natural (NSE) and log-transformed (LNSE) flood quantiles obtained by the three approaches considered in the study.

	GLS blended with TK		GLS		TK	
	NSE	LNSE	NSE	LNSE	NSE	LNSE
Q10	0.91	0.91	0.77	0.73	0.93	0.90
Q50	0.89	0.86	0.76	0.73	0.90	0.88
Q100	0.87	0.82	0.74	0.73	0.88	0.86
Q500	0.75	0.80	0.67	0.72	0.83	0.82

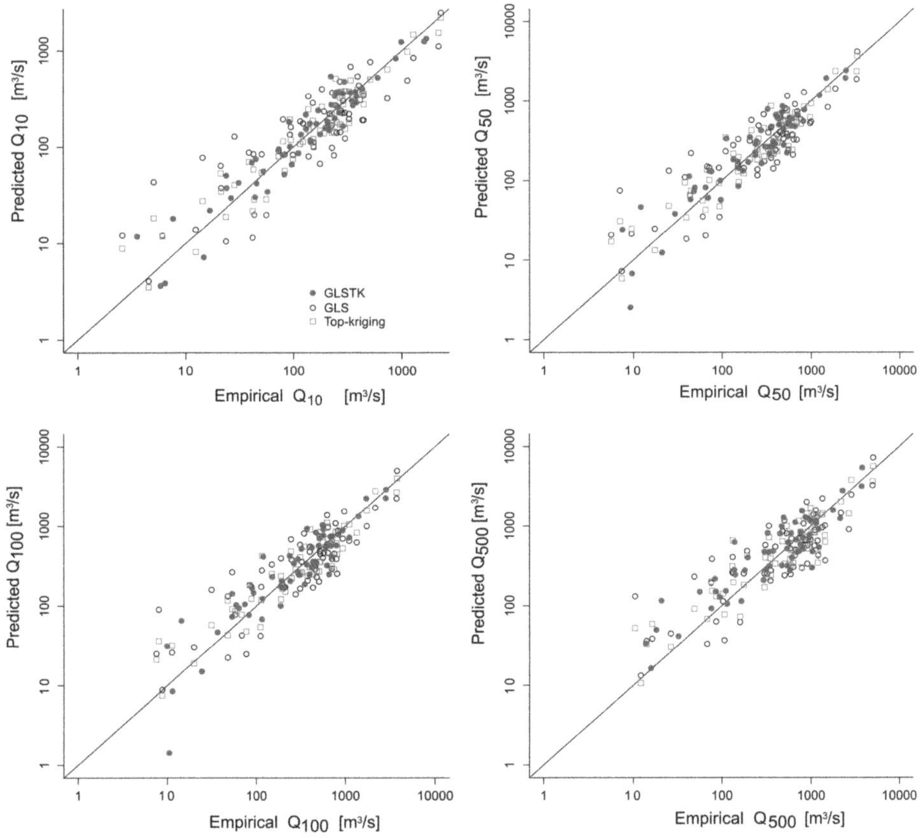

Fig. 12.4. Comparison between empirical and predicted flood quantiles for GLSTK (green filled dots), GLS (black circles) and TK (blue squares).

Nevertheless the blended approach GLSTK never outperforms TK applied directly to the prediction of flood quantiles, the latter showing NSE and LNSE always higher, even though the performances could be considered comparable due to the very limited differences in terms of efficiency (differences in terms of Nash-Sutcliffe efficiency range between 1% and 9%).

Scatter diagrams in Fig. 12.5 compare the results obtained from the three considered models by plotting the absolute value of the prediction error between empirical and predicted quantiles for each site. The *y*-axis

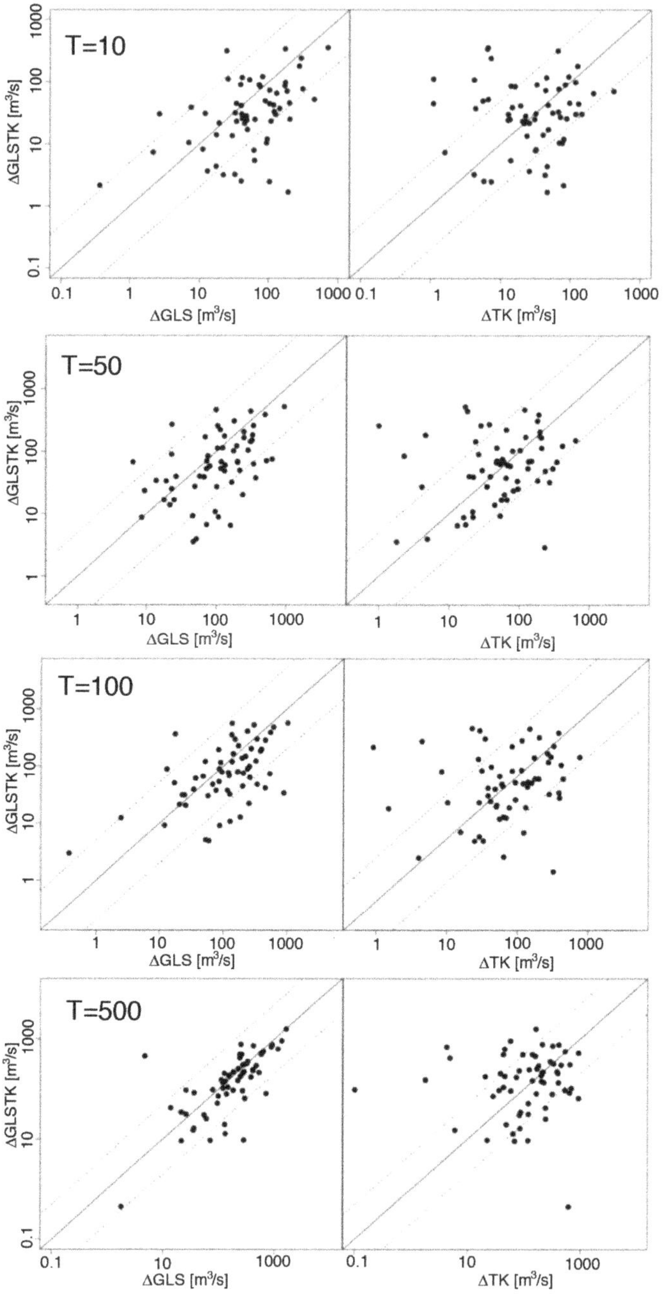

Fig. 12.5. Comparisons between the prediction errors obtained using the proposed GLSTK approach and GLS and TK.

reports errors computed for the proposed blended method (ΔGLSTK) while the *x*-axis reports in turn errors from the benchmark models (ΔGLS and ΔTK, respectively). In this representation the equivalence between model performances is represented by the solid bisecting line, hence if one point falls below the 1:1 line it means that GLSTK performs better than the benchmark model (either GLS or TK) for that site, otherwise if it falls above the 1:1 line. Points below the lower (or above the upper) the dotted lines indicates that GLSTK significantly outperforms (is significantly outperformed by) - i.e. the prediction error is less than one half - the benchmark technique (either GLS or TK). Looking at the left panels, which compare GLSTK with GLS, there is a clear pattern of points shifted down-right, confirming the outcome that was previously highlighted (i.e. GLSTK outperforms GLS). In the right panels, i.e. GLSTK compared with TK, points result scattered around the bisecting line, which means a substantial equality of prediction capabilities (see Table 12.4).

Table 12.4. Number of catchments (out of 61) for which the application of GLSTK results in higher prediction errors (absolute value) relative to GLS or TK .

	GLSTK vs. GLS	GLSTK vs. TK
Q10	18	28
Q50	19	29
Q100	22	28
Q500	25	31
Average	21	29

12.6 DISCUSSION

TK and GLS are fundamentally different by their very nature and by the way they treat spatial correlation among annual flood series observed at different sites. GLS uses observations as training data for fitting a regression model which takes into account the reduction of the total information when the observations are highly correlated. Similarly to GLS, Top-kriging (TK) models spatial variability of empirical data as a function of distance between sites, nevertheless TK handles upstream and downstream sub-catchments differently from neighbouring catchments (i.e. it predicts the streamflow index at the site of interest by assigning a larger weight to the hydrological information coming from upstream/downstream stream gauges - i.e. from the same catchment - relative to the information coming from stream gauges located in neighbouring catchments). This complementarity between the two techniques suggests the possibility to blend them to improve the accuracy of predictions of streamflow indices (i.e. flood quantiles in this study).

On the one hand, the results obtained in this study by blending GLS with TK clearly show that the application of a geostatistical model, such as Top-kriging, to GLS prediction residuals could enhance GLS prediction capability and increase the robustness of the approach. This result is clearly visible if one looks at the improvement of the performances in terms of Nash-Suttcliffe efficiency (see Table 12.3) associated with the blended approach (GLSTK) and at the scatterplots reporting empirical values vs. predictions in Fig. 12.4, albeit both Fig. 12.5 (left panel) and Table 12.3 show a clear dependence of the difference between the performances of the approaches on the recurrence interval (GLS becomes more similar to GLSTK in terms of performances as the recurrence interval increases). On the other hand, by comparing GLSTK overall performances with TK ones one can note that they are practically the same and do not show any dependence on the recurrence interval (see Fig. 12.5 right panels). This suggests that, although blending the geostatistical procedure with GLS skills is associated with a significant additional workload if compared to the application of TK alone, GLSTK does not result in significant improvements if compared with TK.

As a final remark, discussing our results in the light of those obtained by Archefield et al. (2013) one may note that TK is capable of using the hydrological information across sites very efficiently, resulting in better predictions of flood quantiles at ungauged locations if compared to GLS. This characteristic of TK is likely to result from its ability to profit from a correct schematization of the nesting structure of stream networks and the great deal of importance that drainage area plays in the application of the procedure. It is well known that drainage area plays a key role during the generation of extreme flood events, the same events used for predicting flood quantiles in gauged basins, becoming often much more important than land-cover or soils characteristics. These catchment characteristics are not accounted for by TK, while they may be considered in other regression-based approaches (such as GLS). Opposite considerations hold for the predictions of low-flow indices, for which soil types and land-cover characteristics may become very important descriptors of a given basin, reducing the prediction ability and accuracy of TK relative to other regression based regionalization approaches (see e.g. Laaha et al. 2014; Castiglioni et al. 2011).

12.7 CONCLUSIONS

In this study we present an adaptation of the Generalized Least Squares (GLS) regression based model for the prediction of flood quantiles in ungauged locations. In particular, we show how GLS can be blended with Topological kriging, a geostatistical regionalization approach that was recently presented in the scientific literature. Top-kriging is used in this study to geostatistically interpolate GLS residuals, which were computed as the difference between GLS predictions of flood quantiles obtained in cross- validation and the corresponding empirical values of the quantiles for all study catchments over a wide study area located in the south-east United States. In particular, we applied GLS to predict 10-, 50-, 100- and 500-year floods at 61 stream gauges simulating the ungauged conditions at each and every site by applying a leave-one-out resampling strategy. The GLS residuals obtained in cross-validation were then standardized by catchment area and modelled with Top-kriging (TK), still in cross-validation. Modelling GLS residuals through TK enables one to blend the two techniques for predicting flood quantiles in ungauged sites of the study area (i.e. GLS is first applied and the GLS estimate is then adjusted by summing the residual estimated with TK). The final results showed that blending GLS with TK (GLSTK technique) significantly improves the accuracy and reliability of predictions of GLS, resulting flood quantile estimates that are associated with smaller bias relative to GLS estimates. It is worth noting, though, that the proposed GLSTK does not seem to significantly outperform TK, since the application of the GLSTK approach results in flood quantile predictions that are slightly more accurate than predictions obtained by TK in a few instances, and as reliable as TK predictions in the majority of the cases. Future analyses will further investigate the reasons behind TK excellent performances for flood-quantile prediction in ungauged basins of the study area.

***Acknowledgements*:** Armando Famulari is acknowledged for his contribution to a set of preliminary analyses.

12.8 REFERENCES

Archfield, S. A., A. Pugliese, A. Castellarin, J. O. Skøien, and J. E. Kiang. 2013. "Topological and Canonical Kriging for Design Flood Prediction in Ungauged Catchments: An Improvement over a Traditional Regional Regression Approach?" *Hydrol. Earth Syst. Sci.* 17 (4): 1575–88. doi:10.5194/hess-17-1575-2013.

Blöschl, G., M. Sivapalan, W. Thorsten, A. Viglione, and H. Savenije. 2013. *Runoff Prediction in Ungauged Basins: Synthesis across Processes, Places and Scales.* Cambridge University Press.

Burn, D. H. 1990. "Evaluation of Regional Flood Frequency Analysis with a Region of Influence Approach." *Water Resour. Res.* 26 (10): 2257–65.

Castellarin, A., D. H. Burn, and A. Brath. 2001. "Assessing the Effectiveness of Hydrological Similarity Measures for Flood Frequency Analysis." *Journal of Hydrology* 241 (3–4): 270–85. doi:10.1016/S0022-1694(00)00383-8.

Castellarin, A., R. M. Vogel, and N. C. Matalas. 2005. "Probabilistic Behavior of a Regional Envelope Curve." Water Resources Research 41 (6): W06018. doi:10.1029/2004WR003042.

Castellarin, A. 2014. "Regional Prediction of Flow-Duration Curves Using a Three-Dimensional Kriging." *Journal of Hydrology* 513: 179–91. doi:10.1016/j.jhydrol.2014.03.050.

Castiglioni, S., A. Castellarin, A. Montanari, J. O. Skøien, G. Laaha, and G. Blöschl. 2011. "Smooth Regional Estimation of Low-Flow Indices: Physiographical Space Based Interpolation and Top-Kriging." *Hydrol. Earth Syst. Sc.* 15 (3): 715–27. doi:10.5194/hess-15-715-2011.

Ceola, S., and A. Pugliese. 2014. "Regional Prediction of Basin-Scale Brown Trout Habitat Suitability." In *Evolving Water Resources Systems: Understanding, Predicting and Managing Water-Society Interactions*, 26–31. Bologna: IAHS Publ. 364, 2014.

Chokmani, K., and T. B. M. J. Ouarda. 2004. "Physiographical Space-Based Kriging for Regional Flood Frequency Estimation at Ungauged Sites." *Water Resources Research* 40 (12): W12514–.

Cressie, N.A.C. 1993. *Statistics for Spatial Data.* Wiley Series in Probability and Mathematical Statistics: Applied Probability and Statistics. J. Wiley.

Dalrymple, T. 1960. *Flood-Frequency Analyses, Manual of Hydrology: Part 3*. WSP - 1543-A. United States Geological Survey.

Eng, K., Y.Y. Chen, and J. E. Kiang. 2009. "User's Guide to the Weighted-Multiple-Linear Regression Program (WREG Version 1.0)." In, 4:21. U.S. Geological Survey.

Gotvald, A.J., T.D. Feaster, and J. C. Weaver. 2009. *Magnitude and Frequency of Rural Floods in the Southeastern United States, 2006: Volume 1, Georgia*. 2009-5043. Reston, Virginia, USA: U.S. Geological Survey.

Griffis, V. W., and J. R. Stedinger. 2007. "The Use of GLS Regression in Regional Hydrologic Analyses." *Journal of Hydrology* 344 (1–2): 82–95. doi:10.1016/j.jhydrol.2007.06.023.

Hosking, J. R. M., and J. R. Wallis. 1997. *Regional Frequency Analysis: An Approach Based on L-Moments*. Cambridge University Press.

Hrachowitz, M., H.H.G. Savenije, G. Blöschl, J.J. McDonnell, M. Sivapalan, J.W. Pomeroy, B. Arheimer, et al. 2013. "A Decade of Predictions in Ungauged Basins (PUB)—a Review." *Hydrological Sciences Journal* 58 (6): 1198–1255. doi:10.1080/02626667.2013.803183.

Laaha, G., J.O. Skøien, F. Nobilis and G. Blöschl. 2013. "Spatial Prediction of Stream Temperatures Using Top-Kriging with an External Drift." *Environmental Modeling & Assessment*, 1–13. doi:10.1007/s10666-013-9373-3.

Laaha, G., J.O. Skøien, and G. Blöschl. 2014. "Spatial prediction on river networks: comparison of top-kriging with regional regression." *Hydrol. Process.* 28: 315–324. doi: 10.1002/hyp.9578.

Merz, R., and G. Blöschl. 2005. "Flood Frequency Regionalisation-Spatial Proximity vs. Catchment Attributes." *Journal of Hydrology* 302: 283 – 306. doi:10.1016/j.jhydrol.2004.07.018.

Merz, R., G. Blöschl, and G. Humer. 2008. "National Flood Discharge Mapping in Austria." *Natural Hazards* 46 (1): 53–72. doi:10.1007/s11069-007-9181-7.

Ouarda, T.B.M.J., C. Girard, G.S. Cavadias, and B. Bobée. 2001. "Regional Flood Frequency Estimation with Canonical Correlation Analysis." *Journal of Hydrology* 254: 157 – 173. doi:10.1016/S0022-1694(01)00488-7.

Pugliese, A., A. Castellarin, and A. Brath. 2013. "Geostatistical Prediction of Flow-Duration Curves." *Hydrology and Earth System Sciences Discussions* 10 (11): 13053–91. doi:10.5194/hessd-10-13053-2013.

Reed, D.W., D. Jakob, A.J. Robinson, D.S. Faulkner, and E.J. Stewart. 1999. "Regional Frequency Analysis: A New Vocabulary." In *Hydrological Extremes: Understanding, Predicting, Mitigating*, 255:237–43. Birmingham: IAHS.

Skøien, J. O., R. Merz, and G. Blöschl. 2006. "Top-Kriging - Geostatistics on Stream Networks." *Hydrology and Earth System Sciences* 10 (2): 277–87. doi:10.5194/hess-10-277-2006.

Skøien, J. O. 2014. *Rtop: Interpolation of Data with Variable Spatial Support.* http://CRAN.R-project.org/package=rtop.

Stedinger, J. R., and G. D. Tasker. 1985. "Regional Hydrologic Analysis: 1. Ordinary, Weighted, and Generalized Least Squares Compared." *Water Resour. Res.* 21 (9): 1421–32.

Stedinger, J. R., Vogel R.M, and E. Foufula-Georgiou. 1993. "Frequency Analysis of Extreme Events." In *Handbook of Hydrology*, edited by R. Maidment, 18.11–18.66. New York: McGraw-Hill.

Tasker, G. D., and J. R. Stedinger. 1989. "An Operational GLS Model for Hydrologic Regression." *J. Hydrol.* 111: 361–75. doi:10.1016/0022-1694(89)90268-0.

Vormoor, K., T. Skaugen, E. Langsholt, B. Diekkrüger, and J. O. Skøien. 2011. "Geostatistical Regionalization of Daily Runoff Forecasts in Norway." *International Journal of River Basin Management* 9 (1): 3–15. doi:10.1080/15715124.2010.543905.

CHAPTER 13

THE 2010 PAKISTAN FLOOD: HIGH-RESOLUTION SIMULATIONS WITH THE WRF MODEL

Francesca Viterbo[1,2], Antonio Parodi[2], Luca Molini[2], Jost von Hardenberg[1], and Antonello Provenzale[1]

[1]ISAC-CNR, Corso Fiume 4, 10133 Torino, Italy
[2]CIMA Research Foundation, Via Armando Magliotto 2, 17100 Savona, Italy
e-mail: francesca.viterbo@cimafoudation.org

ABSTRACT

High resolution simulations of extreme rainfall events are needed in order to study the response of catchments during floodings and to improve risk mitigation. These needs become even more prominent in mountain areas, where small distances correspond to high spatial and altitudinal variations. The predictability of the WRF (Weather Research Forecast) model, operated at the very fine resolution of 3.5 Km has been explored in reproducing the 2010 Pakistan flood over the complex topography of the Hindu-Kush-Himalaya-Karakorum (HKKH) region. The WRF simulation has been carried on in cloud-permitting mode (d01 14 km and d02 3.5 km) for the period July 26th – July 30th 2010, varying different convective closures (explicit and KF), microphysics parameterizations (WSM6 and Thompson) and days of initialization. These model outputs in terms of daily precipitation estimates have been compared with available observations: radio-sounding data and geostationary/polar satellite images. In our study, the ability of WRF in reproducing the time-scale interactions of clouds and precipitation processes from large to small scales is found to be reasonably satisfying, even if more experiments are needed in order to use the WRF model as a dynamic downscaler over this area characterized by a complex topography.

Keywords: extreme precipitation, 2010 Pakistan flood, WRF simulations, high resolution modelling, complex topography.

13.1 INTRODUCTION

Having a good precipitation estimate (QPE) at a fine resolution is the starting point to address the question of risk mitigation for flood and extreme events forecasting. High resolution precipitation forecasts can be used as input for hydrological models so that the catchment's response to heavy rainfall inputs can be understood and predicted by means of a good estimate of water runoff.

In this study we address the question on how closely a particular meteorological model can predict the true state of a certain severe rainfall scenario as a function of parameterization of the subgrid processes, data availability, model initialization, model coupling, resolution and orography (Molini et al. 2011).

In particular, in the experimental part of the work, we want to gain a deeper understanding, by means of the Advanced Research Weather Research and Forecasting (ARW-WRF) model, of the atmospheric conditions that led to the most intense days of 2010 Pakistan flood (July 26th – 31st, 2010).

The tragic 2010 Pakistan floods began in late July triggered by the persistent heavy monsoonal rains and caused countrywide inundations of the Indus river basin: approximately one-fifth of Pakistan's total land area was covered by water. The more affected regions of Pakistan were the Khyber Pakhtunkhwa, Sindh, Punjab and Balochistan, causing a death toll of about 2000 people.

In order to test the performance of WRF in a complex environment, the outputs of the model have been compared with the observations in hindcast mode, looking for the best simulation of actual rainfall fields.

Precipitation, in fact, is recognized as one of the most difficult parameters to estimate by a numerical prediction system. The accuracy of the estimation is affected by many uncertainties (Litta et al. 2012): the limited knowledge of precipitation processes, the data deficiencies that often restricts the reliability of model's initial conditions and the representation of both resolved and subgrid-scale precipitation processes in a mesoscale model (the convective parameterization issue). A good choice of the convective scheme in terms of closure, cumulus parameterization and microphysics is important to well simulate the physical processes at play.

In the following paragraphs the model is tested, varying different parameters for each run, looking for the setting which is best able to reproduce the event. The WRF simulations are validated, comparing the model output with satellite data, radio-soundings and ground observations to find whether or not the simulations well describe the actual meteorological features. The complex topography (given by the presence of the Tibetan plateau and of the Himalaya range) and the climatic features of a monsoon and westerly influenced environment introduce an higher degree of complexity due to the strong scale interactions, ocean-continent contrast and air-sea interaction in these regions. In addition to that the model performance is expected to be strongly influenced by the steep orography.

The outputs of the model, such as rainfall quantity, columnar water vapor, temperature, wind, pressure and other meteorological features have been compared with observations provided by satellites (Tropical Rainfall Measuring Mission-TRMM, Advanced Scatterometer -ASCAT, Special Sensor Microwave/Imager -SSM/I, The Advanced Microwave Scanning Radiometer-Eos -AMSR-E, CloudSat), radiosoundings and the available ground measures.

Moreover, in addition to comparing maps, statistical tools has been applied to WRF and TRMM rainfall fields (frequency histograms, scatterplots, pixel-by-pixel differences and root mean square error – RMSE, precipitation power spectra) in order to better understand the comparison with the observations and to give a more quantitative information about the model performance.

In the following sections a brief explanation of the model (Section 13.2) and of the observational dataset (Section 13.3) is given. The analysis is explained in Section 13.4, while in Section 13.5 the Pakistan 2010 case study is discussed in detail. Finally, conclusions and future tasks are illustrated in Section 13.6 and Section 13.7.

13.2 THE WRF MODEL

The computational model used for this study is the Advanced Research Weather Research and Forecasting model (ARW-WRF), in its version 3.3.1. It is a fully compressible, 3D, Eulerian, non-hydrostatic model conservative for scalar variables. It is a next-generation mesoscale numerical weather prediction system designed to serve both operational forecasting and atmospheric research needs. The developing of WRF model has been a collaborative partnership, principally among the National Center for Atmospheric Research (NCAR), the National Oceanic and Atmospheric Administration (the National Centers for Environmental Prediction (NCEP) and the Forecast Systems Laboratory (FSL), the Air Force Weather Agency (AFWA), the Naval Research Laboratory, the University of Oklahoma, and the Federal Aviation Administration (FAA). WRF allows researchers the ability to conduct simulations reflecting either real data or idealized configurations (Skamarock et al. 2008). WRF is suitable for a broad spectrum of applications across scales ranging from meters to thousands of kilometers. It uses a terrain following hybrid sigma-pressure vertical coordinate. The grid is the staggered Arakawa E-grid.

A general model description is given in the present paragraph, but for a comprehensive description, the reader is referred to Skamarock et al. (2005) and Skamarock et al. (2008).

A mesoscale model is a numerical weather prediction model with a sufficiently high horizontal and vertical resolution to forecast mesoscale phenomena. The grid spacing is typically less than 30 km. This kind of models are able to provide great detail in order to accurately represent the intensity of the smaller scales phenomena and obtains superior forecasts in coastal and mountainous region, when compared to large scale phenomena.

Non-hydrostatic models are generally applied when the length scale of the phenomena is similar to the height scale, such as convective storms, gust fronts, convergence lines, gravity waves, turbulence and tornadoes. They are used in forecast problems that require very high horizontal resolution and cover relatively small domains. The release of buoyancy in the atmosphere and its detailed effects on the development of deep convection is well represented in the model. Vertical accelerations and vertical motions are taken into account in an additional forecast equation, rather than determining the vertical motion diagnostically solely from the horizontal divergence (as an hydrostatic model does). The main disadvantage of a non-hydrostatic model is the longer computation time. Since this is a limited area model, it requires boundary condition from other external models, such as a global climate model or, as in this case, a reanalysis.

WRF offers multiple physics options in order to better describe the features of the atmosphere and they can be combined in many ways. The model physics permits to set different parameterization regarding the microphysics, cumulus parameterizations, surface physics, planetary boundary layer and atmospheric radiation physics (Skamarock et al. 2008). Cloud microphysics involves all the aspects related to the property of a cloud at the micro-scale, including droplet concentrations and sizes, ice-crystal formation, interaction among droplets and rain drop formation. The cumulus convective parameterized scheme contains different closure assumptions to help the computation of the cumulus scale convection. (Prater et Evans, 2002). It describes the change taking place in atmospheric parameters such as moisture and temperature of the air column with the start of convection and interaction between the process of convection and grid-scale dynamics (Kerkhoven et al. 2006).

The surface physic setting can be multi-layer land surface models ranging from a simple thermal model to full vegetation and soil moisture models, including snow cover and sea ice.

The WRF model supports multiple domain nesting. This kind of technique allows resolution to be focused over a region of interest by introducing an additional grid (or grids) into the simulation. The nested grids are rectangular and are aligned with the parent (coarser) grid within which they are nested. Additionally, the nested grids allow any integer spatial (Δx_{coarse} / Δx_{fine} and temporal refinements of the parent grid. The boundary conditions of the finer grid are provided by the interpolation of the coarse grid. The way of interaction between the two grids can be 1-way nesting or 2-way nesting: in the first way there is an only way information exchange from the coarse grid to the fine grid; in the second way the information exchange between the grids is in both directions where the fine grid solution replaces the coarse grid solution for coarse grid points that lie inside the fine grid (coarse-to-fine for the fine-grid lateral boundary computation and fine-to-coarse during the feedback at each coarse-grid time step).

13.2.1 Model Set Up and Methodology

The WRF model was originally designed for simulations in specific convective environments in America and Europe (Tadross et al. 2006). However, the use of this model has been extended in recent years in other parts of the globe. The climate and the topography of the area of study strongly affects the physical environmental processes and, consequently the different parameterizations do not perform in the same way in different parts of the world. For this reason the appropriate scheme has to be chosen in order to well depict the environmental parameters in some specific regions of the world.

The Indian subcontinent presents the highest mountain range of the world and thus it represents a particularly challenging region for the use of the WRF model and for the selection of an optimal configuration, both in terms of parameterizations and of model resolution.

The runs are performed on the two domains (d01 and d02 at 14 and 3.5 km grid spacing, respectively), covering the upper and lower boundaries of the cloud-permitting range (Arakawa, 2004), in a two-way nesting mode (Fig. 13.1).

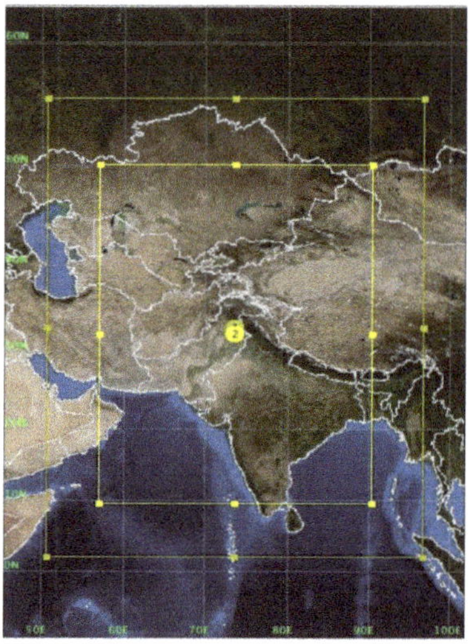

Fig. 13.1. The two nested domains d01 and d02.

The vertical coordinate is discretized in 42 levels, with hydrostatic pressure levels equally spaced, with 10 levels in the lowest 2.5 km. The enhanced number of layers near the surface is intended to help in resolving the complex flow over the orography and its associated convective activity. The turbulent parameterization is the Mesoscale 1D scheme and the planetary boundary layer scheme is the Yonsei University (YSU). The boundary and initial conditions for the experiment are provided by the ERA-INTERIM reanalysis at its native resolution (0.7°, about 78 km).

This study is carried on through two different steps (Fig.13.2). In the first step the best model configuration is explored both in terms of microphysics and convection closures with the same ERA-INTERIM initialization at July 26th at 00 UTC. In the second step we investigate the predictive ability of WRF mode at the different scales, with the variation of the initialization day, in order to find the best initialization.

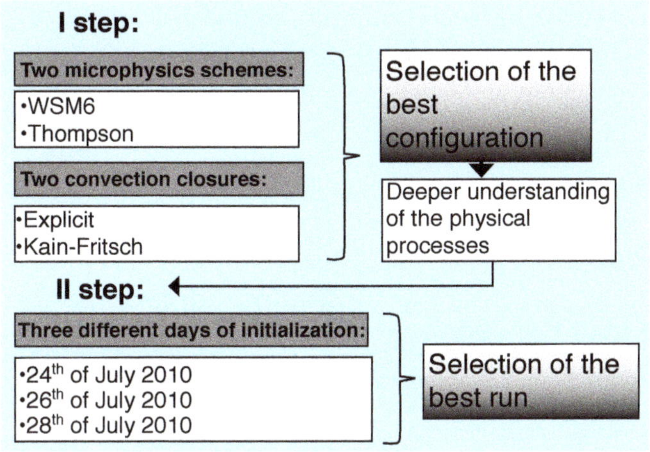

Fig. 13.2. Scheme of the two step development of the experiment.

Four different configurations are tested. The different configuration in terms of convective closures and microphysics schemes are summarized in the following chart (Table 13.1).

Table 13.1. The different configuration sets of the experiments.

Configuration	Convective closure	Microphysics scheme
Ex_WSM6	explicit	WSM6
KF_WSM6	Kain-Fritsch	WSM6
Ex_Thom	explicit	Thompson
KF_Thom	Kain-Fritsch	Thompson

13.2.1.1 Convective closure schemes

Two different single-moment or 'bulk' microphysics parameterizations are evaluated in this work: the WRF Single-Moment 6-Class (WSM6, Hong and Lim, 2006) and the Thompson et al. (2004) microphysics scheme. The six-class scheme extends the WSM5 scheme to include graupel and its associated processes. Some of the graupel-related terms follow Lin et al. (1983), but its ice-phase behavior is much different due to the changes of Hong et al. (2006).

A new method for representing mixed-phase particle fall speeds for the snow and graupel particles is introduced, assigning a single fallspeed to both species, weighted by the mixing ratios, and applying that fallspeed to both sedimentation and accretion processes (Dudhia et al. 2008). Compared with the previous WSMs schemes, the WSM6 scheme is the most suitable for cloud-resolving grids, considering its efficiency and theoretical backgrounds (Hong and Lim, 2006).

The Thompson scheme incorporates a large number of improvements, if compared to earlier single-moment approaches. The main work has been done to better describe the physical processes and reducing the computer coding. Particular attention is given to snow processes: the snow size distribution depends on both ice water content and temperature and is represented as a sum of exponential and gamma distributions. Furthermore, snow assumes a non-spherical shape with a bulk density that varies inversely with diameter as found in observations (Thompson et al. 2004 and 2008).

13.2.1.2 Microphysic schemes

Two different convection closure approaches are evaluated: explicit method and Kain-Fritsch, mass-flux type scheme (Kain and Fritsch, 1990).

Cumulus parameterizations are generally used only for coarse grid sizes, (e.g., greater than 10 km), where they are necessary in order to well depict the convective processes in terms of release of latent heat. Generally, they should not be used at the scale in which the model can resolve the convective eddies itself (e.g., ≤ 5 km grid). If the simulations stay in the "grey zone" between 5–10 km the use of a cumulus parameterization scheme has to be considered: in some cases the assumptions about the convective eddies being entirely sub-grid-scale break down for finer grid sizes, in other cases these schemes have been found to be helpful in triggering convection in grid applications (Skamarock et al. 2008). Even if the highest model resolution in the case study analized is 3.5 km (less than 5 km), we fall close to the grey zone in which it is not obvious if the cumulus parameterization should be used or not. For these reason we decided to test both approaches.

In the explicit simulation no sub-grid closure formulation is used. The high resolution model tries to resolve explicitly, albeit crudely, the convective processes (Kain et al. 2006). The main advantage of the explicit mode is related to the computationally gain related to this choice.

At the contrary, the Kain-Fritsch parameterization is a mass-flux scheme designed to rearrange mass in a column so that CAPE is consumed. Its assumption about consumption of CAPE is appropriate for short time and space scales and it accounts for microphysical processes in convection (Gallus 1999). It uses a simple cloud model with moist updrafts and downdrafts, including the effects of detrainment, entrainment, and relatively simple microphysics. Furthermore, the scheme offers a plausible treatment of convection triggering and of convective inhibition. However, it may tend to leave unrealistically deep saturated layers in post-convective soundings, causing in turn the microphysics scheme to be activated and to simulate post-convective stratiform precipitation (Gallus 1999).

13.2.1.3 Different days of initializations

The WRF model solves a set of partial differential equations, obtaining an approximate solution by means of the finite difference methods. A set of initial conditions and boundary conditions are necessary to solve the problem, given (in this study) by the ERA-INTERIM reanalysis. The starting day of the simulation is very important and its variation may cause consistent differences in the simulation results.

For this reason it has been decided to test how different initialization days could affect the model output and in which terms, considering also the spin up time of the model. After determining the best configuration in terms of parameterizations for experiments initialized on July 26th at 00UTC (IC26), in the second part of the work we used the same configuration to explore experiments initialized on July 24th (IC24) and on July 28th (IC28).

The goal was to assess how the WRF model driven by large scale atmospheric I.Cs and B.Cs was able to enhance the features of meteorological variables at a higher spatial and temporal resolution. We want to investigate if the model develops its own atmospheric circulation and provides a better evaluation of the mesoscale processes characteristic of the extreme event in such a complex orography area.

13.3 OBSERVATIONAL DATASETS

Both the best configuration and the best initialization have been selected through a comparison with observations in terms of daily rainfall, columnar water vapor content, wind speed and direction.

The orographic complexity of the region and the general lack of meteorological station observations represent some of the main difficulties encountered in this model-observation comparison. Ground observations are scarce or of difficult access, both in terms of raingauge networks and of radar data.

Raingauges have not been used in this work because of difficulties in accessing the data and due to their strong bias in a complex topography region like Pakistan (Anders et al. 2006). The existing raingauge network, especially for mountainous areas, is generally not dense enough to capture the fine-grained variability of precipitation, at resolution smaller than tens of square kilometers. Raingauges are subject to several kind of errors including the difficulty to measure the snow water equivalent depth, the deflection of precipitation by winds and the fact that gauges are punctual measures, and do not necessarily represent a good estimator of a widespread precipitation pattern (Anders et al. 2006).

The radiosounding stations from the university of Wyoming database have helped in providing a general idea of the vertical structure of the atmosphere, even if the data quality is not optimal due to the aforementioned issues. (Collins, 2001). Nevertheless the integration of radiosoundings with the remotely-sensed data of TRMM, AMSR-E, SSM/I and ASCAT has provided a satisfactory set of observations for the model comparison.

TRMM satellite precipitation estimates represent the main source of satellite data in that part of the world. The TRMM 3B42 rain products are used for model comparison. The purpose of the 3B42 algorithm is to produce TRMM-adjusted merged-infrared (IR) precipitation and root-mean-square (RMS) precipitation-error estimates. The TRMM rainfall mission was originally developed to observe the tropical and subtropical region of the world: the spatial coverage extends from 50 degrees south to 50 degrees north latitude. The spatial resolution is a 0.25-degree by 0.25-degree grid and the adjusted merged-IR precipitation (mm/hr) and RMS precipitation-error estimates has a 3-hourly temporal resolution.

Because of the fact that the finer is the scale and the more significative is the incidence of the sampling error, the temporal resolution of daily accumulation is preferred to the 3-hour resolution. The reader interested in reading more about the TRMM mission is referred to Kummerow et al. 1998 and by the official NASA product website (http://disc.sci.gsfc.nasa.gov/TRMM).

13.4 ANALYSIS

The meteorological analysis is performed starting from the large scales, down to the mesoscale fields for all the different simulations performed, in order to understand the inter-scale interplay of the phenomena. At larger scales, variables such as wind speed and direction, geopotential height, water vapor mixing ratio and temperature, interpolated on the vertical 500mb isobar level, are compared with the synoptic analysis described in literature. The large scale circulation simulated by WRF is analyzed, investigating the high and low pressure fields distribution, the motion field and the humid air masses transport during the event.

In addition to that, a deeper insight to the mesoscale characteristic of the WRF simulations is gained, mainly considering the daily precipitation fields. Rainfall amount and patterns are, in fact, known to be one of the most difficult variable to be modeled correctly (Litta at al., 2012). Moreover, it is also important to take into account that rainfall quantity and localization are the result of the interplay of other meteorological variables like the distribution and the quantity of columnar water vapor, the wind patterns, etc. and thus a very valuable metric to judge the modeling performances. The model simulations have been compared with the TRMM daily rainfall dataset. To overcome the spatial scale incoherence existing between the two datasets, WRF data have been reaggregated on TRMM's native grid (0.25°).

The study is focused on a geographic window centered on northern Pakistan (23N to 40N in latitude and 66E to 78E in longitude), because this area was the one mainly affected on July 28th and 29th (days of maximum rainfall intensity).

The 2010 Pakistan flood case study will be explained in the next paragraphs. Not all the results of the different steps of the analysis will be shown hereinafter, a more detailed description of the results obtained is given for the final best configuration.

13.5 CASE STUDY: PAKISTAN FLOOD 2010

13.5.1 Synoptic Analysis

The scientific community agrees on the fact that the meteorological conditions that led to the Pakistan 2010 flooding were rather unique and anomalous, with respect to the normal summer monsoonal rainy season (Webster et al. 2011; Houze et al. 2011). The heavy rains caused a countrywide inundation, affecting the economic and human life of Pakistan country both in the long and in the short term. The death toll rose to 1,781 deaths, 2,966 people were injured, and more than 1.89 million homes were destroyed.

In early July, a strong ridge of high pressure began to develop near the Ural Mountains in Russia, creating a blocking pattern known as "omega block" (due to its shape) throughout nearly all western Russia that remain stable in the next two months. Due to this fact, the jet stream assumed an anomalous position, riding around the perimeter of the omega block into the western Pakistan, acting as a carrier of hot and moist air, creating a "supercharged monsoon" and unstable atmospheric conditions (Pakistan Flood Event Recap Report,2010). The study of Hong el al. (2011) demonstrates that the interaction between strong tropical monsoon surges and extratropical disturbance downstream of the European blocking was the key factor leading to the catastrophic Pakistan flood. The moisture flux convergence was anomalous: in normal monsoonal events the low-level moisture flow originates predominantly from the Bay of Bengal, with smaller contributions from the Arabian Sea. In that case the indirect contribution of La Niña phase, introduced a low-level easterly anomaly in south and southeast Asia, which weakened eastward moisture transport and helped enhance moisture transport and convergence in the northern Arabian sea and Pakistan (Hong et al. 2011).

This large scale anomalies created an unusual localization of monsoonal heavy precipitation stratiform patterns (typical of mountains and wetlands of northeastern India and Bangladesh), in the arid mountainous region of northern Pakistan (Houze et al. 2011; Rasmussen et al. 2014). This flow anomaly extended to lower levels, conveyed moisture into a region of Pakistan orthogonal to the Himalayan barrier, and was thus largely responsible for establishing the environment of the mesoscale rain systems that produced the flooding in the Indus region (Houze et al, 2011). The European blocking acted on the persistence of the event, blocking moist air inside an usually dry air region, causing the anticipation of saturation condition, leading to a less convective vertical growing of the cells and to a more stratiform horizontal extension (Houze et al. 2011).

The dynamics was characterized by a strong scale interaction between larger and smaller scales: even though the event at the large scale was predictable even 12 days in advance (Webster et al. 2011), the real differences in producing heavy rain was made by the storm occurring at lower scales (Rassmussen et al. 2014). For these reasons the anomalous precipitating system needs a deeper detail of analysis in order to be understood.

13.5.2 Step 1: Different WRF Configurations

A good configuration of the model is important to well reproduce the physical processes leading to the case study. In this work, the four different settings introduced in paragraph 13.2.1 have been evaluated in the numerical domain d02.

All the results here presented are shown for the most intense days of the event: July 28th and July 29th (Fig. 13.3).

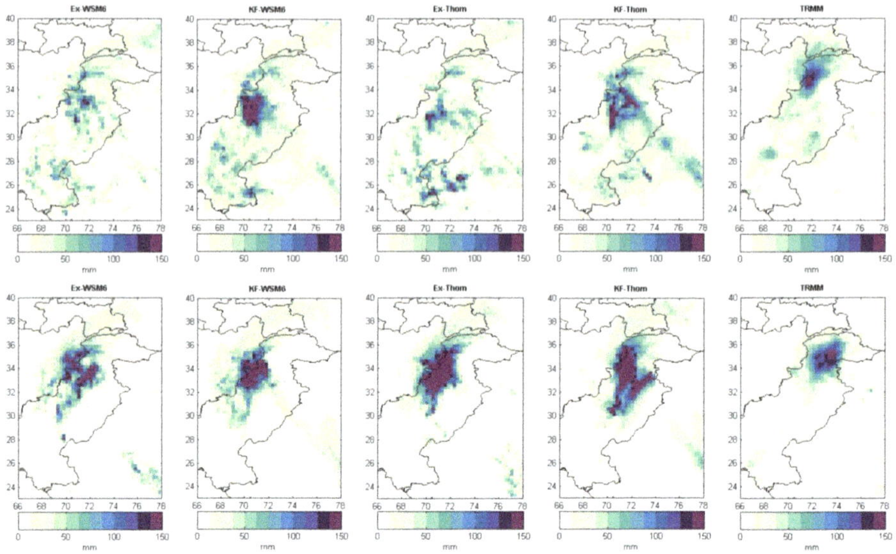

Fig. 13.3. Comparison between WRF QPF (left panels, from left to right: Ex_WSM6 configuration, KF_WSM6 configuration, Ex_Thom configuration, KF_Thom configuration) and TRMM daily rainfall (right panel) at the same 0.25° resolution in the study area, on the 28th of July 2010 (upper panel) and on 29th of July 2010 (lower panel).

Overall, high resolution WRF model simulations seem to reproduce satisfyingly the observed rain depths, the wind direction and the water vapour column content fields, with minor and yet significant differences as parameterizations change. Results turned out to be more sensitive to different convection schemes than to a change of microphysics schemes.

The overestimation of the Thompson microphysics has been highlighted by the frequency distribution histograms, the scatter plots and RMSE. The Thompson Mp8 microphysics scheme tends to overestimate precipitation when used together with both the implicit Kain-Fritsch and with explicit convection.

Focusing on convective closure approaches, the Kain-Fritsch scheme produces more uniform rainfall patterns in the whole domain than the explicit scheme. This latter configuration creates a rainfall localization that was more precise and the field was, on the average, less scattered around the precipitation peak. Although the two convective schemes outputs are really likewise, Kain-Fritsch better reproduced the precipitation patterns.

Both quantitative estimation errors and localization errors are evaluated in the computation of the RMSE: from the summary chart (Table 13.2), the minimum error values on the most intense days of the event have been calculated for the Ex_WSM6 setting.

Considering the results of the map comparison and the RMSE values obtained, the best setting performance in the most intense days of the event of 2010 Pakistan flood was given by the explicit convection WMS6 microphysics scheme (Ex_WSM6).

Table 13.2. Root mean square errors for the different configurations tested, days July 28th and July 29th.

ROOT MEAN SQUARE ERROR		
	July 28th	**July 29th**
Ex_WSM6	24.11	32.38
KF_WSM6	28.78	39.53
Ex_Thom	31.01	45.55
KF_Thom	28.59	42.56

13.5.3 Step 2: Different Initializations

The selection of the best initialization day has been carried out, using the optimal configuration found above, that is using an explicit convection with WSM6 microphysics scheme. Different initialization days were explored, using July 26th (IC26), July 24th (IC24) and July 28th (IC28).

The main meteorological variables have been interpolated on the vertical 500 mb isobar level. They allow to analyze the large scale circulation simulated by WRF, investigating the high and low pressure fields distribution, the motion field and the humid air masses transport during the event. The large scale meteorological fields are shown in Fig.13.4, Fig. 13.5, Fig. 13.6, for the IC26 initialization.

The synoptic features that led to the severe 2010 events on Pakistan are well reproduced by the model in all its different initializations, if compared with the reanalysis maps of the paper of Houze et al. 2011 (Fig. 13.1 and Fig. 13.2, Houze et al, 2011) and the detailed synoptic analysis of the event found in literature (Houze et al. 2011; Hong et al. 2011; Webster et al. 2011; Rasmussen at al., 2014). The geopotential height coupled with wind speed and direction underlines the presence of a strong high pressure block in the northern part of the domain, lasting over a month at that time being (Fig.13.4). This blocking high, associated with the presence of smaller scale troughs in the Arabian sea and in the south of the Afghanistan, led to the formation of a strong wind blowing from the Arabian sea to the northern part of Pakistan.

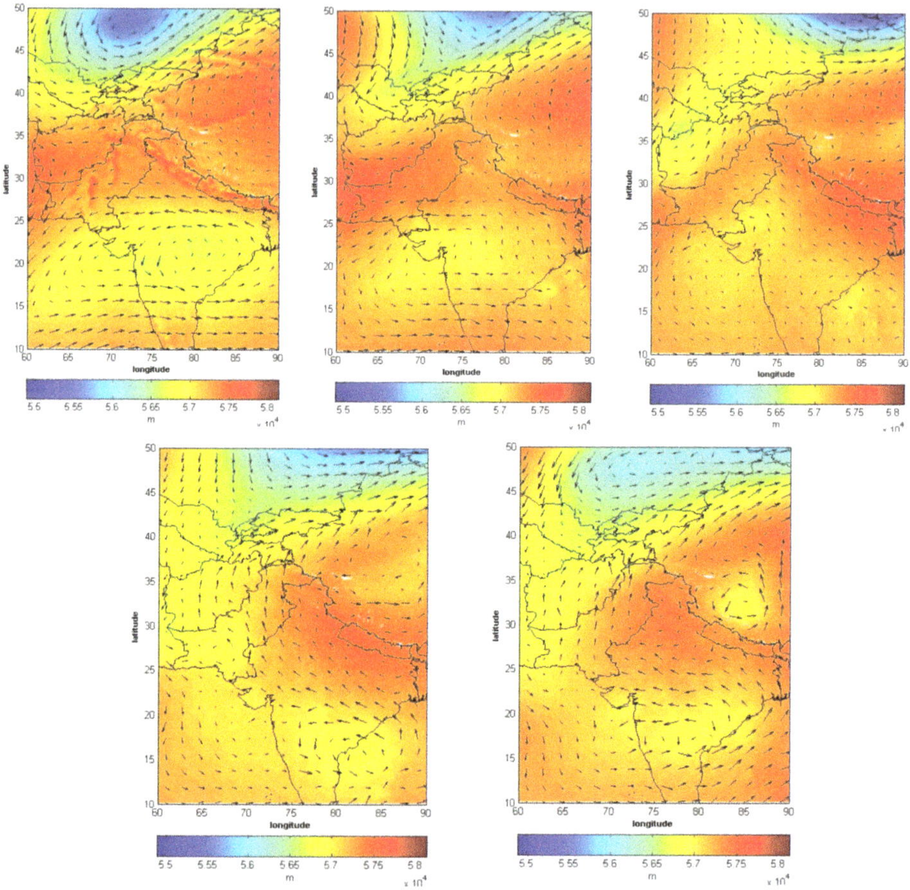

Fig. 13.4. Geopotential height, wind speed and direction at 500 mbar isobar for the 26-30 July 2010 period (from left to right, up to down) in Ex_WSM6 configuration, IC26 initialization.

From the water vapor observation, there is a moisture transport associated to the south-westerly winds that brought a high water vapor quantity up to the northern Pakistan. Another source of vapor is given by the moisture flux approaching from the bay of Bengala, on July 27th and 28th, due to the monsoon circulation (Fig. 13.5). As explained by Reale et al. 2012, it can also be noted that part of easterly moist flow, from the Bay of Bengal flowing across the Gangetic Plain between 21–24 N and 75–90 E, is deflected northwards and becomes entangled in the southerly flow against the northern Pakistan mountains. Both eastern and western contributions suggest a large transport of moisture in the mountainous Pakistan area throughout the days preceding the maximum precipitation (Fig.13.5). Specifically, the moisture flux convergence between July 27th at 00 UTC and July 29th at 00 UTC represents the accumulation of moisture during the two days in which the maximum precipitation occurs.

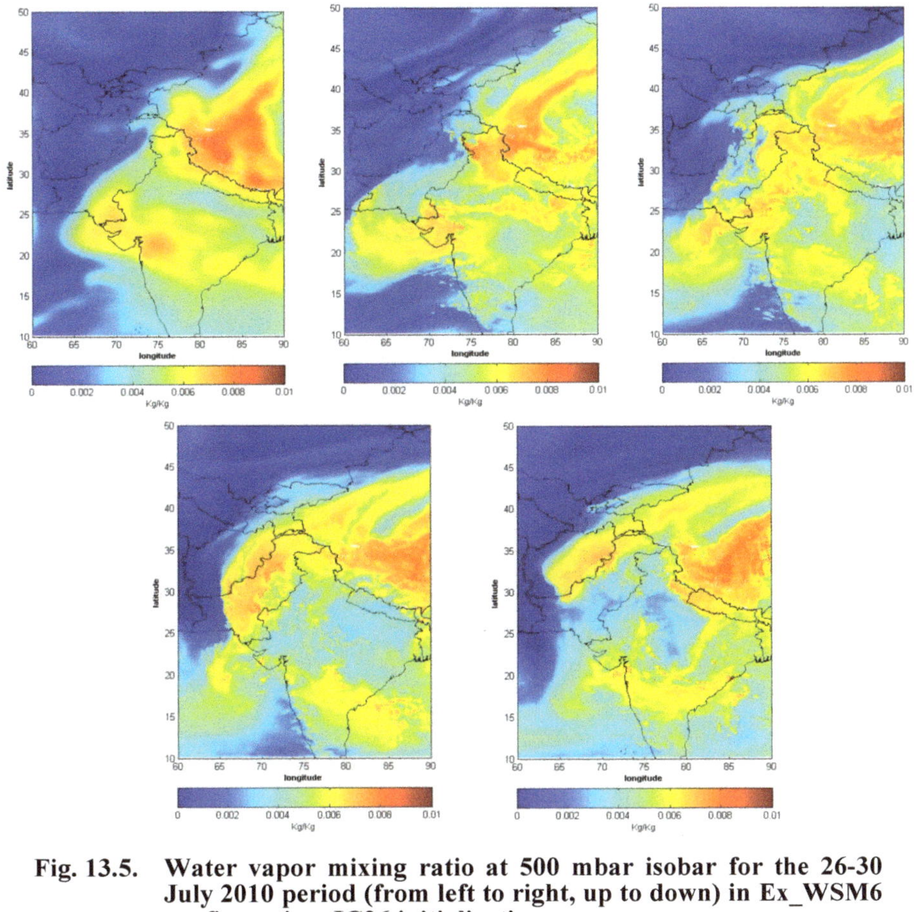

Fig. 13.5. Water vapor mixing ratio at 500 mbar isobar for the 26-30 July 2010 period (from left to right, up to down) in Ex_WSM6 configuration, IC26 initialization.

The large scale cyclonic circulations contribute to the convective development that led to the high rainfall amount that hit Pakistan provinces (Houze et al. 2011; Rasmussen et al. 2014). The temperature at 500hPa underlines the presence of a different heating of the atmosphere in the northern and in the southern part of the domain (Fig.13.6). The temperature gradient matches the behavior of the distribution of high and low pressure. Radiosounding data profiles reveal a good agreement between the observed vertical structure of the atmosphere (blue line) and the simulated diagrams obtained in all the d01 domain (red line). The skew-T log-P diagrams at the available radiosounding stations (Fig. 13.7) are shown in Fig. 13.8, Fig. 13.9 for the IC26 run.

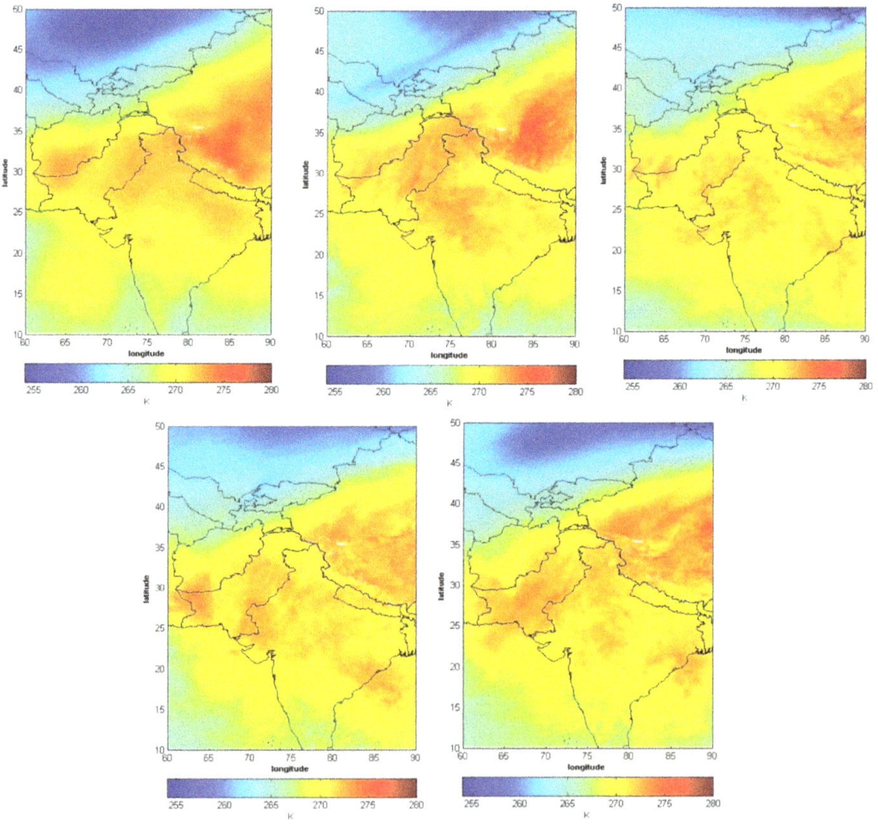

Fig. 13.6. Temperature at 500 mbar isobar for the 26-30 July 2010 period (from left to right, up to down) in Ex_WSM6 configuration, IC26 initialization.

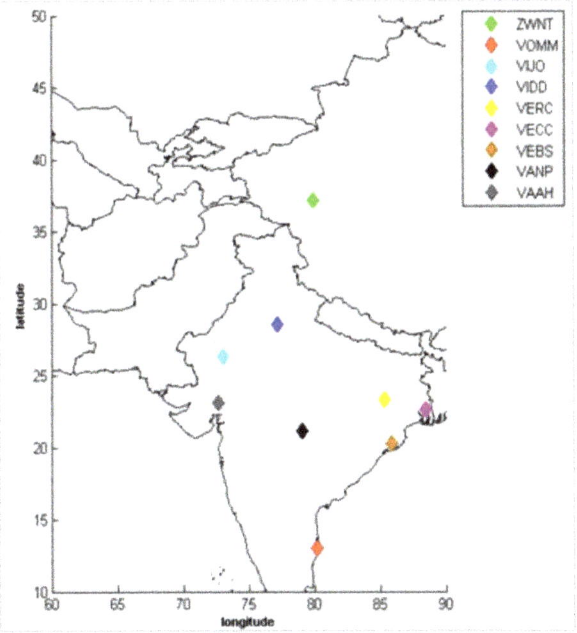

Fig. 13.7. **Radiosounding stations inside the d01 domain, used in this study.**

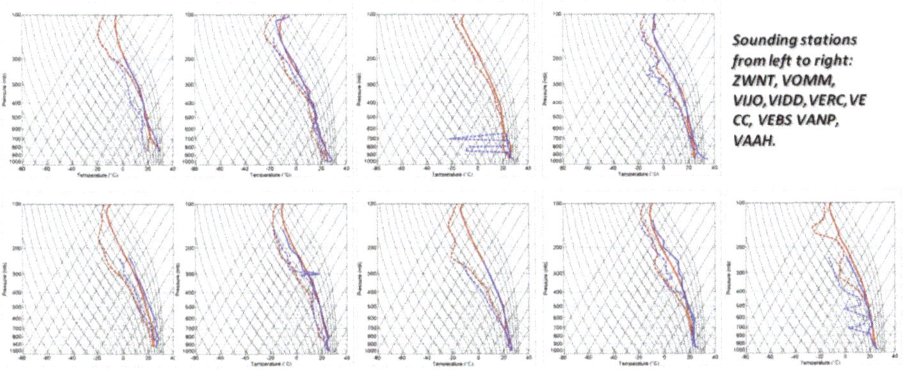

Fig. 13.8. **Skew-T log-P diagrams for July 28th, simulated by WRF IC26 and observed by radiosondes at the radiosounding stations location.**

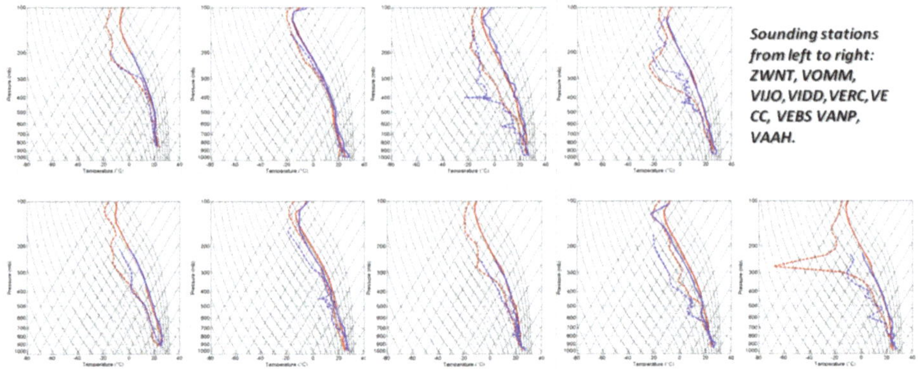

Fig. 13.9.Skew-T log-P diagrams for July 29th, simulated by WRF IC26 and observed by radiosondes at the radiosounding stations location.

Once that the large scale has been considered, the mesoscale results are evaluated comparing the simulated daily rainfall field with the TRMM observations (Fig. 13.10). From the analysis it is evident that the IC24 run performs unreliably in terms of pattern distribution and rainfall amounts.

There are high biases in precipitation position and in total rainfall estimation: The IC24 run well simulated the actual rainfall amounts until July 27th, and then downgraded as the event developed further. The IC26 and IC28 runs better depict the event, on the average. If on the one hand, on July 28th, the pattern is well simulated by the IC26 run, on the other hand IC28 run better matches the daily average. On July 29th, the IC26 run well captured the main rainfall core, even if it is more widespread and localized slightly eastwards (Fig. 13.10). On the same day the IC28 run makes an extremely bad performance, especially in catching the most intense rainfall. This bad result for July 29th is crucial in the final evaluation of the best performance. For this reason, the IC26 run turned out to be the best simulation for the Pakistan flood.

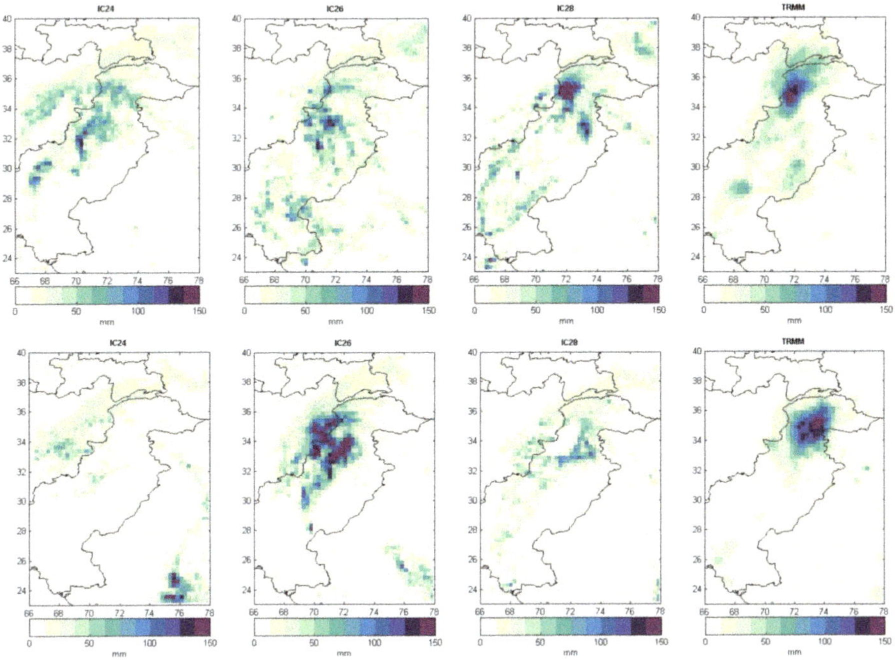

Fig. 13.10. Comparison between WRF QPF (left panels, from left to right: IC24, IC26, IC28) and TRMM daily rainfall (right panel) at the same 0.25° resolution in the study area, on the 28th of July 2010 (upper panel) and on 29th of July 2010 (lower panel) .

13.6 CONCLUSIONS

In this case study the reliability of the WRF model in simulating the extreme Pakistan flood 2010 has been evaluated. A high degree of difficulty is given by the presence of the highest and steepest mountain range of the world and by a strong scale interaction between the atmospheric processes at the different scales. For these reasons high resolution simulations are needed to reproduce the event.

At first different model configurations in terms of convective scheme and of microphysical parameterization have been tested in order to find the best setup in those environmental conditions; then, the influence of the initialization time on the model simulations has been investigated. It has been found WRF model simulations produce results consistent with the observations, to a large extent.

The simulations are found to be more affected by changes in the day of initialization than by the different parameterization schemes that we tested. In the first step, the differences between different configurations were less evident, even if not negligible. Explicit convection together with the WMS6 microphysic scheme produces the best rain estimates and more precise distribution of the main precipitation peak (Fig. 13.1). Once that we have fixed the best configuration, we have varied the different initialization times.

It's commonly known that the nearer is the model initialization to the event, the more reliable is the simulation: in this case the result is quite surprising. The results for July 29th in the IC28 after 48h from the initialization are worse than the results for the same days in the IC26 simulation after 72h (Fig. 13.10). Contrary to what expected, it not always true that the best simulation is the one closest to the day of initialization.

Further analysis has been done considering an additional sets of experiments (even if not shown in this work), performing a sensitivity analysis of the IC28 results to the domain size and initial conditions. Firstly we have reduced the extension of the domain in order to investigate the spin up time of model but no significantly improvements were achieved. Then we have explored the model initial conditions: we have fed the model with the inputs provided by the WRF simulation started on July 26th, setting to zero all the variables related to clouds and vertical velocities. We wanted to observe if the model was able to retrieve a satisfying representation of the event. WRF fine scale processes are better resolved by the new run, than by the ICs given by ERA-INTERIM on July 28th by the IC28 run.

It was found that that the ICs given by ERA-INTERIM at 28 at 00 UTC were the main responsible for the bad simulation of the day 29th offered by IC28. The quality of the initial condition given by the reanalysis deeply affected the WRF simulation and is the prominent factor to justify the better performance of the IC26 run. In conclusion, the IC26 run represent better the event, even if the run starts 72h before the event.

Further investigation will be needed in order to evaluate the performance of WRF in this area also in different study cases.

13.7 FUTURE TASKS

Once that the reliability of the model has been proven over complex orography, high resolution meteorological simulations might be used for understating the occurrence of extreme events, the hydrological processes and flooding generation typical of our target area.

A good estimate of precipitation amounts over complex orography is a difficult but crucial task: mountain areas are the main sources of fresh water, also in terms of snow accumulation and melting. Steepest gradient and ridge-valley phenomena enrich the complexity of processes acting at that altitude. Once that a good precipitation estimate is gained by means of an high resolution model, the future target is to add the hydrological post-processing of the events in order to see the impact of precipitation in terms of runoff, and catchments response.

In this regard, WRF model has a hydro tool called WRF-Hydro, designed to link multi scale processes of the atmosphere and terrestrial hydrology. This tool can analyze the land-atmosphere coupling, considering also the land water feedback over precipitation generation. At the moment some experiments have been carried out in order to getting started with this tool and to test some application of this structure coupled with WRF model in the simulation of extreme events.

13.8 REFERENCES

Anders, A.M., Roe, G.H., Hallet, B., Montgomery, D.R., Finnegan, N.J., and Putkonen, J. 2006. Spatial patterns of precipitation and topography in the Himalaya. In Willett, S.D., Hovius, N., Brandon, M.T., and Fisher, D., eds., Tectonics, Climate, and Landscape Evolution. *Geological Society of America Special Paper* 398, p. 39–53, doi: 10.1130/2006.2398(03).

Collins, W. G. 2001. The operational complex quality control of radiosonde heighs and temperature at the National Centers for Environmental Prediction. Part II: Examples of error diagnosis and correction from operational use. *J Appl. Meteor*. 40, 152-168.

Dudhia, J., S.-Y. Hong, and K.-S. Lim, 2008: A new method for representing mixed-phase particle fall speeds in bulk microphysics parameterizations. *Journal of the Meteorological Society of Japan*, 86A, 33–44, doi:10.2151/jmsj.86A.33.

Hong, S. Y., and Lim, J. O. J. 2006. The WRF single-moment 6-class microphysics scheme (WSM6). *J. Korean Meteor*. Soc 42(2), 129-151.

Hong, S. Y., Noh, Y., & Dudhia, J. 2006. A new vertical diffusion package with an explicit treatment of entrainment processes, *Monthly Weather Review* 134(9), 2318-2341.

Hong, C. C., Hsu, H. H., Lin, N. H. and Chiu, H. 2011. Roles of European blocking and tropical and extratropical interaction in the 2012 Pakistan flooding. *Geophys. Res. Lett. 38*, LXXXXX, doi:10.1029/2011GL047583.

Houze, R. A., Rasmussen, K.L., Medina, S., Brodzik, S.R. and Romatschke, U. 2011. Anomalous atmospheric events leading to the summer 2010 floods in Pakistan, Bul*l. Am. Meteorol. Soc. 92*, 291–298.

Kain, J. S. and Fritsch, J. M. 1990. A One-Dimensional Entraining/Detraining Plume Model and Its Application in Convective Parameterization. *J. Atmos. Sci.*, 47, 2784-2802. doi:http://dx.doi.org/10.1175/15200469(1990)047<2784:AODEPM>2.0.CO;2.

Kain, J. S., Weiss, S. J, Levit, J. J., Baldwin, M. E. and Bright, D. R. 2006. Examination of Convection-Allowing Configurations of the WRF Model for the Prediction of Severe Convective Weather: The SPC/NSSL Spring Program 2004. *Wea. Forecasting*, 21, 167–181. doi: http://dx.doi.org/10.1175/WAF906.1.

Kerhoven, E., Gan, T.Y., Shiiba, M., Reuter, G. and Tanaka, K. 2006. A comparison of cumulus parameterization scheme in numerical weather prediction model for a monsoon rainfall event. *Hydrological processes 20*: 1961-1978.

Kummerow, C., Barnes, W., Kozu, T., Shiue, J. and Simpson, J. 1998. The Tropical Rainfall Measuring Mission (TRMM) Sensor Package. *J. Atmos. Oceanic Technol., 15*, 809–817. doi: http://dx.doi.org/10.1175/1520-0426(1998)015<0809:TTRMMT>2.0.CO;2.

Lin, Y.-L., Farley, R. D. and Orville, H. D. 1983. Bulk parameterization of the snow field in a cloud model. *J. Atmos. Sci.* 22, 1065-1089.

Litta, A.J, Sikka, D.R., Singh, J. 2012. A comparison study of Convective Parametrization Schemes in WRF-NMM model for intense Precipitation Events, International conference on "Opportunities and Challenge in Monsoon Prediction in a Changing Climate (OCHAMP-2012), Pune, India, 21-25 February 2012, Poster Registration ID: OC-000105.

Molini, L., Parodi, A., Rebora, N. and Craig, G. C. 2011. Classifying severe rainfall events over Italy by hydrometeorological and dynamical criteria, *Quart. J. Roy. Meteorol. Soc.*, 137, 148–154.

Rasmussen, K. L., Hill, A. J., Toma, V. E., Zuluaga, M. D., Webster, P. J. and Houze, R. A. Jr. 2014. Three consecutive years of anomalous flooding in Pakistan. *Quart. J. Roy. Meteorol. Soc.*, submitted.

Reale, O., Lau, K. M., Susskind, J. and Rosenberg, R. 2012. AIRS impact on analysis and forecast of an extreme rainfall event (Indus River Valley, Pakistan, 2010) with a global data assimilation and forecast system, *J. Geophys. Res.*, 117, D08103, doi:10.1029/2011JD017093.

Skamarock, W. C., Klemp, J. B. , Dudhia, J., Gill, D. O., Barker, D. M., Wang, W., Powers, J. G. 2005 A description of the Advanced Research WRF version 2. NCAR Tech. Note NCAR/TN-468+STR, 88 pp.

Skamarock, W. C., Klemp. J. B., Dudhia, J., Gill, D.O., Barker, D. M., Duda, M. G., Huang, X., Wang, W. and Powers, J. G. 2008. A description of the Advanced Research WRF version 3. NCAR Tech. Note NCAR/TN-475+STR.

Tadross, M. A., Gutowski, W. J., Hewitson, B. C., Jack, C. and New, M. 2006. MM5 simulations of interannual change and the diurnal cycle of southern African regional climate, *Theoretical and Applied Climatology 86*: 63-80.

Thompson, G., Rasmussen R. M., and Manning, K. 2004. Explicit Forecasts of winter precipitation using an improved bulk microphysics scheme. Part I: Description and sensitivity analysis. *Monthly Weather Rev*iew, vol. 132, 519–542, 2004.

Thompson, G., Field, P. R., Rasmussen, R. M. and Hall W. D. 2008. Explicit forecasts of winter precipitation using an improved bulk microphysics scheme. Part II: implementation of a new snow parameterization. *Monthly Weather Review*, vol. 136, no. 12, pp. 5095–5115.

Webster, P.J., Toma, V.E., Kim, H.M. 2011. Were the 2010 Pakistan floods predictable? *Geophys. Res. Lett. 38*, L04806, http://dx.doi.org/10.1029/2010GL046346.

CHAPTER 14

SPECIFIC CALIBRATION AND UNCERTAINTY EVALUATION FOR FLOOD PROPAGATION MODELS BY USING DISTRIBUTED INFORMATION

Stefania Camici[1], Giuseppe Tito Aronica[2], Angelica Tarpanelli[1], Tommaso Moramarco[1]

[1] CNR - Istituto di Ricerca per la Protezione Idrogeologica, Via Madonna Alta n. 126, 06128 Perugia, stefania.camici@irpi.cnr.it

[2] Dipartimento di Ingegneria Civile, Informatica, Edile, Ambientale e Matematica Applicata. Università di Messina, C.da di Dio, 98166 S. Agata (ME), garonica@unime.it

ABSTRACT

This Chapter provides a new approach for the calibration and uncertainty evaluation of flood propagation models by using distributed information. Specifically, different water depth measurements, recorded during a well monitored flood event occurred in the Central Italy in November 2012, were used for calibrating and evaluating the uncertainty of a two-dimensional hydraulic model (Multi Level Flood Propagation 2-D, MLFP-2D). The results show the importance of including uncertainty analysis in flood risk mapping respect to the deterministic approach.

14.1 INTRODUCTION

Hydraulic models are an essential tool in many fields, e.g. civil engineering, flood hazard and risk assessments, evaluation of flood control measures, etc. Nowadays there are many models of different complexity regarding the mathematical foundation and spatial dimensions available, and most of them are comparatively easy to operate due to sophisticated tools for model setup and control. However, the calibration of these models is still underdeveloped in contrast to other models like e.g. hydrological models or models used in ecosystem analysis. This has basically two reasons:

1. the lack of relevant data necessary for the model calibration. Indeed, flood events are very rarely monitored due to the disturbances inflicted by them and the lack of appropriate measuring equipment.
2. the choice of a suitable performance measure for calibrating, evaluating model predictions in a suitable and consistent way as well as reducing the uncertainty.

Concerning the first point, *in-situ* gauging networks help to quantify the movement of water in river channels but provide few information about the spatial dynamics of surface water extent, such as floodplain flows and the dynamics of wetlands. The water surface measurement is appropriate in situations where the flow is clearly confined in a well-defined channel boundary, but not in more complex riverine environments that involve movement of water over wetlands and floodplains and where the process can be classified as two-or three-dimensional. In these cases, a distributed information can be very useful for improving the hydraulic model calibration, but it is quite rare especially during flood event.

The choice of suitable performance measure is a more complex issue, at the center of interesting debates along with the most correct method to produce flood risk maps, i.e., by using deterministic or probabilistic approach (Di Baldassarre et al. 2009; Aronica et al. 2002; Pappenberger et al. 2005). The probabilistic approach would provide information for making rational decisions, offering potential for additional economic benefits of forecasts to every decision maker. In particular, probabilistic models are based on the idea that there are many different combinations of parameters or model structures that may fit equally well observed data, named equifinality (Beven and Binley 1992). The most common estimation technique used to produce probabilistic maps is the Generalized Likelihood Uncertainty Estimation (GLUE, Beven and Binley 1992) which allows to simply estimate probability flood risk maps by conditioning against different sort of information, i.e., water levels

(Romanowicz and Beven, 1998), maximum inundation extent observations (Romanowicz and Beven, 2003) and satellite imagery data (Schumann, et al. 2009; Bates et al. 2004).

In this study, a procedure to obtain a flood hazard map including uncertainty is presented. Hazard map has been obtained incorporating parametric uncertainty (related to roughness parameter of the propagation model) using the GLUE procedure. Four objective functions have been used to describe the posterior distribution of roughness parameters, and a comparison of resulting hazard maps has been carried out. In order to test the procedure different water depth measurements, recorded during flood event occurred in the Chiani basin (Central Italy) in November 2012, were used, for calibrating and evaluating the uncertainty of the two-dimensional hydraulic model.

14.2 METHODOLOGY

14.2.1 Flood Propagation model

The MLFP-2D (Multi Level Flood Propagation 2-D) model (Aronica et al. 1998) was used. It is a hyperbolic model based on de Saint-Venant (DSV) equations when convective inertial terms are neglected. The conservative mass and momentum equations for two-dimensional shallow-water flow can be written as

$$\frac{\partial H}{\partial t}+\frac{\partial p}{\partial x}+\frac{\partial q}{\partial y}=0 \quad ; \tag{14.1}$$

$$\frac{\partial p}{\partial t}+gh\frac{\partial H}{\partial x}+ghJ_x=0\,; \qquad \frac{\partial q}{\partial t}+gh\frac{\partial H}{\partial y}+ghJ_y=0 \tag{14.2}$$

where $H(t, x, y)$ is the free surface elevation; $p(t, x, y)$ and $q(t, x, y)$ are the x- and y-components of the unit discharge (per unit width); h is the water depth; g is the gravitational acceleration; J_x and J_y are the hydraulic resistances in the x- and y-directions.

If Manning's formula is adopted, the last terms in eq. (14.2) can be expressed as:

$$J_x = \frac{c^2 p \sqrt{p^2 + q^2}}{h^{10/3}}; \qquad J_y = \frac{c^2 q \sqrt{p^2 + q^2}}{h^{10/3}} \tag{14.3}$$

where c is Manning's roughness coefficient.

These equations were solved by using a finite element technique with triangular elements. The free surface elevation is assumed to be continuous and piece-wise linear inside each element, where the unit discharges in the x and y directions are assumed to be piece-wise constant.

14.2.2 Calibration and Uncertainty Analysis

In the traditional approach the purpose of calibration is 'to find an optimum model parameter set that optimize the numerical value of the objective function'. Given that flood models are highly non-linear, it is possible to obtain the same or very similar values of the objective function for different parameter sets (non-uniqueness of the solution of the inverse problem). In this case, the idea that, for a given model structure, there is some optimum parameter set that can be used to simulate the system loses credibility. There is rather an equifinality of parameter sets (Beven, 2001a), where different parameter sets, and therefore model structures, might provide equally acceptable simulators of the system. This leads to a significant uncertainty in the model predictions which should be taking in account when these predictions are for practical application (i.e., water depths in flood risk evaluation).

Here, the calibration and uncertainty analysis were carried out using the Generalized Likelihood Uncertainty Estimation (GLUE) procedure (Beven, 2001b; 2006; Beven & Binley, 1992). The GLUE procedure is a method which transforms the problem of searching for an optimum parameter set into a search for the sets of parameter values which would give reliable simulations for a wide range of model inputs. No requirement is needed to minimize any objective function, but the GLUE methodology provides a simple and flexible approach to parameter sensitivity analysis, model conditioning and uncertainty estimation.

The calibration and uncertainty analysis procedure comprises different steps. The first step is to recognize the parameter to be included in the uncertainty analysis. In this study, friction due to floodplain and river channel roughness were assumed the most important parameters controlling the inundation extent.

The second step is to decide the range of the parameter space to be examined, which relies upon 'expert knowledge' of the system.

In the third step the GLUE procedure is implemented through an ensemble of simulations of the hydrodynamic model ran on a high throughput computing system. Particularly, GLUE methodology is a Monte Carlo based technique that allows to randomly sample an ensemble of parameter sets from prior distributions by a Latin Hypercube Sampling (LHS). The model is run with each set and the acceptability of each run is assessed by comparing predicted with observed values using a chosen index of goodness-of-fit (likelihood measure) (Aronica et al. 2012, 2002).

Various likelihood functions have been proposed in the literature as measures that quantify the closeness between model simulations and observations.

Meaningful indexes for the comparison of the different calibrations used here are: the mean absolute error (MAE), the root mean square error (RMSE) and the average error (BIAS) of the simulation results from the measured maximum inundation depths calculated as:

$$MAE\left(\theta_k \middle| \mathbf{Y}\right) = \frac{1}{m}\sum_{i=1}^{m}\left|\varepsilon_{i.k}\right| \tag{14.4}$$

$$RMSE\left(\theta_k \middle| \mathbf{Y}\right) = \sqrt{\frac{1}{m}\sum_{i=1}^{m}(\varepsilon_{i,k})^2} \tag{14.5}$$

$$BIAS\left(\theta_k \middle| \mathbf{Y}\right) = \frac{1}{m}\sum_{i=1}^{m}\varepsilon_{i,k} \tag{14.6}$$

where $MAE(\theta_k|Y)$, $RMSE(\theta_k|Y)$, $BIAS(\theta_k|Y)$ are the goodness-of-fit measures for the *k-th* model conditioned on the observations $\boldsymbol{Y}$ and $\varepsilon_{i.k}$ is the error between the observed and simulated maximum water depths at the *i-th* site for the *k-th* model given the parameters vector θ.

Most of these likelihood functions are considered pseudo-likelihood functions because they do not adhere to formal Bayesian statistics and, hence, they are not the best choice for uncertainty analysis (Mantovan and Todini, 2006).

To overcome these limitations the following likelihood function which shows a major statistical appeal was used in this study :

$$L\_RMSE(\theta_k|\mathbf{Y}) = \frac{max\left[RMSE(\theta_k|\mathbf{Y})\right] - RMSE(\theta_k|\mathbf{Y})}{max\left[RMSE(\theta_k|\mathbf{Y})\right] - min\left[RMSE(\theta_k|\mathbf{Y})\right]} \quad (14.7)$$

where $L\_RMSE(\theta_k|\boldsymbol{Y})$ is the likelihood measure for the *k-th* model conditioned on the observations and $RMSE(\theta_k|\boldsymbol{Y})$ is the root mean square error as above defined.

In order to obtain hazard maps including uncertainty in model prediction the results of each simulation for each nodes where weighted according to the likelihood L associated to each simulation and the probability of each node having a water depth has been computed by:

$$h_i^{flood} = \frac{\sum_{k=1}^{n\_sim} h_{i,k} \cdot L\_RMSE(\theta_k|\mathbf{Y})}{\sum_{k=1}^{n\_sim} L\_RMSE(\theta_k|\mathbf{Y})} \quad (14.8)$$

where the likelihood L_$RMSE(\theta_k|\boldsymbol{Y})$ is represented by the probability of the *k*-pair (n_{fl}, n_{ch}) in the likelihood surface (Pappenberger et al. 2007, Aronica et al. 2012).

14.3 CASE STUDY

14.3.1 Study Area

The Chiani river is the main tributary of the Paglia river, right tributary of the Tiber River in Central Italy (Fig. 14.1). The drainage area of the Chiani basin is about 460 km^2 representing one third of the total area of Paglia basin.

The climate is Mediterranean with mean annual precipitation of about 950 mm, ranging from 700 mm (at lower altitudes) to 1600 mm along the ridge. The streamgauge network operating along the Chiani river includes

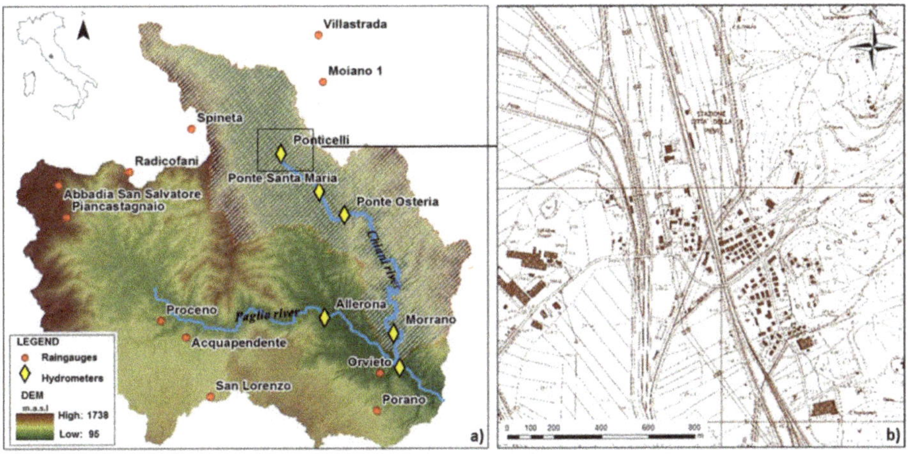

Fig. 14.1. **a) Investigation area overview and Chiani basin location (dashed area). b) Technical map of Ponticelli.**

4 hydrometers that provide sub-hourly records of water level from more than 20 years. For the knowledge of river discharge, Q, at each gauged stations, rating curves obtained by local recorded water levels and velocity measurements, v, occasionally collected for different flow conditions, were employed. For each gauged site, at least 25 pairs of h-v values are available, thus allowing one to obtain a reliable stage-discharge relationship. Concerning the flow regime, the values of maximum and minimum water level of the period 1995–2010 are reported in Table 14.1 for each gauged site.

Table 14.1. **Maximum and minimum water level recorded along the Chiani river in the period 1995–2010.**

Streamflow gauging station	**Water level (m) 1995–2010 period**	
	Maximum	Minimum
Ponticelli	3.68	0.00
Ponte Santa Maria	5.19	0.21
Ponte Osteria	4.18	0.00
Morrano	3.97	-0.04

The study area is a reach of Chiani river of about 3 km between the Città della Pieve station and the Ponticelli gauged station (basin area equal to 100 km^2). The topographical map of the study area is shown in Fig. 14.2 along with the location of the Ponticelli hydrometer.

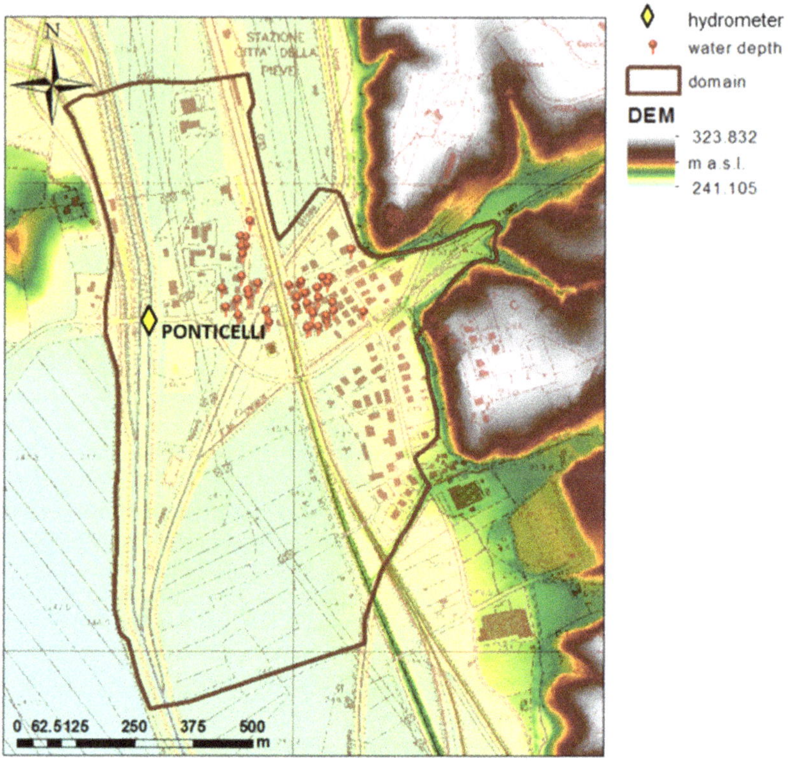

Fig. 14.2. Topographical map of the study area.

14.3.2 The Flooding Event

In the last decade the Paglia River basin has been affected by different flood events, the most severe of which occurred on November 2005, December 2008, January 2010, November 2010 and November 2012.

The event that strucked the Paglia-Chiani rivers from 12th to 16th of November 2012 was the consequence of a widespread, high-intensity storm. Data from the rainfall station of Piancastagnaio in the Paglia basin, indicated an intense precipitation of 328 mm in 48 h, and 350 million m^3

water were estimated to have precipitated in the whole Paglia River basin following this event (Costantini et al. 2012).

The storm caused very large flooded zones affecting the urbanized areas along the Paglia and Chiani rivers (Orvieto, Ciconia Fabro Scalo, Ponticelli), with extensive damages to buildings, roads, commercial and industrial activities, agricultural production fortunately without casualties, but with significant socio-economic damages.

Interesting information was obtained regarding the duration of the flood: while at Orvieto the flood was very rapid, and lasted just a few hours (people were able to return to their businesses by the early afternoon of 12 November 2012), at Ponticelli (in the Municipality of Città della Pieve) the high water lasted for at least a couple of days.

Fig. 14.3a) shows the hydrograph recorded during the flood event at the gauged station of Ponticelli. In order to correctly simulate the flooded areas in the study area also the hydrographs for two streams, Fosso Nuovo and Fosso Osteria, are needed. Due to lacking of more information, the hydrographs for both the rivers are obtained by assuming:

1) hydrological similarity between the basins;
2) the discharges and drainage areas are linearly related.

The resulting hydrographs are shown in Fig. 14.3b and 14.3c for Fosso Nuovo and Fosso Osteria, respectively.

Additional distributed information were available for this case study. Specifically, water depths in 35 points within the floodplain areas were recorded for the post-event analysis (about 10 days later) around the urban area of Ponticelli (see Fig. 14.2). The Fig. 14.4 shows the maximum water level gathered from researcher/regional teams in the study area.

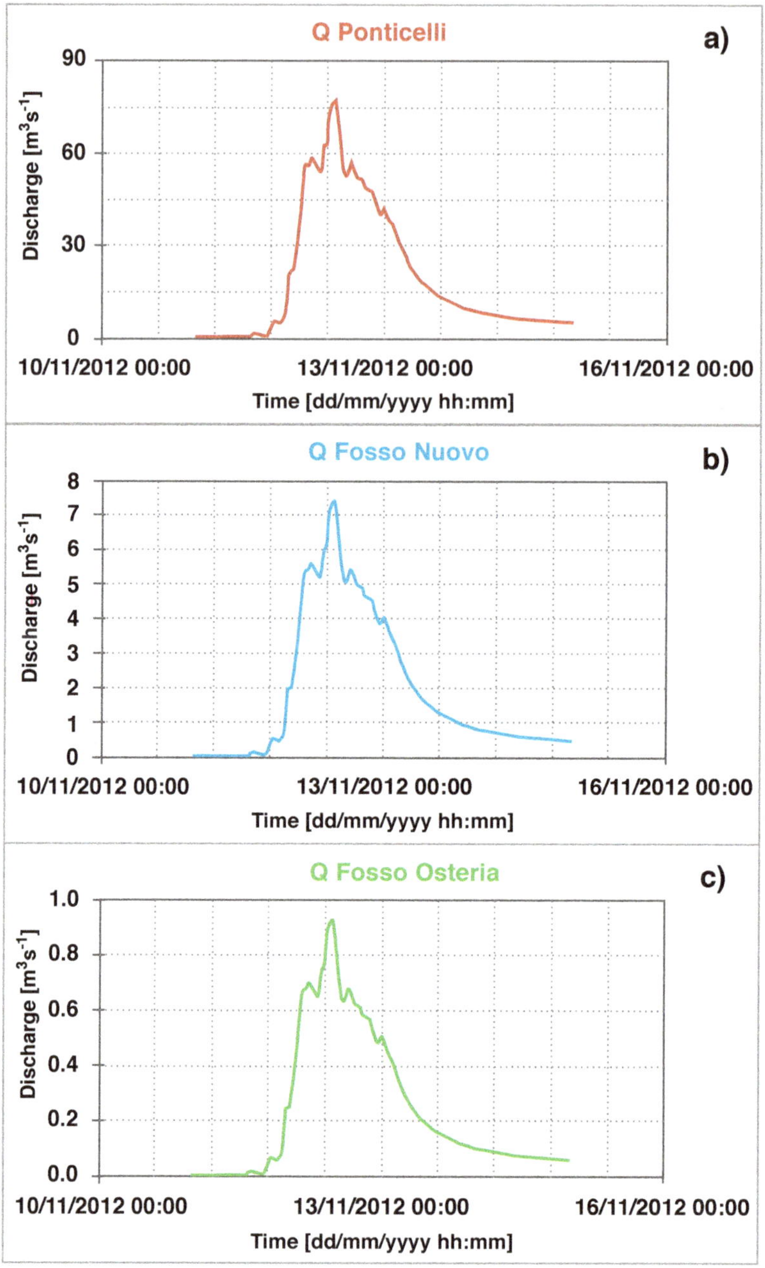

Fig. 14.3. Flood hydrographs recorded at Ponticelli gauged station a) and reconstructed for the Fosso Nuovo b) and Fosso Osteria c) rivers.

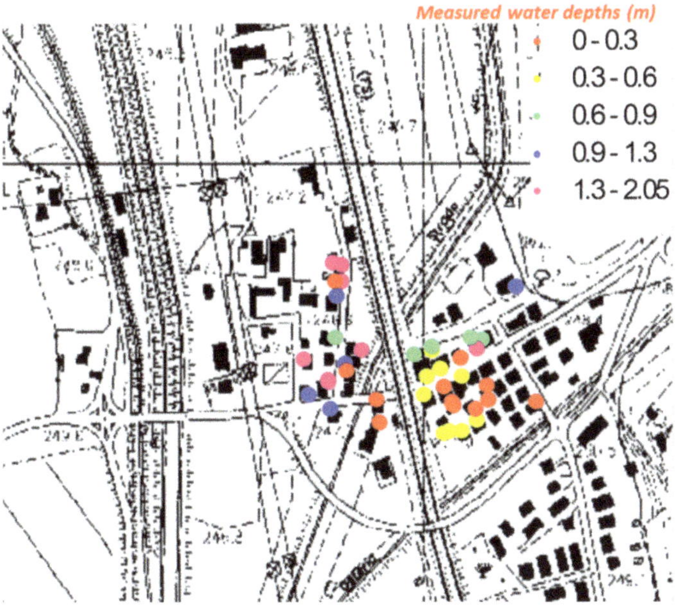

Fig. 14.4. **Maximum water depths measured from post-event survey in the Ponticelli area.**

A detailed mapping of the levee failures and flooding areas were reconstructed for the post-event analysis. The dots in Fig. 14.5a) highlight the levee failure occurred in the study area producing the flooding of the surrounding area as shown in the Fig. 14.5b), c), d).

Moreover, photos and testimonials provided by the technicians of the regional authorities enabled also to reconstruct the actual flooded areas during the flood event as shown in Fig. 14.5a.

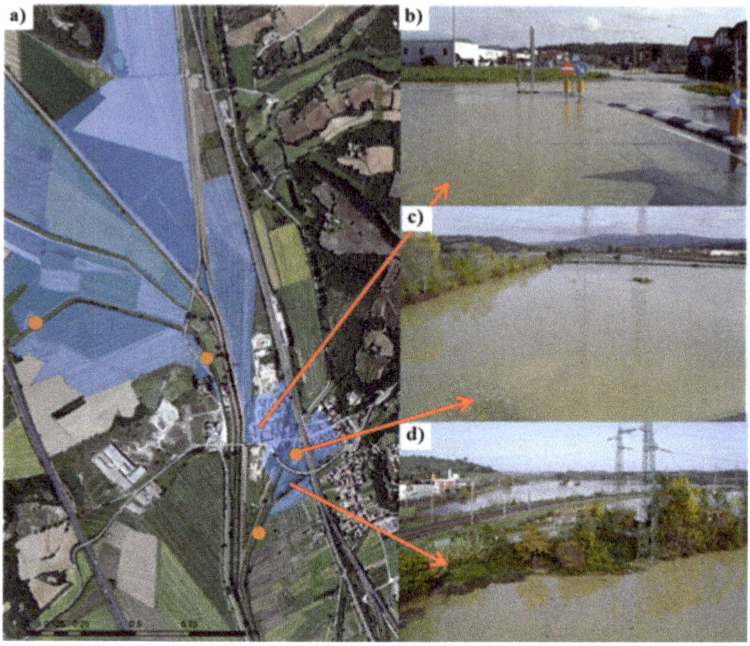

Fig. 14.5. Urban area of Ponticelli during the flood event caused by levee failure (orange dots).

14.4 RESULTS

In the following, the results about the MLFP-2D model set-up, calibration and uncertainty analysis are presented.

14.4.1 MLFP-2D Model Set-up

In the application of the MLFP-2D model, the topography is represented as a smooth surface and elevations are assumed piecewise linear inside triangular elements.

The limits of the finite element mesh were based upon the basin morphology in order to cover the inundated surface and to leave the upper areas out of the domain. The total domain area is about 0.79 km^2, discretized in 39619 triangular elements. The geometric features (x,y,z coordinates) of the 20900 nodes were extracted from a DEM at 2 m resolution available for the study area. The mesh was created from these nodes by using the GIS contouring package TIN that builds triangular irregular networks from point data. The final mesh is shown in Fig. 14.6b, where also two principal regions, floodplain and channel are highlighted (see Fig. 14.6a).

As concerns the boundary conditions, the flood hydrographs showed in Fig. 14.3 were used as inflow conditions to models in three different points as illustrated in Fig. 14.6.

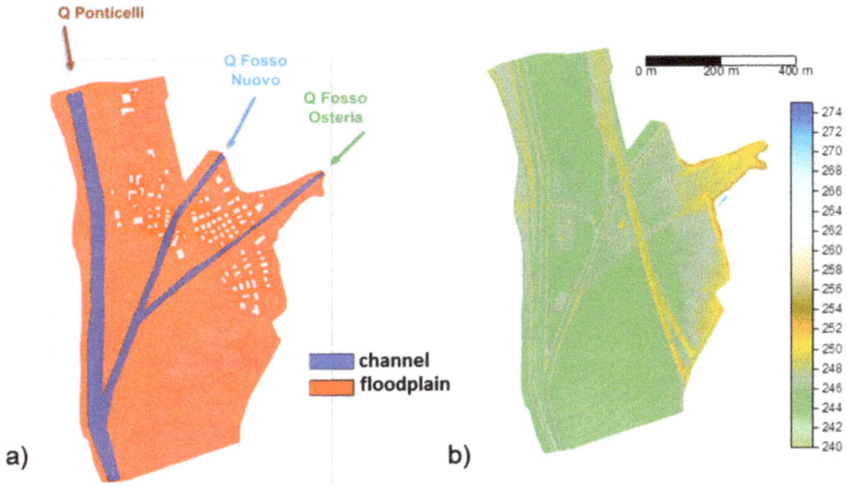

Fig. 14.6. Triangular FE mesh a) and DEM used for model runs b).

14.4.2 MLFP-2D Calibration

In the MLFP-2D model, the Manning's roughness coefficient in Eq. 14.3 is the unique parameter involved in the calibration. The model structure allows one roughness coefficient for each triangular element to be used but, lacking a detailed knowledge of the land use in the study area for allowing the roughness coefficient to vary, the domain was divided into two principal regions, floodplain and river (Fig. 14.6a), and for both of these an ensemble average roughness coefficient was assumed.

The roughness coefficients for these two regions were varied between two limits, resulting a range of values between 0.02 $m^{-1/3}/s$ and 0.1 $m^{-1/3}/s$ for channel and between 0.01 $m^{-1/3}/s$ and 0.5 $m^{-1/3}/s$ for floodplain.

An ensemble of 300 simulations of the hydrodynamic model was run on a high throughput computing system. Each simulation has a separate pair roughness coefficients (channel and floodplain) randomly selected from a uniform distribution by a Latin Hypercube Sampling (LHS) (see Fig. 14.7). Considering maximum water depths residuals, likelihood surfaces were generated using the different performance measures of eq. (14.4-14.7). These surfaces, illustrated in Fig. 14.8, map the parameter space to a continuous space as function of Manning's roughness coefficients for the channel n_{ch} and the floodplain n_{fl}.

From inspection of the different figures, for all the different goodness-of-fit measures, it becomes clear that the roughness values for the river with the greater likelihood measures are in the approximate range n_{ch} = 0.09 $m^{-1/3}/s$ to 0.1 $m^{-1/3}/s$. Since the 'ridge' of relatively more possible parameter combinations exhibits an insensitivity to the floodplain roughness for most of the likelihood surfaces, then we should conclude that the truncated ridge of more possible floodplain

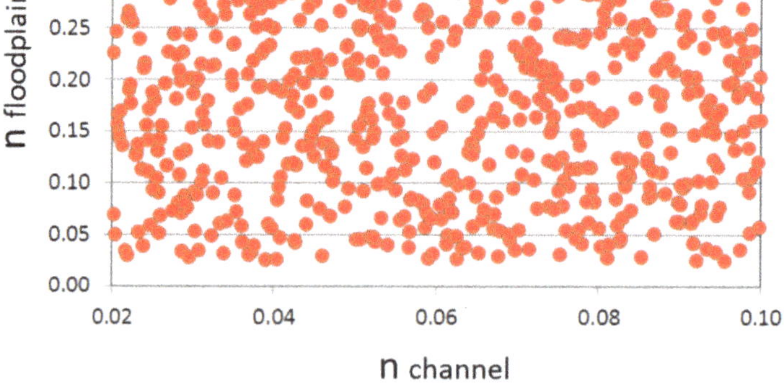

Fig. 14.7. Random uniform values of roughness coefficients used for the simulations.

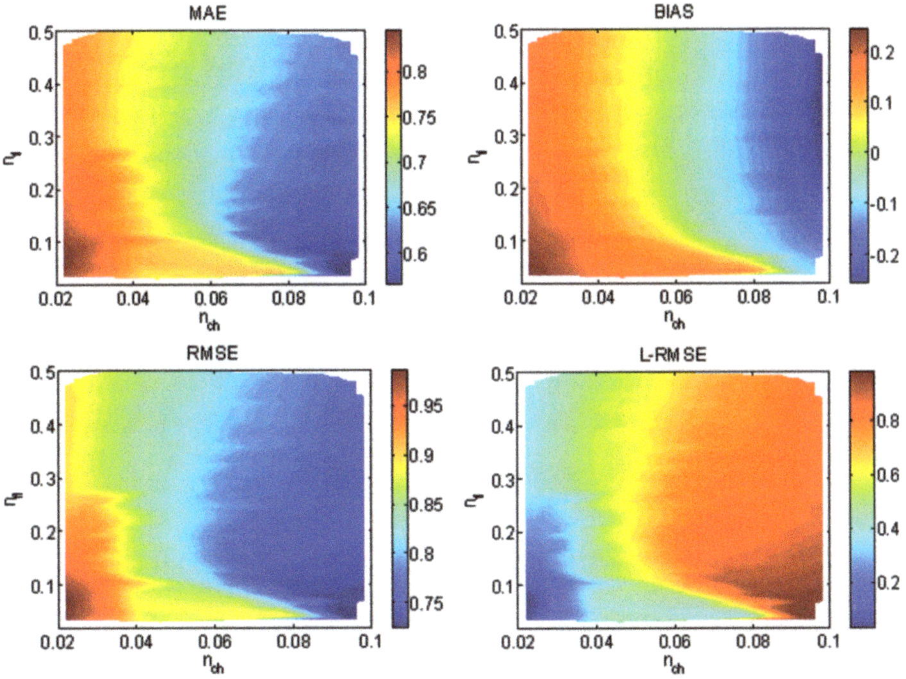

Fig. 14.8. Likelihood surfaces for different performance measures.

roughness values (n_{fl} = 0.05 $m^{-1/3}/s$ - 0.15 $m^{-1/3}/s$) in L_RMSE measure can be interpreted as representing the region of more feasible Manning coefficients. Thus, in this case, the concept of equifinality is accepted.

14.4.3 Uncertainty analysis

In order to obtain hazard maps including uncertainty in model prediction the results of each simulation for each nodes where weighted according to eq. (14.8). Then, two distinctive flood hazard maps were produced. The first map (Fig.14. 9a) has been obtained via flood propagation modelling using the "optimal" values of the roughness parameters corresponding to the maximum value of L-RMSE, i.e. no parameter uncertainty has been considered. Instead, the second map in Fig. 14.9b include the parameter uncertainty treated using the equation above.

The two maps of Fig. 14.9 look similar, but they are not identical. Taking into account the uncertainty due to the roughness, the hazard maps

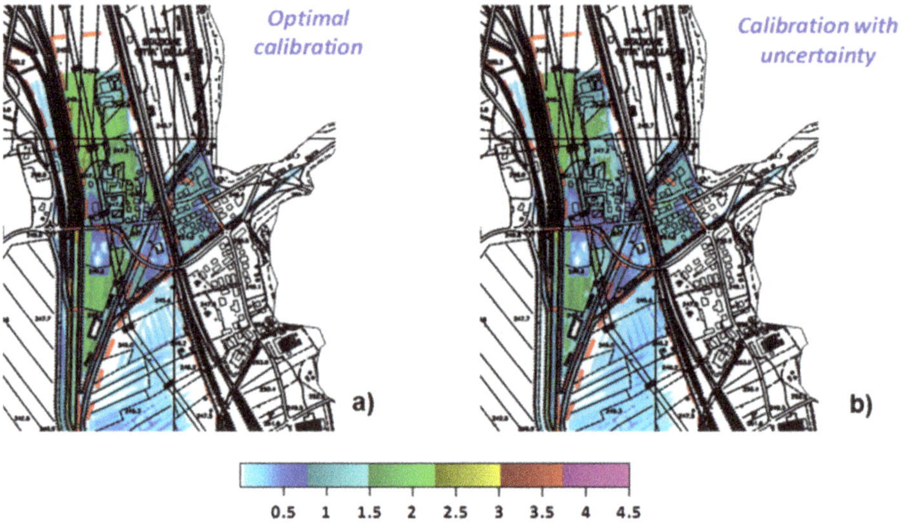

Fig. 14.9. Flood inundation maps including uncertainty (depths in meters).

cover the same area but with a different spatial distribution of maximum water depths. Hence, the comparison between the two maps confirm how crucial is the choice of including uncertainty in model calibration.

14.5 CONCLUDING REMARKS

In this paper a procedure to derive flood hazard maps incorporating uncertainty concepts is presented. The procedure is based on a Monte Carlo simulation procedure to derive flood hazard maps with uncertainty analysis, focusing on uncertainty due to the hydraulic model (roughness coefficients). A new definition of hydrodynamic indexes for flood hazard assessing based on water depth has been used. The results show the importance of including uncertainty analysis in flood risk mapping and more research should be carried out in this field, being these maps essential tool to aid planning decisions and for flood warning purposes.

14.6 REFERENCES

Apel, H., Aronica, G.T., Kreibich, H., and Thieken, A.H. 2009. Flood risk analyses—how detailed do we need to be?, *Nat. Hazards*, 49: 79–98,

Aronica, G.T., Fabio, P., Candela, A., Santoro, M. 2012. Estimation of flood inundation probabilities using global hazard indexes based on hydrodynamic variables, *Journal of Physics and Chemistry of Earth*, 42–44: 119-129

Aronica, G.T., Bates, P.D. & Horritt, M.S. 2002. Assessing the uncertainty in distributed model predictions using observed binary pattern information within GLUE, *Hydrological Processes*, 16: 2001-2016.

Aronica, G.T., Tucciarelli, T., and Nasello, C. 1998a. 2D multi-level model for flood wave propagation in flood-affected areas, *Journal of Water Resources Planning and Management*, 124(4): 210-217.

Aronica, G.T., Hankin, B., Beven, K.. 1998b. Uncertainty and equifinality in calibrating distributed roughness coefficients in a flood propagation model with limited data. *Advances in Water Resources*, 22 (4): 349–365.

Bates PD, Horritt MS, Aronica G, Beven K. 2004. Bayesian updating of flood inundation likelihoods conditioned on flood extent data. *Hydrological Processes,* 18(17): 3347–3370.

Beven, K.J. 2001. How far can we go in distributed hydrological modelling? *Hydrology and Earth System Science*, 5 (1): 1–12.

Beven, K.J., 2000b. Rainfall–Runoff Modelling. *the Primer Wiley*, 360 pp.

Beven, K.J., 2006. A Manifesto for the equifinality thesis. *Journal of Hydrology*, 320: 18–36.

Beven, K.J. , Binley, A. 1992. The future of distributed models: model calibration and uncertainty prediction, *Hydrological Processes*, 6: 279–298.

Chow, V 1959. *Open Channel Hydraulics*, McGraw-Hill, New York.

Di Baldassarre G, Castellarin A, Montanari A, Brath A. 2009. Probability-weighted hazard maps for comparing different flood risk management strategies: a case study. *Natural Hazards* 50(3): 479–496.

Fabio P., Aronica G. T., and Apel H. 2010. Towards automatic calibration of 2-D flood propagation models, *Hydrol. Earth Syst. Sci.*, 14: 911–924.

Mantovan, P., and Todini, E.: Hydrological forecasting uncertainty assessment: Incoherence of the GLUE methodology, *Journal of Hydrology*, 330, 368-381, 2006.

Pappenberger F, Beven KJ, Horritt MS, Blazkova S. 2005. Uncertainty in the calibration of effective roughness parameters in HEC-RAS using inundation and downstream level observations. *Journal of Hydrology*, 302(1-4): 46–69.

Pappenberger, F., Beven, K., Frodsham, K., Romanowicz, R. & Matgen, P. 2007. Grasping the unavoidable subjectivity in calibration of flood inundation models: A vulnerability weighted approach, *Journal of Hydrology*, 333: 275-287.

Romanowicz R, Beven KJ. 1998. Dynamic real-time prediction of flood inundation probabilities. *Hydrological Sciences*, 43(2): 181–196.

Romanowicz R, Beven KJ. 2003. Estimation of flood inundation probabilities as conditioned on event inundation maps. *Water Resources Research*, 39(3): 1–12.

Schumann G, Di Baldassarre G, Bates PD. 2009. The utility of space-borne radar to render maps of observed possibility of inundation. *IEEE Transactions on Geoscience and Remote Sensing*, 45(6): 1715–1725.

Tarpanelli, A, Brocca, L., Melone, F. , Moramarco, T.2012. Hydraulic modelling calibration in small rivers by using coarse resolution synthetic aperture radar imagery. *Hydrological Processes*, DOI: 10.1002/hyp.9550.

CHAPTER 15

HYDROLOGICAL IMPACTS OF AGRICULTURAL MANAGEMENT PRACTICES ON FLASH FLOOD WAVE CHARACTERISTICS, SEDIMENT LOAD TRANSFER AND RESULTING ECONOMICAL LOSSES

George P. Karatzas and Nektarios N. Kourgialas

Technical University of Crete – School of Environmental Engineering, Polytechneioupolis, 73100 Chania, Greece.
e-mail: nektarios.kourgialas@enveng.tuc.gr; karatzas@mred.tuc.gr

ABSTRACT

A modelling system for the estimation of flash flood propagation characteristics, sediment transport load and the resultant damages is developed in this study. This modelling framework can be applied to a river reach of any hydrogeological river basin. In this modeling system four components were employed: a) a modelling framework based on the hydrological model HSPF, b) the hydrodynamic module of the hydraulic model MIKE 11 (quasi-2D), c) the advection-dispersion module of MIKE 11 as a sediment transport model, and d) a flood loss estimation model. An important parameter in hydraulic modelling for the determination of the flood inundation is the Manning's coefficient, an indicator of the channel resistance, which is directly depended on riparian vegetation changes. Thus, in this study, based on the developed modeling framework, the role of riparian vegetation on water depth (inundation), discharge, flow velocity, sediment transport load and the consequent damages is investigated. The obtained results show that when the weed cutting percentage is increased, the flood wave depth and flood damages decrease while flow discharge, velocity and sediment transport load increase. Based on this, a well-balanced selection of the most appropriate agricultural cutting practices of riparian vegetation was performed. Ultimately, the proposed modelling system is used to evaluate and illustrate the flood hazard for different cutting riparian vegetation scenarios. The proposed methodology was applied, tested and validated at the downstream part of a small Mediterranean river basin in Crete, Greece.

Keywords: Ecohydraulics; water depth; discharge; flow velocity; sediment transport; flood damages; flood inundation; flood hazard maps; Riparian vegetation.

15.1 INTRODUCTION

The frequency and magnitude of flash flood events are expected to increase in the near decades, due to climate change, especially in the Mediterranean areas (de Moel et al. 2009; Kourgialas and Karatzas 2014). Additionally, the economic damages caused by flash flood events in Europe have been increasing dramatically during the past years, mainly due to the increase in population and economic activities in flood hazard zones (Plate, 2002; Georgakakos, 2006). In these zones, inappropriate flood management measures may lead to considerable loss of property and life.

Hydrodynamic modeling, providing the knowledge of hydrodynamic parameters in time and space, can play a significant role for establishing a proper floodplain management plan. Commonly, these parameters are flood inundation, discharge and flow velocity (Pramanik et al. 2010). Several numerical models, using different computational algorithms, have been developed in order to solve the Saint Venant equations for river and floodplain flow simulation (Chatterjee et al. 2008). These models are divided into three categories: a) one-dimensional (1D), b) two-dimensional (2D), and c) one-dimensional river flow models coupled with two-dimensional floodplain flow models (1D-2D). Software tools, such as the MIKE 11 hydraulic model developed at the Danish Hydraulic Institute (DHI, 2007) and the HEC-RAS (HEC River Analysis System) model from the US Army Corps of Engineers (USACE1, 2002), have been used extensively for the dynamic one-dimensional flow simulation in rivers. One-dimensional models are simple to use but usually fail to provide detailed information regarding the river flow. This can be overcome by applying two-dimensional modelling. The most widely used software packages for 2D modelling are the MIKE 21 model (DHI, 2007) and the FLOW 2D model (O'Brien, 2006). However, 2D models have some disadvantages, for example they require high computational time and detailed topographic data. Thus, nowadays, modelling packages have been generated that couple 1D river flow models with 2D floodplain flow models. A characteristic example of these models is the software package MIKE FLOOD, developed at the Danish Hydraulic Institute. The MIKE FLOOD model links dynamically the MIKE 21 model to the MIKE 11 model (DHI, 2007). The combined 1D-2D models are generally computationally expensive, and also have the disadvantage of maintaining two models instead of one. For these reasons, an alternative approach has been developed namely the quasi-two-dimensional (quasi-2D) models, which combine a simplified 2D representation of the flood inundation characteristics with a fast computation process. MIKE 11 (quasi-2D) is one of these models (Kourgialas and Karatzas 2013). Tuteja (2008) compared the MIKE 11 (quasi-2D) model with the MIKE 21 and MIKE FLOOD models and concluded that all three models exhibited similar accuracy in the estimation of the flood inundation area for low and medium flooding flows.

The sediment transport load is a very important factor that affects the flooding flow and can lead to the phenomenon of siltation in waterways and erosion of river banks (Etemad-Shahidi et al. 2010). Simulation river flood sediment transport load is a very difficult task and up to now has not been widely studied. Sediment in rivers is usually divided based on its physical properties into two categories: i) cohesive fine sediment, with particles which can cluster together (typically wash load and suspended load), and ii) non-cohesive gravel sediment, behaving as individual particles, which can be suspended load or bedload. The non-cohesive sediment dominates the main channel and affects the overall channel shape (e.g., width, depth, slope), while fine cohesive sediment is important for floodplain sedimentation or siltation of river waterways (DHI, 2011). Due to the erosion, deposition and flocculation processes the simulation of sediment transport is a very difficult task. Thus, considerable effort has been put into the modelling of hydrodynamics and cohesive sediment transport because the equations of sediment transport require current velocity components that are normally obtained by a hydrodynamic model (Etemad-Shahidi et al. 2010; Kourgialas and Karatzas 2014).

It is well known that, in flood hydrodynamic simulation, flow resistance is a critical parameter in determining flow propagation characteristics and sediment load. In numerical flood simulations, flow resistance is commonly defined by Manning's coefficient (*n*), whose high values indicate high flow resistance. In addition, high resistance is mainly caused by the existence of vegetation in channel bed/bank and floodplain area. Riparian vegetation, which consists of the vegetation on the land immediately adjacent to a stream, is often the dominant factor affecting channel roughness (Tabacchi et al. 2000; Gurnell et al. 2006).

It is important to identify to what extent the riparian vegetation affects the magnitude of a flood event. Anderson et al. (2006) and Tabacchi et al. (2000), show the smaller floods with low or mid-sized flood peaks (high probability of occurrence) are more sensitive to the riparian vegetation conditions than larger floods with high discharges (low probability of occurrence). Therefore, in the case of flash floods, which are high probability events, occur suddenly, and with limited spatial extent, riparian vegetation management can control the consequences of floods. Moreover, the importance of riparian vegetation on flood control depends on the length and width/depth ratio of the river channel. According to Anderson et al. (2006) when the smaller the channel, the higher the relative effect of riparian vegetation. Specifically, riparian vegetation is likely to have a significant effect on flood control for channel width/depth ratio less than 17. Thus, the role of riparian vegetation in controlling overbank flows in small channels with extensive riparian vegetation can be very important. A common agricultural practice to overcome flooding is the annual mowing of the riparian vegetation by mechanical means. However, this practice restores discharge capacity but has negative ecological impact (Vereecken et al. 2006). Thus, the best method is to use an optimal riparian vegetation cutting plan, taking into account the balance between flood control minimizing the flood damages, and protect riparian ecology (Leu et al. 2008; Kourgialas and Karatzas 2013).

Regarding the estimation of flood losses, different approaches can be used for estimating the potential flood damages using flood parameters obtained from: a) historical databases, b) GIS modelling, and c) hydrodynamic physically based models (Kourgialas and Karatzas 2013). The accuracy in the estimation of flood damages depends on the accuracy in the estimation of various flood parameters. Taking into consideration that flood wave parameters are the most important inputs to a flood loss estimation model, using hydrodynamic models can provide an adequate representation of these. Recently, works that combine a physically based flood inundation simulation model with a loss estimation model have been presented (Dutta et al. 2003; Dutta et al. 2006; Jonkman and Vrijling, 2008; Kourgialas and Karatzas, 2013).

Based on the European Floods Directive (CEC, 2007), all EU members should prepare flood hazard maps and flood risk management plans by the end of 2015. These flood hazard maps should include various flood scenarios and the corresponding damages that can be generated. Out of several flood parameters, flood hazard maps usually show only water depth information. However, in order to evaluate the overall flood hazard, flow parameters such water depth and flow velocity should be aggregated into qualitative classes and combined to calculate a measure for the flood danger (de Moel et al. 2009).

The main objective of this study is to develop an integrated modelling system for the estimation of flood propagation characteristics (water depth, flooded area, flow discharge, and flow velocity), the sediment transport load, and the resulting losses for an oncoming flash flood event. Consequently, utilizing the above, flood hazard maps can be generated based on the requirements of the European Floods Directive. The modelling system has a modular structure that combines several submodels: a) a modelling framework based on hydrological model HSPF, b) the distributed MIKE 11 hydrodynamic module (quasi-2D), c) the MIKE 11 suspended sediment transport module and d) a grid-based flood loss estimation model based on the empirical stage-damage curves of Dutta et al. (2003). Moreover, this system is used to investigate the relationships among riparian vegetation, flood propagation characteristics, sediment load and resulting damages of an oncoming flash flood event. Therefore, the proposed modeling system can be used to study the effect of different riparian vegetation cutting scenarios minimizing the degree and extend of flood hazard, especially in small river basins which have short response times. This methodology is applied to the downstream area of the Koiliaris River basin in Crete-Greece, where flash flood phenomena occur from time to time.

15.2 STUDY AREA

The Koiliaris River basin is a typical Mediterranean watershed located in the eastern part of the Chania Prefecture in Crete, Greece (Fig. 15.1). The basin extends from the White Mountains (highest altitude 2041 m AMSL)

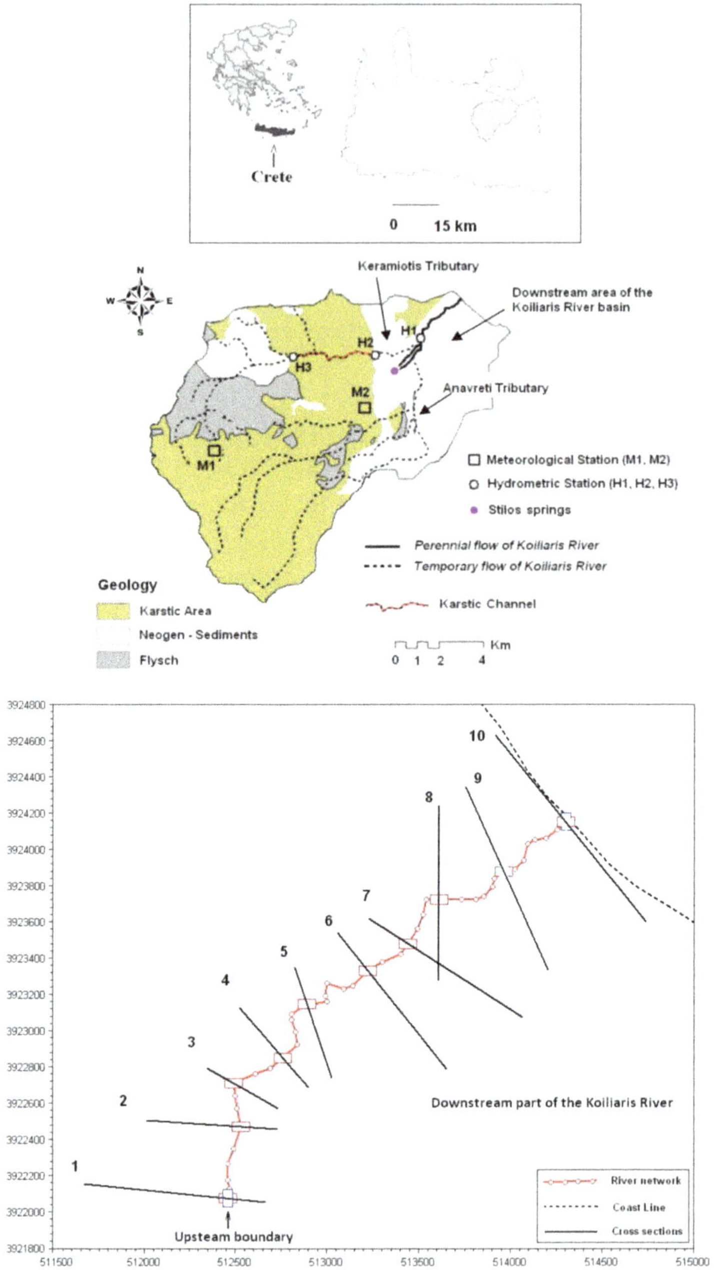

Fig. 15.1. The Koiliaris River basin, the downstream river part and the cross sections.

to the coastline and covers a total area of 130 km^2. The Koiliaris River has four tributaries (two permanent and two temporary). In this basin, there are three hydrometric stations (H1, H2, H3) and two meteorological stations (M1, M2), all telemetric, providing hourly real-time data (Kourgialas and Karatzas, 2011) (Fig. 15.1). The study area consists of carbonate formations, flysch, quaternary-neogenic and alluvial deposits. Rangeland accounts for 58% (101 km^2) of the total watershed area, cultivated areas cover 29.4% (51km^2), urban areas 2.8% (5 km^2), forests 8.5% (14,8 km^2), and aquatic areas 0.6% (1km^2). The downstream area of the watershed is agricultural with main products being oranges, lemons, olive oil, pears, and vegetables.

In the past, several extreme flash flood events have occurred in the downstream area of the basin (Fig. 15.1). The downstream part of the basin, starting at the location where all tributaries meet and ending to the outflow point, has a small length (3.6 km), and a very small width/depth ratio, less than 3.5, and is characterized by dense riparian vegetation. Moreover, the topography of the study downstream part is relative smooth with a low topographic slope of 6 %. In addition, when the cross-section of the hydrometric station H1 (downstream area) is completely covered with water, the hydraulic depth is equal to 3.4 m and the flooding flow (Q_{max}) is relatively small and equal to 92.6 m^3/sec (Kourgialas and Karatzas, 2011). Taking into consideration the above channel characteristics, it is concluded that appropriate riparian vegetation management can assist in flash flood prevention plans. In order to investigate the role of riparian vegetation on flash flood propagation characteristics and the resulting damages, the flash flood events of December 7, 2000 and December 10, 2003 that took place in the downstream area of the Koiliaris River and caused several damages was considered.

15.3 METHODOLOGY

15.3.1 The Concept of the Modelling System

The developed modeling system is a modular design model which connects four submodels. The first one is a framework based on hydrological model HSPF. The time series output from HSPF becomes input to the second submodel, the hydrodynamic module of MIKE 11, for simulating flood propagation characteristics at river reach scale. The hydrodynamic module output is used as input to the third submodel, the advection-dispersion (AD) module of MIKE 11 which has the capability of estimating the suspended sediment load during a flash flood event. In addition, the estimating flash flood characteristics are used as input to a grid-based flood loss model. Apart from the above, the present work investigates the role of riparian vegetation on water depth, flooded area, discharge, flow velocity, sediment transport and flood losses. The changes on the riparian vegetation based on different agricultural cutting practices were simulated in 1D and 2D. Using the results of the proposed modeling approach, two-dimensional flood hazard maps were generated for different agricultural cutting practices. Such maps can be used by decision makers (hydro-environmental agronomists and engineers) to compare different management scenarios that can affect the degree and extend of flood hazard in flood-prone areas. A graphical representation of this integrated modeling system is presented in Fig. 15.2.

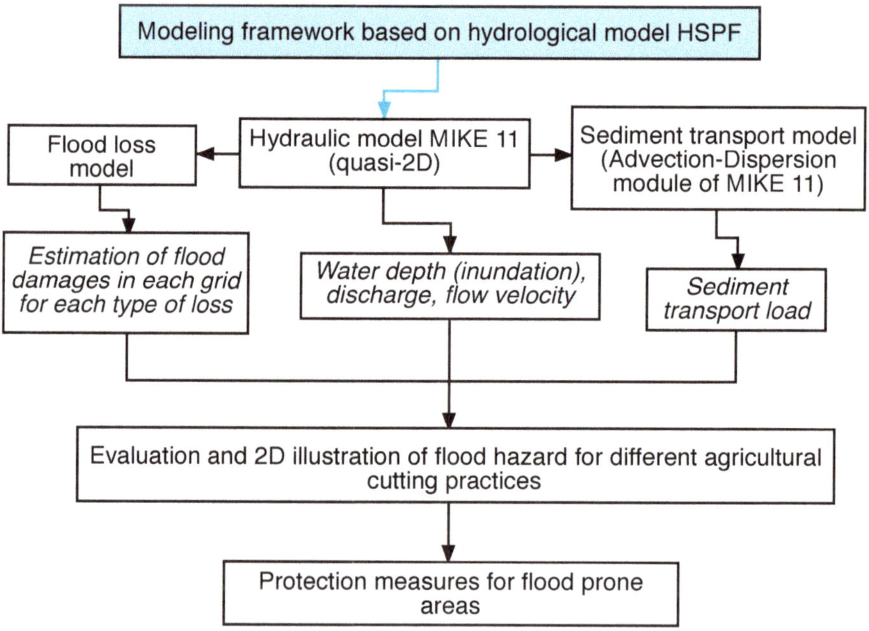

Fig. 15.2. Flow chart of the proposed modeling system.

15.3.2 A Framework Based on HSPF

This modeling framework consists of the following sub-models: a) a Karstic model for estimating spring discharge, b) a GIS based Energy Budget Snow Melt model that simulates and computes the snow melt flow that enters the karstic system of the high mountainous area, c) an empirical surface model developed to simulate the flow in karstic channels, and d) the hydrological model HSPF. This framework is capable of simulating the surface and groundwater flow that generates flash flood events in the downstream area of any complex hydrogeological basin. Such basins can consist of perennial and temporary rivers (tributaries), high mountainous karstic areas, springs, and downstream karstic channels. Analytical information regarding the framework based on HSPF model, the matlab code that incorporates all these, and the calibration and validation process can be found in Kourgialas et al. 2010.

15.3.3 The Hydrodynamic Module of MIKE 11

The hydrodynamic module of MIKE 11 is an implicit finite difference model for unsteady channel flow simulation and quasi two-dimensional for floodplain flow simulation. The model has been widely used to simulate water levels and flow in river systems and floodplains (Mishra et al. 2001; Hammersmark et al. 2005; Kourgialas and Karatzas 2013). It solves the full dynamic and width integrated Saint-Venant equations of continuity and conservation of momentum (Eqs. 15.1 and 15.2).

$$\frac{\partial A}{\partial t}+\frac{\partial Q}{\partial x}=q \tag{15.1}$$

$$\frac{\partial Q}{\partial t}+\frac{\partial\left[\gamma\frac{Q^2}{A}\right]}{\partial x}+gA\frac{\partial H_1}{\partial x}+\frac{gQ|Q|}{n^2AR}=0 \tag{15.2}$$

where:

Q = discharge (m^3/s),
A = flow area (m^2),

q	=	lateral inflow (m^2/s),
H_l	=	stage above datum,
n	=	Manning's coefficient ($m^{1/2}/s$),
R	=	hydraulic radius (m),
γ	=	momentum distribution coefficient,
g	=	acceleration due to gravity (9.81 m^2/s),
x	=	longitudinal distance in the direction of the flow (m),
t	=	elapsed time (s).

Before running simulations using the MIKE 11 (quasi-2D) model, input data such as the river network, the cross-section, the boundary parameter file, and hydrodynamic parameter file must be processed (DHI, 2007). In our case, the river network of the studied downstream part (3.6 km) of the Koiliaris River was digitized using topographic maps (1:5000). Moreover, 10 cross-sections were used with widths ranging from 500 to 1300 m. The geometry of these cross-sections was extracted, as a first step, from DEM and afterwards it was corrected with measurements from field-surveyed data in order to improve their accuracy (Fig. 15.1). The boundary parameter file includes the hourly time series boundary conditions that were defined using the time series output of the HSPF-based framework (Kourgialas et al. 2010). The hydrodynamic parameter file includes the bed and floodplain resistance data for the river channel. In this file the Manning's coefficient (*n*) is used as the resistance coefficient. In addition, in this file the river cross-sections were divided into three riparian vegetation zones with different bed resistance values: a) the bottom zone for the description of low to moderate flows (Riparian Zone 1), b) the bank zone for the description of moderate to high flows (Riparian Zone 2), and c) the overbank and floodplain flow zone (Riparian Zone 3). Based on the above division into river cross-sections, agricultural cutting scenarios were applied to investigate their effects on the flood propagation characteristics, sediment load transport and on the resulting damages. The grid cell size of the output 2D maps was defined as 1 m, while the computational time step was set to 1 minute. The detailed description of the development of this submodel is provided in Kourgialas and Karatzas (2013).

15.3.4 The Advection-Dispersion (AD) Module of MIKE 11

The advection-dispersion module of MIKE 11 for transport modelling is based on the one-dimensional equation of conservation of mass of dissolved or suspended material. The module requires the output from the hydrodynamic module of MIKE 11. The basic equation in this module is the advection-dispersion equation (Kourgialas and Karatzas 2014):

$$\frac{\partial AC}{\partial t}+\frac{\partial QC}{\partial x}-\frac{\partial}{\partial x}\left(AD_L\frac{\partial C}{\partial x}\right)=-kCA+C_2q+Ew \qquad (15.3)$$

where:

A	=	cross-section area,
C	=	the concentration,
D_L	=	the horizontal dispersion coefficient,
k	=	the linear decay coefficient,
C_2	=	source/sink concentration,
q	=	the lateral inflow,
E	=	the net deposition /erosion,
x	=	the space coordinate,
t	=	the time coordinate.

In our case study, C_2 is the tributary concentration, q the tributary inflow per unit length and w the river width.

The above equations are solved numerically using an implicit finite difference scheme (DHI, 2011). The AD module includes a relatively simple description of the erosion and deposition as source/sink terms in the AD equation. In the case of deposition, the net deposition term E is given by:

$$E=\frac{W_S C}{H^*}\left(1-\frac{v^2}{v_{cd}^2}\right) \qquad v<v_{cd}\ \ (deposition) \qquad (15.4)$$

where:

v	=	the flow velocity,

v_{cd} = the critical deposition velocity,
W_s = the fall velocity,
H^* = the average depth (of the section)

through which the particles settle.

In the case of erosion, the net erosion term E is expressed as:

$$E = \frac{M}{h}\left(1 - \frac{v^2}{v_{ce}^2}\right) \qquad v \geq v_{ce} \ (erosion) \qquad (15.5)$$

where:

v_{ce} = the critical erosion velocity,
M = the erodibility coefficient,
h = the flow depth.

15.3.5 The Flood Loss Estimation Model

The next sub model concerns a flood loss model, in this submodel, at the end of each time step, the simulated flood parameters from the hydrodynamic module of MIKE 11 model are passed for the estimation of direct flood damage. The flood loss estimation model is based on the concept of unit loss for the different categories of flood loss (Smith, 1994; Dutta et al. 2006; Kreibich et al. 2010). By using the dynamic interaction between the hydraulic model and the flood loss estimation model, the total flood damage can be estimated in each grid cell of the area of study.

The hydrodynamic module of MIKE 11 model 'feeds' the flood loss estimation model with the flood inundation parameters (water depth, duration) and, in turn, the loss estimation model uses the stage-damage functions to estimate the flood damage for each category. Depending on the category, a stage-damage function should be created to estimate the damage for each object in the specific category. The stage-damage functions for residential structures and crops by Dutta et al. (2003) were used in this study (Kourgialas and Karatzas, 2013). Utilizing these functions, the flood loss model estimates the damage for each category in each grid cell. The total damage is estimated as the summation of all the damages in each cell. The stage-damage functions should be calibrated to account for the specific characteristics of each study case. This has emphasized the need for reliable data collection regarding flood damages or each category, in order to accurately specify the shape of the loss-probability curve (Merz et al. 2010). The mathematical models for the residential buildings (structure), non-residential buildings and agricultural damage categories considered in this study can be expressed as follows (Dutta et al. 2006; Kourgialas and Karatzas 2013).

Structure damage for residential buildings:

$$D_{str} = \sum_{k=1}^{rt} NR(k)FA(k)UP_{str}(k)C_{str}(k) \tag{15.6}$$

Structure damage to non-residential buildings:

$$D_{nrb} = \sum_{n=1}^{NI} NU(n)UP_{nrb}(n)C_{nrb}(n) \tag{15.7}$$

where:

rt = the number of types of residential buildings based on building materials,

NR(k) = the number of residential buildings of type k,

FA(k) = the unit residential floor area for building type k,

UPm = the unit price for the respective category m in current condition (m = str, nrb),

Cm = the depth–damage function for the respective category m,

NI = the total number of industry types,

NU(n) = the number of units for industry of type n.

Agricultural damages:

$$AD = \sum_{k=1}^{n} DUA(k)TAC(k)CUW(k) \tag{15.8}$$

where:

n	=	the total number of crops,
DUA (k)	=	the damage to crop k per unit area (damage as a proportion of flood-free gross returns),
TAC(k)	=	the total area of cultivation of crop k,
CUW(k)	=	the estimated cost per unit weight of crop k.

15.3.6 Calibration and Validation of the Modeling System

The modeling framework based on HSPF model was successfully calibrated and validated for an extended time period of one year and two years, respectively (Kourgialas et al. 2010). Nevertheless, in this study emphasis was given to the other three components of the proposed modeling system. The hydrodynamic module of MIKE 11 was calibrated based on the flash flood event of December 10, 2003, and validated for the flash flood event of December 7, 2000 (Kourgialas and Karatzas, 2013). The advection-dispersion module was calibrated and validated for the time period from 2011-2013, when a sampling campaign took place at the downstream part of the Koiliaris River. Specifically, fifteen (15) samples of suspended sediment were collected at hydrometric station H1 and river bank soil samples were also collected at nine (9) different locations along the downstream part of the river. The soil samples were used to determine the sediment characteristics experimentally. The calibration process was based on the dispersion coefficient because it plays a significant role in sediment transport. Nine (9) of the collected suspended sediment samples were used for calibration, while the remaining six (6) were used for the validation process. The calibrated and validated advection-dispersion module was then used for the study the sediment load of the high flash flood event that took place on December 7, 2000. Due to lack of sediment load measurements for this specific event, in order to determine the boundary time series of suspended sediment, a regression equation was developed based on the suspended sediment samples collected at station H1. This regression equation is related to discharge and rainfall, factors controlling the production and delivery of sediment, and is described as follows (Kourgialas and Karatzas 2014):

$$SSC = aD + bP + c \tag{15.9}$$

where:

SSC = the suspended sediment concentration (mg/l),
D = the discharge (m^3/s),
P = the accumulated rainfall (mm),
a, b, c are coefficients.

The discharge measurements from station H1 and the rainfall data from station M1 were used in the above equation.

The flood damage for the validation flash flood event that took place in the study area on December 7, 2000 was estimated utilizing the flood loss model which was tested with historical data (damages).

15.3.7 Riparian Vegetation Cutting Scenarios and the Flood Hazard Rating

In hydraulic modeling, the component that corresponds to weed cutting is Manning roughness coefficient (Leu et al. 2008). The determination of Manning roughness coefficient, based on weed-cutting practices, is a very difficult task as it varies constantly due to the different river geomorphological characteristics and the variation in riparian vegetation cutting patterns (Nandalal, 2009). Thus, due to lack of field data, the relationship between Manning roughness coefficient and the weed-cutting percentage was determined, with relative low uncertainty, utilizing literature values from laboratory and field experiments from channels with similar characteristics to the studied downstream part. Next, three different weed cutting scenarios were considered in order to investigate the role of riparian vegetation on flood propagation modelling and the resulting damages in an integrated way for the validation flash flood event under study (December 7, 2000) (Doncker et al. 2009; Kourgialas and Karatzas, 2013):

A) No cutting scenario (using the calibrated Manning coefficient),

B) 40% weed cutting corresponding to a 27% reduction in Manning's coefficient, and

C) 57% weed cutting corresponding to a 62% reduction in Manning's coefficient.

From the results of the authors' prior work (Kourgialas and Karatzas, 2013), the effect of weed-cutting on the flood wave characteristics is minor in Riparian Zones 1 and 2 while is significant in Riparian Zone 3. Thus, in this study, weed cutting only in Riparian Zone 3 was investigated. The developed modeling system allows for the demonstration of the changes in the flash flood propagation characteristics, sediment load transport and the resulting damages due to the different weed-cutting scenarios. These characteristics were simulated in one-dimension for discharge and sediment load and in two-dimensions for water depth, flooded area, and flow velocity. The water depth and flow velocity results for the different weed-cutting scenarios were utilized to determine the rating of flood danger for each scenario, using the UK flood hazard rating (van Alphen and Passchier, 2007; de Moel et al. 2009). The equation of this measure of flood danger is described as follows:

$$FHR = W(V + 0.5) + DF \tag{15.10}$$

where:

FHR	=	the flood hazard rating,
W	=	the water depth (m),
V	=	the flow velocity (m/s),
DF	=	the debris factor.

The debris factor is equal to 1 for $W > 0.25$ and equal to 0.5 otherwise. Using the above flood hazard rating and taking into consideration the sediment load and the consequent damage results the well-balanced weed cutting scenario was selected.

15.4 RESULTS AND DISCUSSION

15.4.1 The Hydrodynamic and Sediment Transport Modules of MIKE 11

The hydrodynamic MIKE 11 module was successfully calibrated and validated in the study downstream river part, based on historical flash flood event that took place on December 10, 2003 and December 7, 2000, respectively. Manning's coefficient (*n*) was selected as a model calibration parameter (Pramanik et al. 2010) at different locations (riverbed and banks) along the downstream river reach. The initial values for coefficient *n* were selected based on Chow (1959) and Sturm (2001), together with land-use information from the area of study. The most suitable values for Manning's coefficient (*n*) were determined by trial and error from the spatial distribution of the historical flooded points reported on December 10, 2003. Thus, its value was found equal to 0.035 in Riparian Zone 1, 0.04 in the Riparian Zone 2, and between 0.04 and 0.05 in Riparian Zone 3 (floodplains).

The comparison between the observed historical flooded points and the model simulation shows that the values used for Manning's coefficient (*n*) were appropriate, since all the historical flooded points are located inside the area of the 2D maximum flood inundation map determined by hydrodynamic MIKE 11 module, (Fig. 15.3a). Additionally, information related to the values of the maximum flood water depth that was available only for five of these was compared with the model inundation water depth. Statistical analysis of this comparison show very good fit between simulated results and field data, thus for the calibration flash flood event the values of RMSE and R^2 were 0.14 and 0.96, respectively. For validation process (simulation flash flood event on December 7, 2000) the hydrodynamic MIKE 11 module was run by changing only the boundary conditions, as determined by the framework based on the HSPF model, and keeping all other model parameters and river cross-sections the same, as in the calibration. Figure 15.3b represents a spatial comparison between the observed historical flooded points (located on the perimeter of the flooded area) and the simulated 2D maximum flood inundation map for the validation flash flood event. Moreover, the simulated results of the maximum flood water depth were compared with existing historical data at four historical flooded points (Fig. 15.3b). This comparison shoes a very good agreement with RMSE = 0.18 and R^2 index = 0.91. Thus, the values of Manning's coefficient (*n*) used for the calibration of the model are also appropriate for the validated historical flash flood event.

Regarding the sediment load, a boundary condition time series of suspended sediment concentration had to be created for the advection-dispersion module. This was accomplished by using a regression equation

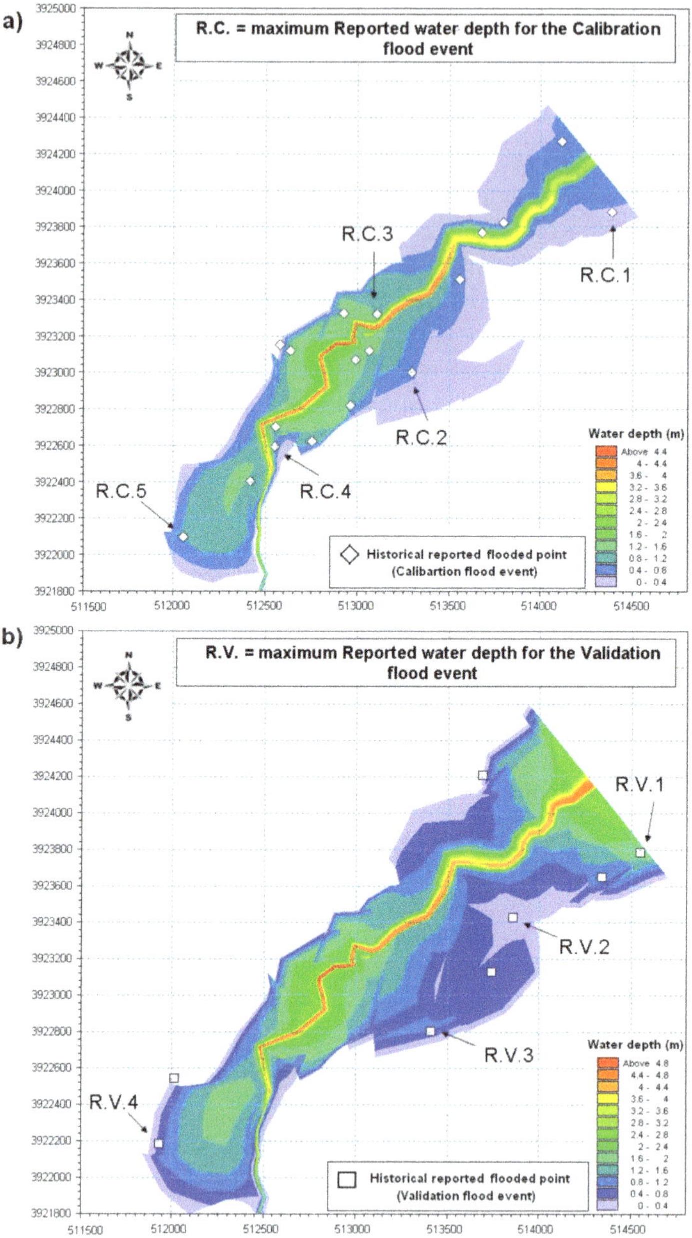

Fig. 15.3. Calibration and Validation flood events (Kourgialas and Karatzas, 2013).

based on the recent suspended sediment concentration measurements, on the discharge measurements, and the rainfall data as follows (Kourgialas and Karatzas, 2014):

$$SSC = 0.064D + 0.06P + 2.112 , \qquad R^2 = 0.85 \qquad (15.11)$$

where:

SSC = the suspended sediment concentration (mg/l),
D = the discharge (m^3/s),
P = the accumulated rainfall (mm).

The sediment transport model calibration was performed using the dispersion coefficient as calibration parameter, as the rest of the input data were available from experimental results. A statistical comparison between the available sediment transport load data and the corresponding simulation results for the calibration period; show that the determination index (R^2) was equal to 0.93, while for the validation period it was equal to 0.86. This good fit was achieved for both processes using a dispersion coefficient equal to 0.32 m^2/s.

15.4.2 Flood Damage Estimation

Using the simulation results for the validation flash flood event that took place in the study area on December 7, 2000, the damages were estimated. This was achieved by using the dynamic combination of the hydrodynamic MIKE 11 module and the flood loss model. The 2D grid based inundation maps of hydrodynamic MIKE 11 simulation with the relative information (flooded area, depth and duration of the flood wave) were used in order to determine the economic impact of this flood event. For this specific event the inundation depth varies from 0 to 4.8 m (Fig. 15.4).

The range of the damage that can influence the agricultural and structure areas depends on the dynamic of the flood inundation in each grid cell of the flooded area. The total flood damages were computed by using Eqs. (15.6-15.8) for structure and agricultural damages. The

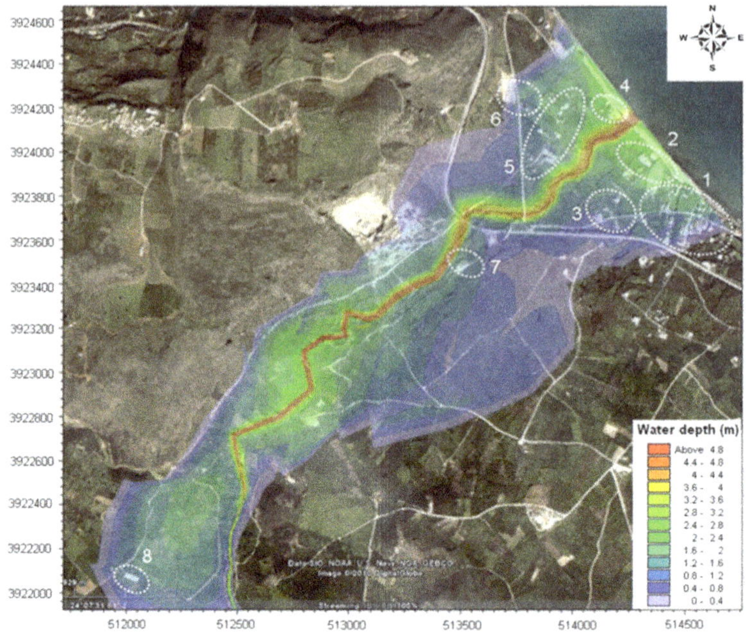

Fig. 15.4. Validation flood event and the affected areas (Kourgialas and Karatzas, 2013).

inputs data in the above equations were the flooded area (structure/agricultural land), the economic characteristics (building/crop), and the stage-damage functions for each grid cell. Due to the lack of stage-damage historical data in the study area, the stage-damage functions after Dutta et al. 2006, were used (Kourgialas and Karatzas, 2013).

Figure 15.4 shows the flood-affected structure areas, which were determined by using remote sensing methods. The same methods were used to determine the floodplain agricultural areas. Specifically, the main cultivations that were flooded were olive trees with a flooded area about 1.497 km², citrus trees about 0.555 km² and vegetation about 0.109 km². Based on the above, the total structural and agricultural damages were estimated at 776,435€ for structural damages and 156,177€ for agricultural damages, while the total damages were estimated at 932,612€. Regarding the reliability of the flood loss model, the only available damage data for the validation flash flood event were recorded by the Hellenic Agricultural Insurance Organization (ELGA).

Specifically, the reported agricultural damage was equal to 153,131€, which is in very good agreement with the modeled value of 156,177€.

At this point, it will be useful to mention that apart from inundation depth and flood duration the extent of damage can be affected by flow velocity. Nevertheless, the duration of a flood event usually plays a significant role for the estimation of agricultural damage; however in the case of flash floods this role is limited. Moreover, damage due to high flow velocities has a considerable influence only on road infrastructure, near the dike breach, with relatively high topography slope (Kreibich et al. 2009; Bouwer et al. 2009). Taking into consideration that the study downstream area is characterized as agricultural with low slopes and limited infrastructures, the impact of high flow velocities is minimal and can be ignored. Thus, the model was used in order to estimate total damages only for the structural and agricultural loss categories.

15.4.3 Validation Flash Flood Event and the Role of Riparian Vegetation Cutting Scenarios

Riparian vegetation affects flood propagation characteristics and its cutting is part of the water management strategy. Nevertheless, the riparian vegetation has an essential role in water quality control which contributes to maintenance of a sustainable ecosystem along the river and also controls the stability of stream banks by controlling erosion (Vereecken et al. 2006; Doncker et al. 2009). Therefore, it has several advantages from an ecological point of view, and it would be desirable to limit the mowed area of the riparian vegetation reducing the risk of flooding while at the same time retaining the ecosystem. The effect of weed cutting scenarios A, B and C on water depth, flooded area, discharge, flow velocity, damages and sediment transport was investigated in this section. The results show that the higher the weed cutting percentage, the lower the flood wave depth and the higher the flow discharge. For instance, 40% weed cutting (scenario B) reduces the flood wave peak from 4 to 3.6 m, while 57% weed cutting (scenario C) reduces it from 4 to 3.4 m. Moreover, the total damages were reduced by 21.41% (40% weed cutting) and 30.83% (57% weed cutting) compared to no weed cutting. Furthermore, the total flooded area was reduced by 12.2% and 37.32%, respectively. The simulation results for flow velocity at a cross-section located 2100 m from the study area's upstream boundary (Fig. 15.1) show that the higher the weed cutting percentage, the higher the velocity. Specifically, the maximum velocity is 0.14 m/s for scenario A, 0.18 m/s for scenario B, and 0.33 m/s for scenario C. The above values correspond to a 22.2% and 57.6% increase in flow velocity peak for scenarios B and C, respectively, compared to the no cutting scenario A.

The simulation results for the sediment transport load were generated for the same cross-section that was used for the velocity results. As observed, the higher the weed cutting percentage, the higher the sediment transport load. The maximum sediment transport load is 0.110 kg/s for scenario A, 0.127 kg/s for scenario B, and 0.179 kg/s for scenario C. These values correspond to a 13.4% and 38.5% increase in the sediment transport peak for scenarios B and C, respectively, compared to the no cutting scenario. Furthermore, as the cutting percentage increases, changes in the sediment transport load as a function of time are steeper. Moreover, the sediment transport peak as well as flow velocity peak appear earlier in time as the weed cutting percentage increases. The same trends sediment load and flow velocity parameters were verified for all river cross-sections examined.

15.4.4 Flood Hazard Maps Based on Agricultural Cutting Practices

In this section, hazard rating map for each of the three weed cutting scenarios was generated using the two-dimensional maps of the maximum flood inundation and flow velocity in the river and floodplain of the hydrodynamic module of MIKE 11. These maps were combined according to Eq. 3.10 in order to derive the final hazard rating map for scenarios A, B and C, respectively (Fig. 15.5). In this figure, the flood hazard ratings were divided into four classes of different degree (low, moderate, significant and extreme) (Defra/Environment Agency, 2006).

Table 15.1 shows the flooded area for each flood hazard class for each scenario. As shown, the total flooded area was reduced by 5.95% and 23.9% for scenario B and C respectively, compared to scenario A. Specifically, the area of the low hazard class increases slightly for scenarios B and C. This is viewed as a positive effect of weed cutting since low hazard zones correspond to flood zones with shallow flowing

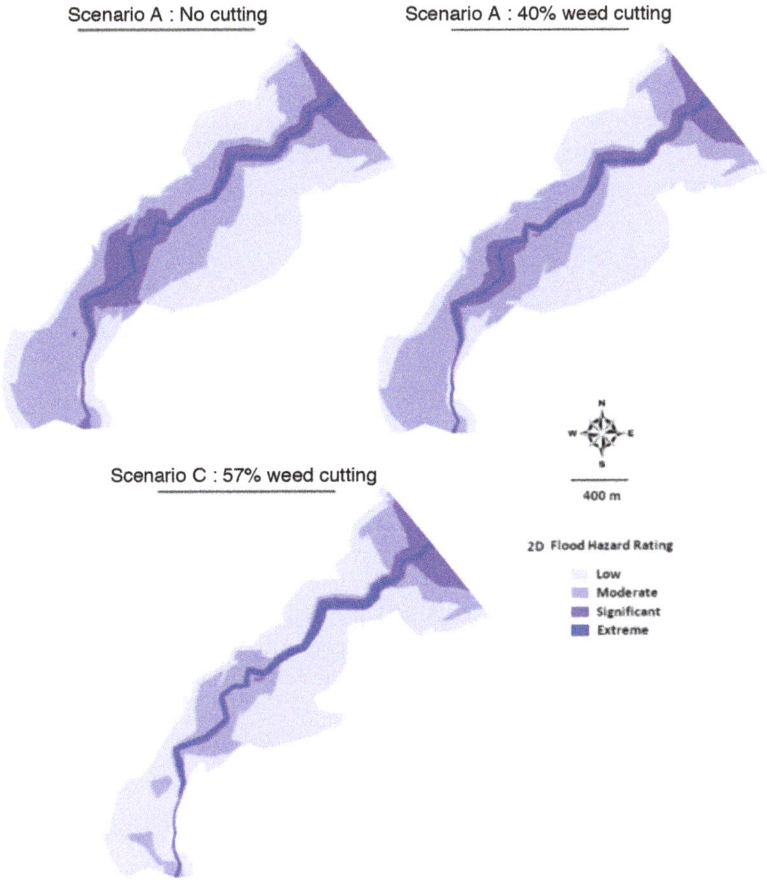

Fig. 15.5. Flood hazard rating maps for scenarios A, B, and C.

Table 15.1. Flooded area and hazard degree (Kourgialas and Karatzas, 2014).

Thresholds for Flood Hazard Rating (FHR)*	Degree of flood hazard	Flooded area (km^2)		
		scenario A	scenario B	scenario C
< 0.75	Low	1.105	1.161	1.240
0.75 - 1.25	Moderate	1.051	0.970	0.492
1.25 - 2.5	Significant	0.374	0.259	0.140
> 2.5	Extreme	0.087	0.090	0.117
	Total flooded area	2.617	2.480	1.989

* Defra/Environment Agency, 2006

water. On the other hand, the moderate and significant hazard zones decrease significantly when the weed cutting percentage is increased. Regarding the extreme hazard zones, they increase when weed cutting is increased. The increase of the hazard level for the areas close to the main channel is a negative effect of weed cutting. This is more intense in the areas where the channel line meanders (Kourgialas and Karatzas, 2014).

The simulation results indicate the necessity for an optimal weed cutting plan which takes into account the fact that flood water depth and flow velocity are affected differently by the riparian vegetation cutting agricultural practices. Moreover, according to the simulation results for the sediment transport load and flood loss model the higher the weed cutting percentage the heavier the transport load and the lower the total damages. Specifically, regarding the total flood damages, they were reduced by 17.64% for scenario B compared to scenario A, while the reduction for scenario B compared to scenario C was relatively small and equal to 7.2%. According to the above and given that heavy sedimentload leads to more pronounced riverbank erosion and has a negative impact on riparian ecology and human health point of view, scenario B seems to be preferable, as it provides the best balance among the flood characteristics (water level, discharge, velocity, flooded area), sediment load and flood damages in regards to their intensity.

15.4.5 Limitations and Uncertainty of the Proposed Modelling System

One of the main limitations of the proposed modeling system is related to the hydrodynamic MIKE 11 module, which does not consider the dynamic water exchanges between the river channel and the floodplain. Specifically, evapotranspiration losses or interactions between the river water and groundwater are ignored. Although, these processes can play an important role in flood simulation of long rivers, in case of relatively short river reaches of less than 5 km and for flood events with short duration (flash floods) the proposed hydrodynamic module is particularly suitable (Kourgialas and Karatzas, 2013). Regarding the AD module of MIKE 11, this can simulate wash and suspended load but not bed load particles (rolls or jumps) along the riverbed, thus this sub model is not appropriate for simulating significant long-term morphological changes in a river reach. In the downstream part of the Koiliaris River, no significant morphological changes have been marked during the last 50 years, thus the AD module can be used with increased accuracy.

The main source of uncertainty for the proposed modeling system corresponds to the flood loss module. Generally, a shortcoming of damage modeling is that model validation is scarcely performed and thus quality assessment of damage estimates can hardly be achieved. In the present study, the lack of flood damages data led us to an attempt to estimate the consequent flood damages based on the empirical damage function model of Dutta et al. (2003), using the corresponding hydro-economical characteristics of the study area as input parameters. It should be noted, that for areas of limited damage data, damage models can be transferred from another situation/area with caution in order to avoid extra uncertainty in the modeling process (Merz et al. 2004). Based on the above, in order to minimize the uncertainty in the proposed flood loss model, in the study area, the estimation of flood damage was limited to the structural and agricultural sectors. Analytically, regarding the agricultural damages, apart from the very good absolute validation of the agricultural damages, this sector has small variability for the study area (three types of crops). Thus, for the estimation of agricultural damage, the stage-damage curves presented by Dutta et al. (2003) for the specific types of crops were used. On the other hand, the structural sector has small variability and high homogeneity, since all the flooded structures were made of the concrete. Thus, the stage-damage curve presented by Dutta et al. (2003) for concrete structures was chosen to ensure a relative similarity between the two areas. In addition, for both cases, agricultural and structural sectors, the curves of stages (water depth) were determined using the output from the calibrated and validated hydrological model. The damage percentage calculated from this curve was then multiplied with the unit price for the current condition in the study area in order to determine the final damage. In addition, the total structure damage as well as the total agricultural damages was estimated based on the aggregation of the elements at risk (flooded structures) that reduces the uncertainty compared to the micro-scale (single elements at risk) approach.

15.5 CONCLUSIONS

This study focuses on the development and evaluation of a modeling system for the simulation of flash flood inundation characteristics, sediment load transport and the resulting flood damages that can be applied in any river basin. Moreover, this study highlights the significant impact of riparian vegetation on flood propagation modelling, focusing on the estimation of flood propagation characteristics (water depth, flooded area, discharge, and flow velocity), sediment transport load and the resulting damages. Thus, the proposed modeling system offers the capability of flash flood risk management analysis at river reach scale and in accordance to various mitigation options based on the European Floods Directive. In addition, the aim of this study was to introduce and investigate a new modeling approach that involves riparian vegetation, flash flood inundation, sediment transport load and the resultant damages. The system was applied and tested in the downstream area of the Koiliaris River basin under different agricultural cutting scenarios. Three different weed cutting scenarios were considered in order to investigate the role of riparian vegetation on flood propagation modelling. Based on the derived simulation results an appropriate weed cutting scenario can be planned for the riparian vegetation in order to reduce flood danger and guarantee ecological quality. Moreover, it is concluded that when the weed cutting percentage is increased, the flood wave depth and the resulting damages decrease while flow discharge, velocity and sediment transport load increase. Additionally, the proposed modeling system can incorporate the water depth, flow velocity, sediment transport load, and the consequent damages in order to estimate flood hazard from an oncoming flash flood event. Using the above approach, a well-balanced agricultural cutting practice with respect to the level of flood hazard can be selected.

15.6 REFERENCES

Anderson, B.G. Rutherfurd, I.D. and Western, A.W. 2006. An analysis of the influence of riparian vegetation on the propagation of flood waves. *Environmental Modelling & Software* 21: 1290-1296.

Bouwer, L.M. Bubeck, P. Wagtendonk, A.J. and Aerts, J.C.J.H. 2009. Inundation scenarios for flood damage evaluation in polder areas. *Nat. Hazards Earth Syst. Sci.* 9(6): 1995-2007.

CEC (Commission of European Communities). 2007. *Directive 2007/60/WE of the European Parliament and of the Council 23 October 2007 on the Assessment and Management of Flood Risk*, Brussels: Official Journal of the European Communities 6.11.2007 L 288, 2007.

Chow, V.T. 1959. *Open channel hydraulics*, New York: McGraw-Hill Book Co.

de Moel, H. van Alphen, J. and Aerts, J.C.J.H. 2009. Flood maps in Europe - Methods, availability and use. *Natural Hazards and Earth System Science* 9:2: 289-301.

Defra/Environment Agency. 2006. The Flood Risks to People Methodology, Flood Risks to People Phase 2, FD2321 Technical Report 1, March 2006. (http://sciencesearch.defra.gov.uk/Document.aspx?Document=FD2321_3436_TRP.pdf).

DHI. 2007. Danish Hydraulic Institute (DHI) w.a.e., *MIKE 21 Environmental Hydraulics, Advection–Dispersion Module Reference Manual*, Horsholm, Denmark.

DHI. 2011. Danish Hydraulic Institute (DHI), MIKE 11 Sediment transport, cohesive and non-cohesive sediment transport model, Short Description, Horsholm, Denmark.

Doncker, L.D. Troch, P. Verhoeven, R. Bal, K. Meire, P. Quintelier, J. 2009. Determination of the Manning roughness coefficient influenced by vegetation in the river Aa and Biebrza river. *Environmental Fluid Mechanics* 9: 549-567.

Dutta, D. Herath, S. Musiake, K. 2006. An application of a flood risk analysis system for impact analysis of a control system in a river basin. *Hydrological Processes* 20: 1365-1384.

Dutta, D. Herath, S. Musiake, K. 2003. A mathematical model for flood loss estimation. *Journal of Hydrology* 277:1-2: 24-49.

Etemad-Shahidi, A. Shahkolahi, A. and Liu, W.-C. 2010. Modeling of hydrodynamics and cohesive sediment processes in an estuarine system: Study case in Danshui river. *Environmental Modeling and Assessment* 15:4: 261-271.

Georgakakos, K.P. 2006. Analytical results for operational flash flood guidance. *Journal of Hydrology* 317:1: 81–103.

Gurnell, A.M. van Oosterhout, M.P. de Vlieger, B. Goodson, J.M. 2006. Reach-scale interactions between aquatic plants and physical habitat: River Frome, Dorset. *River Research and Applications* 22:6: 667-680.

Hammersmark, C.T. Fleenor, W.E. Schladow, S.G. 2005. Simulation of flood impact and habitat extent for a tidal freshwater marsh restoration. *Ecological Engineering* 25: 137-152, doi:10.1016/j.ecoleng.2005.02.008.

Hellenic Agricultural Insurance Organization (ELGA). 2000. Statistical Bulletin for flood damages (www.elga.gr).

Jonkman, S. Vrijling, J. 2008. Loss of life due to floods. *Journal of Flood Risk Management* 1:1: 43-56.

Kleinen, T. and Petschel-Held, G. 2007. Integrated assessment of changes in flooding probabilities due to climate change. *Climatic Change* 81:3-4: 283-312.

Kourgialas, N.N. Karatzas, G.P. and Nikolaidis, N.P. 2010. An integrated framework for the hydrologic simulation of a complex geomorphological river basin. *Journal of Hydrology* 381:3-4: 308-321.

Kourgialas, N.N. and Karatzas, G.P. 2011. Flood management and a GIS modelling method to assess flood-hazard areas - a case study. *Hydrological Sciences Journal* 56:2: 212-225.

Kourgialas, N.N. and Karatzas, G.P. 2013. A hydro-economic modelling framework for flood damage estimation and the role of riparian vegetation. *Hydrological Processes* 27:4: 515-531.

Kourgialas, N. N. and Karatzas, G. P. 2014. A hydro-sedimentary modeling system for flash flood propagation and hazard estimation under different agricultural practices. *Nat. Hazards Earth Syst. Sci.* 14: 625-634, doi:10.5194/nhess-14-625-2014.

Kreibich, H. Piroth, K. Seifert, I. Maiwald, H. Kunert, U. Schwarz, J. Merz, B. Thieken, A.H. 2009. Is flow velocity a significant parameter in flood damage modelling?. *Nat. Hazards Earth Syst. Sci.* 9: 1679-1692.

Kreibich, H. Seifert, I. Merz, B. Thieken, A.H. 2010. Development of FLEMOcs - a new model for the estimation of flood losses in the commercial sector. *Hydrological Sciences Journal* 55:8: 1302-1314.

Leu, J.M. Chan, H.C. Jia, Y. He, Z. Wang, S.S.Y. 2008. Cutting management of riparian vegetation by using hydrodynamic model simulations. *Advances in Water Resources* 31:10: 1299-1308.

Merz, B. Kreibich, H. Schwarze, R. Thieken, A. 2010. Review article "Assessment of economic flood damage". *Nat. Hazards Earth Syst. Sci.* 10:8: 1697-1724.

Merz, B. Kreibich, H. Thieken, A. and Schmidtke, R. 2004. Estimation uncertainty of direct monetary flood damage to buildings. *Nat. Hazards Earth Syst. Sci.*, 4:153–163, doi:10.5194/nhess-4-153-2004.

Mishra, A. Anand, A. Singh, R. Raghuwanshi, N.S. 2001. Hydraulic modeling of Kangsbati main canal for performance assessment. *J. Irrig. Drain. Eng.* 127:27-34, doi:10.1061/(ASCE)0733-9437(2001)127,1(27).

Nandalal, K. 2009. Use of a hydrodynamic model to forecast floods of Kalu River in Sri Lanka. *Journal of Flood Risk Management* 2:3: 151-158.

O'Brien, J.D. 2006. *FLO-2D user's manual, Version 2006.01*, FLO Engineering, Nutrioso.

Plate, E. 2002. Flood risk and flood management. *Journal of Hydrology* 267:1-2: 2-11.

Smith, D.I. 1994. Flood damage estimation - a review of urban stage - damage curves and loss functions. *Water SA* 20: 23-238.

Sturm, T. 2001. *Open channel hydraulics*, New York: McGraw-Hill Science.

Tabacchi, E. Lambs, L. Guilloy, H. Planty-Tabacchi, A.M. Muller, E. and Decamps, H. 2000. Impacts of riparian vegetation on hydrological processes. *Hydrological Processes* 14: 2959-2976.

Tuteja, N.K. 2008. *Koondrook-Perricoota Forest Flood Enhancement Project - Hydraulic Modelling.* Dept. of Environment and Climate Change NSW. Sydney South. ISBN:9781741226720.

USACE1. 2002. US-Army Crops of Engineers, *HEC-River Analysis System, Hydraulic Reference Manual*, Version 3.1.

van Alphen, J. and Passchier, R. 2007. Atlas of Flood Maps, examples from 19 European countries, USA and Japan, Ministry of Transport, Public Works and Water Management, The Hague, Netherlands, available at: http://ec.europa.eu/environment/water/flood_risk/flood_atlas/index.htm.

Vereecken, H. Baetens, J. Viaene, P. Mostaert, F. and Meire, P. 2006. Ecological management of aquatic plants: effects in lowland streams.

CHAPTER 16

FLOOD HAZARD DETECTION AT THE CATCHMENT SCALE FROM REMOTE SENSED INFORMATION

Giorgio Roth[1], Massimiliano Degiorgis[1], Angela Celeste Taramasso[1]

[1] Department of Civil, Chemical and Environmental Engineering University of Genoa, Via Montallegro 1, 16145 Genoa, Italy. giorgio.roth@unige.it

ABSTRACT

This contribution deals with the classification of the flood hazard at the catchment scale. The aim is to recognize flood-exposed areas, extending available information on the flood hazard up to the complete catchment coverage. Threshold binary classifiers, based on quantitative morphological features derived from data stored in Digital Elevation Models, are used to investigate the relationships between morphology and the flooding hazard. Results show that threshold binary classifier techniques should be taken into account when either one is interested in a first low-cost detection of flood-exposed areas, or flood hazard estimation is needed in specific areas, for which limited information are available. The method has been tested on two basins, pertaining to different environments: the Tanaro (Italy) and the Niger (Africa). In the first case, we proceed on the basis of a wide flood hazard database and results present a high degree of accuracy. For the Niger case study, the flood hazard database is limited to non-existent. Available data have been therefore used, with good results, to qualitatively validate a single-feature procedure based on relative elevation coupled with the interpretation of remote sensed images of the floodplain.

16.1 INTRODUCTION

Within a flood prone area, the land, usually not covered by water, is occasionally flooded as a consequence of natural storms, high tides, cyclones, tsunamis, and so on. In some areas, flood risks are forced by a particular action of the man, such as an inappropriate land use, or they are consequence of climate change. Important civilizations were borne near rivers, because water is one of the fundamental aspects needed to support life. As a consequence, people usually accepted the risk to live near the coast, rivers or lakes. Nowadays, floods are the more dangerous natural events, both in terms of human life and occurrence. Where an appropriate flood risk management was adopted, the amount of losses has decreased. In rural areas, or in less developed countries, the number of casualties as a consequence of floods is still increasing, even if this is not the primary problem of these countries.

The goal of this work is to estimate, throughout a geomorphological approach, the extension of flood-prone areas, as different morphologic and hydrologic characters are able to characterize them. The two proposed approaches are both based on the geomorphologic approach described in detail in Degiorgis et al. (2013). The first, relevant to basins for which official flood hazard data based on historical hydrologic observations are available, is applied to the Tanaro basin, in Italy. The second, intended for less monitored basins, is applied to the Niger basin, in Africa.

To apply the geomorphologic approach, data are needed to describe the shape of the land and flood risk sources. In this work, this is performed throughout a Digital Elevation Model (DEM), associated with the HydroSHEDS (Hydrological data and maps based on SHuttle Elevation Derivatives at multiple Scales) river network. Beyond the approach originally proposed by Degiorgis et al. (2012), here we exploit the use of Google Earth data and images, potentially available for the entire globe, to both draw transversal sections along the river and infer altimetric information related to floods areal extension from vegetation state during dry and wet seasons, or inside flooded area before and after an event.

16.2 THE GEOMORPHOLOGIC APPROACH

In the context of the present work, all streams that constitute the drainage network should be assumed as potential hazard sources. For a given basin, the hazard is delineated when flood-prone areas are identified for all the streams that constitute the drainage network.

16.2.1 Data Sources

The DEM used in this work is obtained from the HydroSHEDS dataset (see: hydrosheds.cr.usgs.gov/index.php), a remotely sensed elevation data product originated from the NASA Shuttle Radar Topography Mission (SRTM) at 3 arc-second resolution. The Hydrosheds DEM is characterized by two sub-products: the VOID-filled (DEM–VOID) and the hydrologically CONditioned (DEM–CON) elevation model. In DEM–VOID the no-data cells are filled, and main elevation inconsistencies removed. DEM–CON is further conditioned to produce a river network coherent with the actual network (stream burning by decreasing the elevation of DEM cells along digitized river channels). Note that, since the conditioning process of the DEM-CON significantly alters the original elevation data, its use is limited only to the drainage network identification procedures, while the quantitative measures of morphologic characteristics (e.g., local slope, curvature or elevation difference between points) should be derived from original DEM–VOID data.

The reference drainage network is derived from the hydrologically conditioned DEM (DEM-CON) adopting the stream network delineation procedure proposed by Giannoni et al. (2005). Such algorithm stems from the classical slope-area method (Montgomery and Dietrich, 1988) and takes into account both the contributing area, A, and the local slope, S, for the identification of channel initiation. In particular, channel heads are located at cells where the quantity AS^k exceeds a given threshold, while the channel path to the outlet is identified by following the steepest direction (Jenson and Domingue, 1988; Nardi et al. 2008).

In this work, simple morphologic features are taken into account. The selected features, specified for each DEM cell, are: distance from the nearest stream, denoted by D, elevation to the nearest stream, H, Laplacian of the local elevation, ΔH, contributing area, A, local slope, S, and the modified Topographic Index, TI_m.

16.2.2 Identification of Areas Subject to a Marginal Hazard

The calibration of a binary classifier finalized to the identification of flood-prone areas require the prior knowledge about the DEM cells associated to a low flood risk level, i.e. those areas actually pertaining to a marginal hazard class. In the present work, the flooding hazard for a given location can be defined as marginal if i) all the potential flood risk source for the specific location have been taken into account and ii) for each single risk source the location is outside flood-prone areas, as introduced by Degiorgis et al. (2013).

16.2.3 Identification of Areas Subject to the Flood Hazard

Where available, official data should be used to identify and rank the flood hazard. For the Tanaro basin, data produced for different return periods by the Italian Po River Basin Authority (BA) will be used in the following. For each point in the dataset, i.e. for each DEM cell, the following features are stored: latitude, longitude, D, H, ΔH, A, S, IT_m. Moreover, on the basis of BA information, points are catalogued into two classes: marginal hazard (Class 0) and flood-prone ones (Class 1).

If different relevance is assigned to the two events (marginal hazard and flood-proneness), as it happens in some applications, achieving a small average misclassification error on the more relevant class is more meaningful than getting a small error on the other class. Assigning different weights to the average misclassification errors on the two classes during the training procedure is an option to take this into account, but one has to face the difficulty of choosing suitable weights.

A more convenient way of proceeding, followed in our analysis, is based on the so-called Receiver Operating Characteristics (ROC) curve (see, e.g., Fawcett, 2006), defined as the set of pairs (r_{fp}, r_{tp}) obtained by varying the threshold of a binary classifier, where r_{fp} and r_{tp} are the false positive rate and the true positive rate, respectively. For a given binary classifier, the false positive rate, r_{fp}, is the probability that a sample from Class 0 is erroneously classified as a sample from Class 1, i.e., in our case, a marginal-hazard area is classified as flood-prone. Similarly, the false negative rate, r_{fn}, is the probability that a sample from class 1 is erroneously classified as a sample coming from class 0. The true positive rate, r_{tp}, and the true negative rate, r_{tn}, are given by $r_{tp}=(1-r_{fn})$ and $r_{tn}=(1-r_{fp})$, respectively.

Introducing the ROC curve allows one to compare different kinds of binary classifiers, by comparing the corresponding values of the Area Under the ROC Curve (AUC). This approach does not require either to choose the weights assigned to the average misclassification errors on the two classes or to specify their a-priori probabilities. Note that the AUC takes on values between 0 and 1. The larger its value, the better the classifier. A random classifier (which corresponds to recognizing true positives by flipping a coin) has an AUC equal to 0.5. So, classifiers performing better than chance are those with AUC values greater than 0.5.

16.2.4 Hazard Identification with Limited Data

For those areas for which hazard data are not available, we propose to define the extension of the hazard areas by applying geomorphological and vegetation filters to the information provided by the Google Earth application. In fact, through this application it is possible to both obtain the transversal sections along the river reach under study, and to recognize the geomorphologic and vegetation limits that define the boundaries of frequently flooded areas.

From these data, significant threshold values to flood prone areas indicators – for the whole basin or for a part of it – can be estimated. One indicator is the difference in elevation between the DEM cell under analysis and the nearest stream cell, selected by following the maximum slope direction. Moreover, using satellite images, one could recognize rapid vegetation changes in areas near the river under study; these changes can be assumed as indicators of past floods. From the limits of this area, the elevation of the boundary and its distance from the river can be assumed as input for the geomorphologic approach above summarized.

In this second approach, results are expected to be strongly influenced by the accuracy of data extracted from Google Earth. Nevertheless, this is expected to increase in the near future. In fact, different projects aims to increase their reliability.

16.3 THE TANARO CATCHMENT: A WELL DOCUMENTED AREA

In this chapter, the approach to infer flood-prone areas in well-documented areas is summarized, and results briefly presented. The reader is referred to the original work by Degiorgis et al. (2012; 2013) for further information. The Po Basin Authority reports, for the Tanaro catchment, the hazard map for a drainage network of limited extension (Fig. 16.1a). The drainage network identification procedure described in Giannoni et al. (2005), when applied to this catchment with a target drainage density equal to D_d=0.74 m^{-1}, produces the network depicted in Fig. 16.1b, with a significant increase in the extension of the network. The network of Fig. 16.1b is therefore assumed as the risk source for the catchment. In the procedure proposed by Degiorgis et al. (2013), the first step is to produce the marginal hazard areas for the official drainage network. Results of this step are shown in Fig. 16.1c. The resulting dataset is made up of 187,306 labeled data points, each belonging either to the Tanaro basin flood-prone areas, identified by the Po Basin Authority, or to adjoining areas that are subject to a marginal flood hazard. Each point in the dataset contains latitude, longitude, and values for six simple morphologic features, used to delineate flood-prone areas. Out of the six selected features, the maximization of the Area Under ROC Curve (AUC), used as a guide of the optimal feature set, indicate D and H as the best performing couple. The final hazard map is shown in Fig. 16.1d.

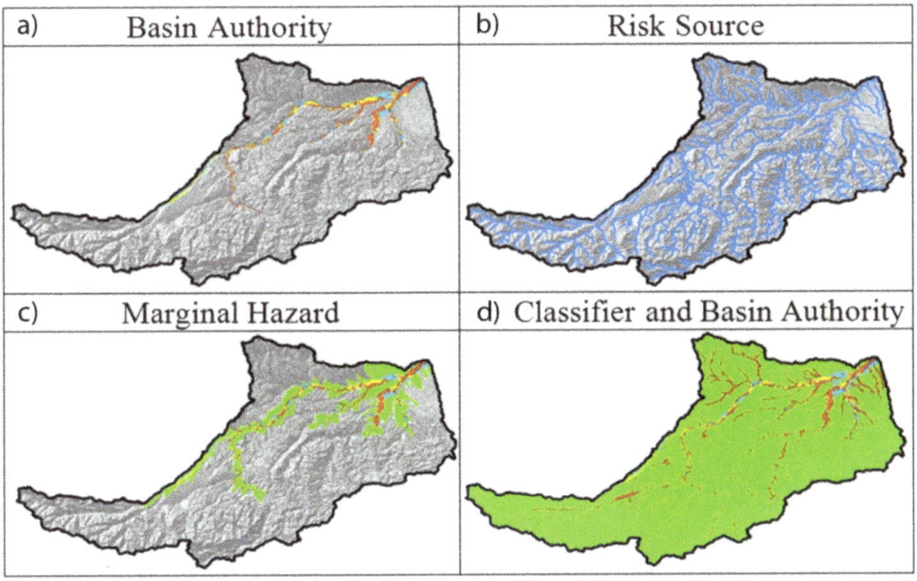

Fig. 16.1. Tanaro Basin: Basin Authority) Flood prone areas of Po Basin Authority studies. Risk Source) Reference drainage network derived from DEM. Marginal Hazard) Marginal hazard areas (green) identified according to the Po Basin Authority hazard studies. Classifier and Basin Authority) Composite of Po Basin Authority predictions and results of approach proposed. Red, yellow, blue and green respectively indicate high, medium, low and marginal hazard areas.

16.4 THE NIGER CATCHMENT: A LIMITED DATA AREA

In this chapter, the above outlined procedure is adapted and applied to a catchment without official data relevant to flooded areas. The Niger catchment was selected for the purpose, due to its importance for a number of African Countries. Its drainage network is particular. Along its path from the source to the sea it realizes two deltas: one in an inland area, near Timbuktù. Moreover, the first part of the river is characterized by the presence of rich vegetation, while the last part form a significant humid area, important for ecologic and economic aspects.

Most of the data commonly used to evaluate hydrologic and hydraulic processes relevant to the identification of flood prone area are not easily available for this catchment. This aspect is common for large areas in the world. We therefore follow a different path, that mimic the one above outlined for the Italian case study. Accordingly, the first step is to define the drainage network, i.e. the hazard source. Then, using Google Earth data and images, we try to identify flooded area boundaries. These steps are depicted in Fig. 16.2.

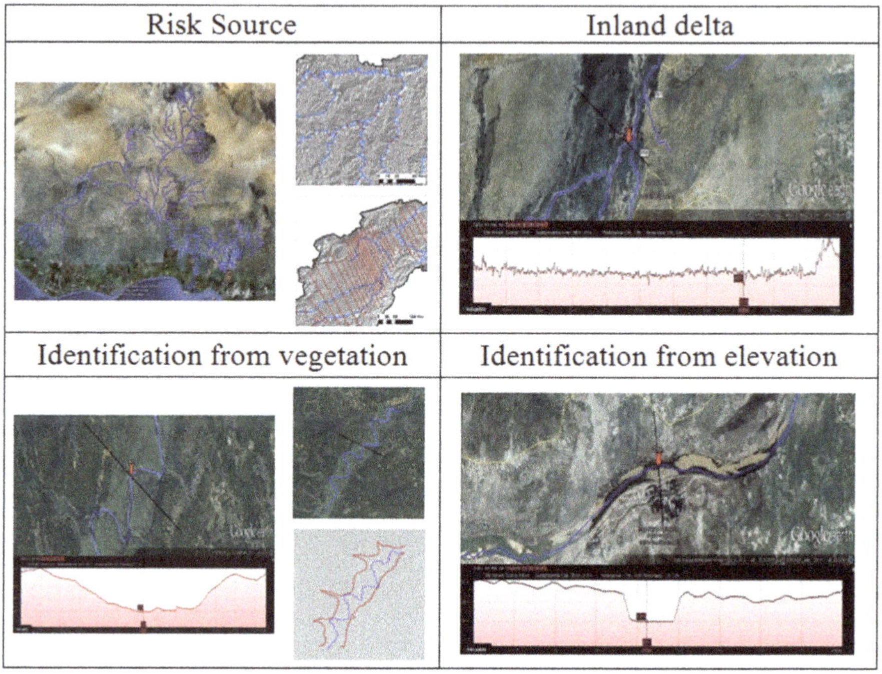

Fig. 16.2. Niger Basin: Risk Source) Reference drainage network and cross sections of calibration datasets. Inland delta) Difference in elevation of Inland delta. Identification from vegetation) Calibration datasets for an area recognized by vegetation. Identification from elevation) Calibration datasets for an area recognized by elevation.

A total amount of 300 sections were elaborated along the above defined river network with specific reference to two areas: the upper Niger and the inner delta. The value of the difference in elevation from the river to the boundary limit identified through satellite images vary in a wide range. Moreover, the exact location of the inner delta and the extension of flooded area rapidly change for different periods along the year, according to seasonality and rain amount. This influences the position of the sections in this area. For this reason, two threshold values are defined: one is used to identify flood-prone areas for the upper Niger, the second is specifically used for the inner delta. Results are summarized in Table 16.1 and Fig. 16.3.

Table 16.1. Niger Basin: Threshold values to identify flood-prone areas.

	Upper Niger	Inner delta
Mean Delta H	9.1 [m]	12.2 [m]
Standard deviation	3.1 [m]	1,8 [m]
Coefficient of variation	0.3	0.1

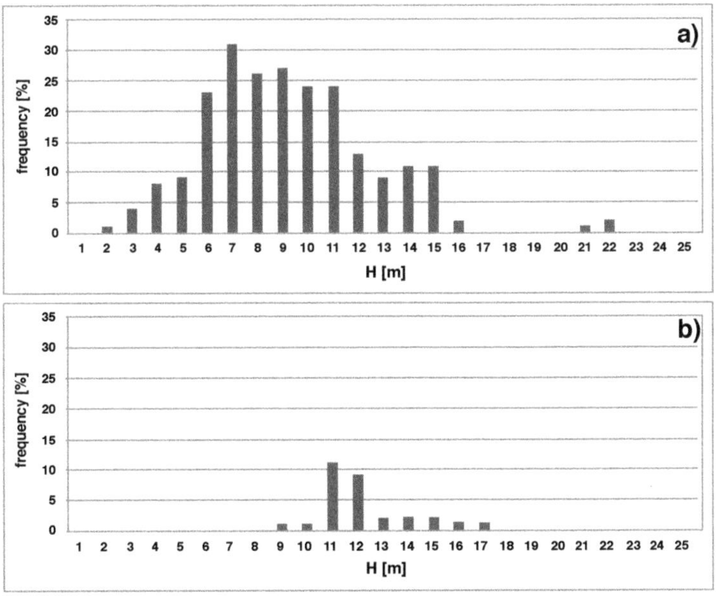

Fig. 16.3. Niger Basin: Frequency distribution of the difference in elevation from each flood-area boundary-cell to the drainage network (H), a) for the upper basin; and b) for inland delta.

Figure 16.3 shows the frequency of occurrence of the difference in elevation from each flood-area boundary-cell to the drainage network. This is used to define the value of the threshold needed to discriminate flood-prone from marginal hazard areas. Results here presented assume the 80% of the cumulative frequency as good representation of this index. For the upper Niger and the inner delta, these threshold values are 11 and 13 m, respectively.

Figure 16.4 shows the results of the application of the approach here presented for the Niger catchment: one image for the upper basin and one for the inner delta. Figure 16.5, by using satellite images of flooded area for the 12th August 2012 event, represents a first tentative of reliability evaluation for the approach here presented. The blue line is the Niger drainage network; blue represents the limits of flooded area for the specific event, and red lines bound the expected limits of flooded area for the specific event, and the blue area are the expected flood-prone areas, as derived from the present approach. One could notice that areas flooded by the specific event are well inside – but not far away from –predicted flood-prone areas.

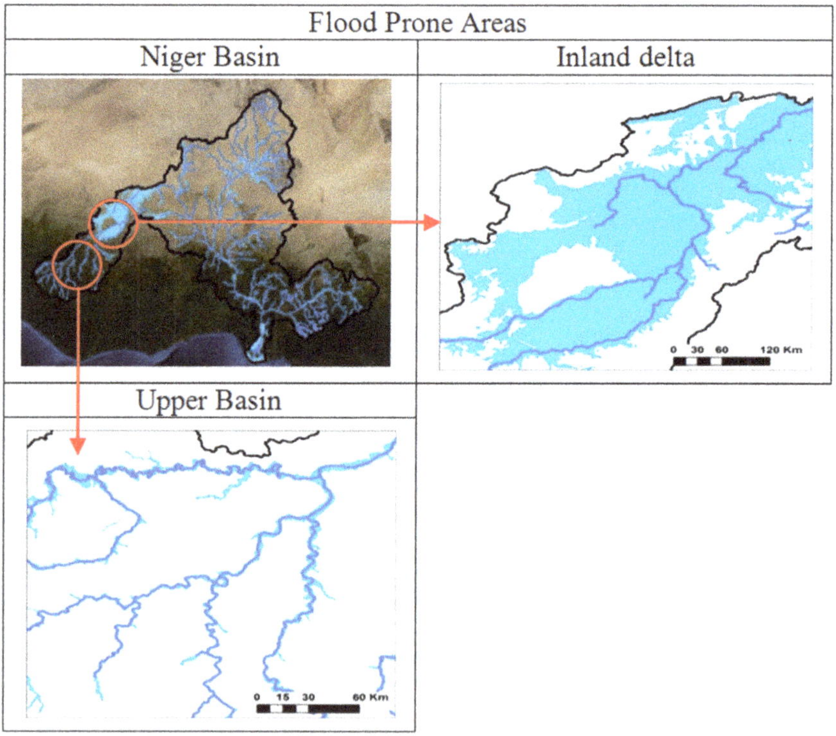

Fig. 16.4. Niger Basin: Results of the application of the methodology, focused on inland delta and the upper basin.

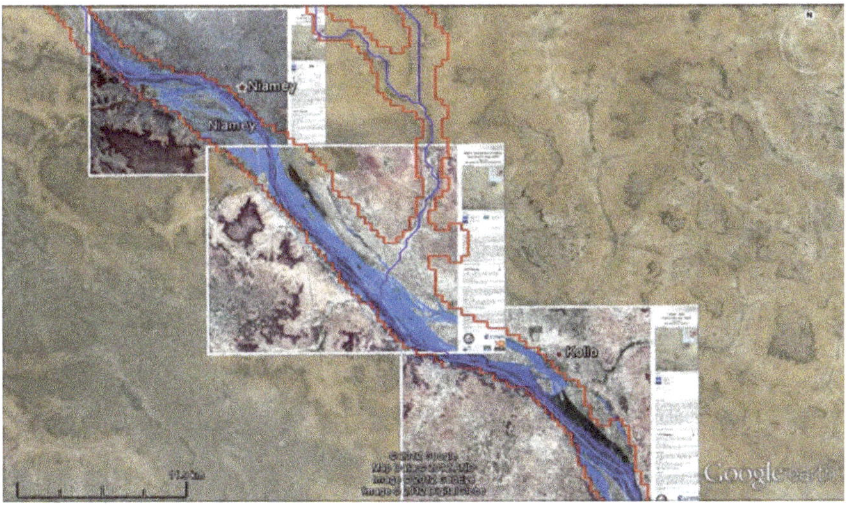

Fig. 16.5. Comparison between satellite image (Source: SERTIT) of the August 2012 event (blue area) and the approach here proposed (red line).

16.5 CONCLUSION

The aim of this contribution is to recognize flood-exposed areas at the catchment scale. To this purpose, threshold binary classifiers are used to extend available information on the flood hazard up to the complete catchment coverage. Based on the classification of quantitative morphological features derived from data stored in Digital Elevation Models, they allow investigating the relationships between morphology and the flooding hazard. The proposed method has been tested on two basins, pertaining to different environments: the Tanaro (Italy) and the Niger (Africa). While in the first case we proceed on the basis of a wide flood hazard database, for the Niger case study the flood hazard database is limited to non-existent. Available data have been therefore used to qualitatively validate the single-feature procedure based on relative elevation coupled with the interpretation of remote sensed images of the floodplain. Results show that threshold binary classifier techniques should be taken into account when either one is interested in a first low-cost detection of flood –exposed areas, or flood hazard estimation is needed in specific areas, for which limited information are available.

16.6 REFERENCES

Degiorgis, M., G. Gnecco, G. Gorni, G. Roth, M. Sanguineti and A. C. Taramasso. 2012. Classifiers for the detection of flood-prone areas using remote sensed elevation. *Journal of Hydrology* 470–471:302–315

Degiorgis, M., G. Gnecco, S. Gorni, G. Roth, M. Sanguineti, and A.C. Taramasso. 2013, Flood Hazard Assessment via threshold binary classifiers: case study of the Tanaro river basin. *Irrigation and Drainage 62* (Suppl. 2): 1–10 (2013)

Fawcett, T. 2006. An introduction to ROC analysis. *Pattern Recognit. Lett*. 27, 861–874.

Giannoni, F., G. Roth and R. Rudari. 2005. A procedure for drainage network identification from geomorphology and its application to the prediction of the hydrologic response. *Adv. Water Resour 28*, 567–581.

Jenson S.K. and Domingue J.O. Extracting Topographic Structure from Digital Elevation Data for Geographic Information System Analysis. *Photogrammetric Engineering and Remote Sensing 1988*, 54 (11): 1593-1600.

Montgomery, D.R. and W. E. Dietrich. 1988. Where do channels begin? *Nature 336*, 232–234.

Nardi, F., S. Grimaldi, M. Santini, A. Petroselli and L. Ubertini. 2008. Hydrogeomorphic properties of simulated drainage patterns using digital elevation models: the flat area issue. *Hydrol. Sci. J. 53*, 1176–1193.

CHAPTER 17

GEOHAZARDS IN KARST, AND AMPLIFICATION OF THE RELATED EFFECTS DUE TO LAND MISMANAGEMENT

Mario Parise

CNR-IRPI, Bari; m.parise@ba.irpi.cnr.it

ABSTRACT

Karst is a highly fragile setting, due to peculiar geological and hydrological features, and the strict connection between the surface and the sub-surface. It presents spectacular landforms, often exploited as natural parks and show caves, and hosts very important natural resources. Karst covers 20% of the Earth's surface and provides 30-35% of drinking water, with high-quality groundwaters. However, the direct connection between surface morphology and the underlying aquifer make karst waters extremely vulnerable to pollution.

Occurrence of geohazards in karst strongly differs from other natural settings. Subsidence and sinkholes originated by presence of cavities, slope movements favoured by conduits and caves, flash floods related to inability of the system to manage water from heavy rainstorms, are the main types of geohazards observed. Even though mostly related to natural processes, their occurrence and consequent damage to the anthropogenic environment are often caused or exacerbated by human activities.

Management of karst territories cannot be carried out without taking into account the peculiar features and behaviour of karst terrains and aquifers. The delicate and fragile equilibrium reached by nature in the course of millennia can be easily threatened by man, with catastrophic consequences to the society in terms of casualties and economic losses.

17.1 INTRODUCTION

Karst territories are among the world's most fragile settings, due to a number of peculiar geological and hydrological features, and to the strict connection between the surface and the sub-surface (White 1988; Palmer 2007). At the same time, they present spectacular surface and underground landforms (Figs. 17.1 and 17.2), which are often exploited as natural parks and show caves, and host very important natural resources.

Fig. 17.1. Karst landscapes: a) folded carbonate ridges in the Sierra Madre Mountains, northern Mexico; b) Devil's sinkhole, Texas, USA; c) a view of the Plitvice Lakes, Croatia, a UNESCO World Heritage Site; d) view of the Grahovsko polje, Montenegro, near the border with Bosnia Herzegovina.

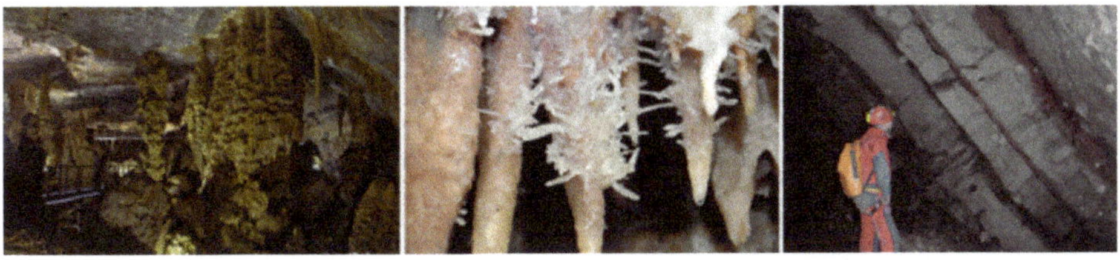

Fig. 17.2. The beauty and importance of underground karst. From left to right: a view within the Grutas de Bustamante (or Gruta del Palmito), a show cave in Nuevo Leòn, Mexico (photo courtesy CaptNemo, at http://chilembwe.com/nemo/); speleothems in a cave in Apulia, Italy; bauxite levels interlayered within the carbonate sequence at the Pertosa Cave, Campania, Italy.

At this latter regard, it must be reminded that karst covers 20% of the Earth's surface and provides 30-35% of the world's drinking water (Fig. 17.3), with high-quality groundwaters (Ford and Williams 2007). FAO (the Food and Agriculture Organization of the United Nations) forecasts that before 2025 at least 80% out of the demand of drinkable water in the Mediterranean Basin will be provided by karst aquifers. However, the direct connection between surface morphology and the underlying aquifer make underground karst waters extremely vulnerable to pollution. Water circulation in karst, as a matter of fact, is greatly different from what occurs in other types of terrain, due to the limited surface runoff, that is balanced by rapid infiltration of water underground through the complex network of discontinuities within the rock mass, and regulated by the role of epikarst (Brancelj and Culver 2005; Williams 2008).

Fig. 17.3. Importance of water resources in karst: a) Syri i kalter (Blue Eye), a beautiful resurgence in southern Albania; b) a spring in the Tampa area, Florida, USA; c, d) entrance (c) and inner view (d) of the Sotop spring, one of the several springs coming out in the Kotor Bay, Montenegro.

In karst, occurrence of geohazards strongly differs from other geological, morphological and hydrological settings (Gutierrez 2010; Parise 2010a; De Waele et al. 2011; Gutierrez et al. 2014). Subsidence and sinkholes originated by presence of underground cavities, slope movements favoured by development of karst conduits and caves, flash floods related to inability of the karst system to manage huge amount of water from heavy rainstorms, are the main types of geohazards observed. Even though mostly related to natural processes, their occurrence and the consequent damage to the anthropogenic environment are often caused and/or exacerbated by several types of human activities (Santo et al. 2007; Parise 2008).

Sinkholes are by far the most common and typical geohazard in karst (Waltham et al. 2005). They can be produced also as a consequence of human activities, from water leakages, or due to overcharge and vibrations in areas interested by presence of underground voids. In the last decades, there has been an increasing number of events related to man-made cavities, especially below and/or in the proximity of built-up areas (Waltham and Lu 2007; Hermosilla 2012; Parise 2012); this points out to the need for detailed and careful geological studies, addressed in particular to the identification of possible voids in the subsoil. Sinkholes may interfere with the urban development at the surface, as regards both expansion of urbanized areas or development of resort areas for touristic purposes (Brinkmann et al. 2008; Bruno et al. 2008; Margiotta et al. 2012).

Gravity-related processes occur underground as a natural evolution of caves, once water moved at greater depths, thus leaving unsupported the cave walls. Breakdown deposits are very common in caves (Klimchouk and Andrejchuk 2002; Iovine et al. 2010), and their presence and distribution within underground systems should be properly faced when the development of karst caves interferes with the built-up environment, with possible consequences for the infrastructures above.

Another hazard that might be very dangerous in karst is represented by floods, and in particular by flash floods related to very intense and clustered precipitation (White and White 1984). These often affect wide areas as poljes (Fig. 17.4), typically dry, but that once or more frequently during a year may be flooded by water. The answer of karst settings to heavy rainstorms is highly dependent upon the correct management of the territory: clogging the natural swallow holes, or covering with

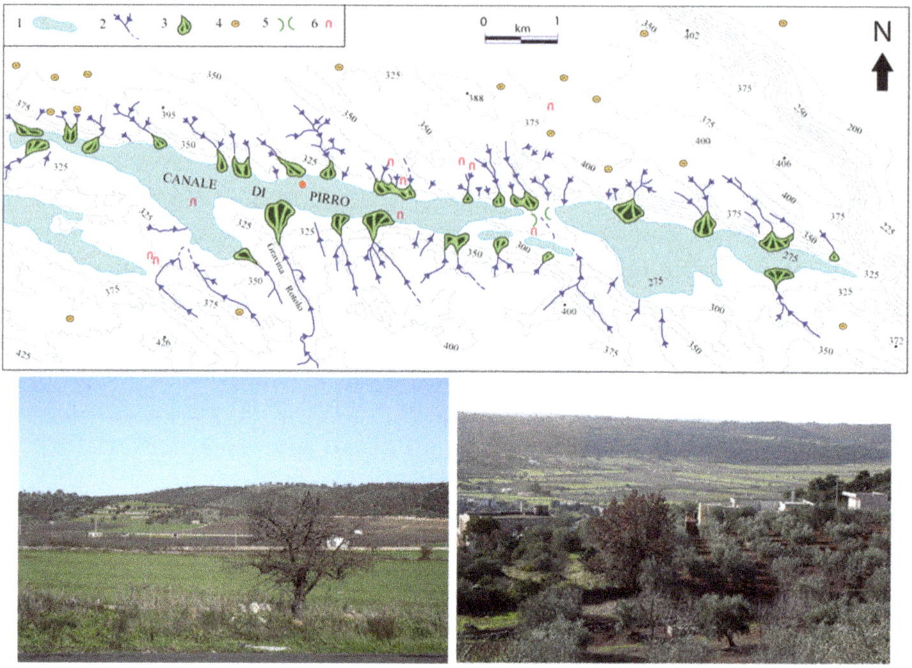

Fig. 17.4. Canale di Pirro, a remarkable polje in the Apulian territory, southern Italy. Above: geomorphological map. Key: 1) polje; 2) water course, dashed where subdued; 3) fan; 4) doline; 5) saddle; 6) cave. In the pictures below, two views of the polje.

road and asphalt the natural infiltration of water underground, may be at the origin of serious problems on the occasion of intense rainfall events, with severe consequences and huge costs to society (Parise 2003; Delle Rose and Parise 2010).

Given the high fragility of karst, in such setting the planning and realization of engineering works (Milanovic 2002) is particularly delicate, and often result in environmental problems and in damage to the natural landscape. Building dams, in particular, may require high costs, and often the results are not those expected, with serious problems in filling the planned reservoirs, due to leakage through karst conduits and fissures in the rock mass.

Management of karst territories cannot be carried out without taking into the due account the peculiar features and behaviour of karst terrains and aquifers. The delicate and fragile equilibrium reached by nature in the course of millennia can be easily threatened by man, with catastrophic consequences to the society in terms of casualties and economic losses.

17.2 MAN AS A GEOMORPHOLOGIC AGENT

Karst terrains present at the surface typical landforms such as closed depressions, dolines and poljes (Nicod 1972; Ford and Williams 2007); further, they are characterized by a complex underground drainage system, and by direct interaction between water flowing at the surface and groundwater. The hydrological and hydrogeological features of karst terrains allows storage of great amount of high-quality water, a natural resource of high value that, however, may be easily threatened by anthropogenic actions (Parise and Pascali 2003). Differently than for the great majority of natural environments on Earth, in karst there is not necessarily correspondence between the surface watersheds, identified by means of topography, and the groundwater circulation (Gunn 2007). Water can infiltrate at a certain site, and, due to complex underground systems, be transported in another, nearby watershed. Consequently, the only way to be sure of the course of water in karst is following it underground, when possible, or using dye tracers (Goldscheider and Drew 2007) to move along its path whenever man is not able to enter narrow or flooded passages.

Locating catchment boundaries in karst represents a highly complex goal, also because of the great variability that may be recorded in time. Further, it has been repeatedly proved that in conditions of high groundwater levels also fossil and inactive karst conduits become activated, thus originating overflow from one catchment to those nearby (Bonacci et al. 2006; De Waele and Parise 2013).

In the last centuries, with an increasing pace in the last decades, man has become among the main agents shaping the landscapes, often causing heavy changes in the original environment. In fragile settings, such as karst, the transformations deriving from man's activities may have a great impact, and cause negative effects to the overall ecosystem (Parise and Gunn 2007), as well as inducing the loss of karst features and natural resources.

Mining and quarrying produce severe impacts on the environment, causing loss of natural landscapes and frequently favoring events of degradation and pollution (Ryka and Werner 1985; Ekmekci 1993; Gunn

1993, 2004; Andriani and Walsh 2003). Sites for extraction of stones developed not only at the surface, as open quarries, but also at depth, which can be at the origin of the development of sinkholes, with further problems and costs to society (Parise, 2010b, 2012). Mining and related human activities, caused repeated episodes of environmental degradation, both during the phase of work and after the end of the extraction. In karst, opening of quarries inevitably is in contrast with the presence of natural caves: finding a cave means typically trouble, because the working activity is delayed, if not stopped. Even where specific rules exist to safeguard karst caves (Quinlan 1986; Assad and Jordan 1994; LaMoreaux 1997; LaMoreaux et al. 1997), the common practice is to rapidly proceed to destroying the cave or filling it with waste materials (Formicola et al. 2010), without allowing cavers to explore the cave, and document it.

Further, abandoned mines and quarries, at the surface and underground, may easily become the sites where to illegally discharge wastes and pollutants (Delle Rose et al. 2007). This represents one of the highest negative impacts to the environment, and for this reason any mining and quarrying activity should necessarily include a final recovery plan of the site, aimed at avoiding such negative circumstances.

17.3 MISMANAGEMENT ACTIONS IN KARST

The fragility of karst environment depends upon its main hydrological features, consisting of a very limited water runoff, rapid underground infiltration through the network of fissures and conduits in the rock mass, turbulent flow, and the direct connection between the surface and the water table (White 2002; Bakalowicz 2005). The latter point, in particular, poses high environmental problems, since a likely pollutant cannot be diluted by passing through different types of rocks in its downward movement, but rather it may reach the karst aquifer with all its potential of pollution, strongly impacting the water quality (Zwahlen 2004; Goldscheider 2005; Farfan et al. 2010).

Dealing with karst, it is nowadays necessary to take into account the several human actions and activities therein occurring. This requires to perform analysis also as concerns the social and economic effects in karst, broadening the usual approach and referring to concepts from other disciplines, in order to carry out a more complete evaluation of the effects caused by man. The term resilience, for instance, was initially proposed in the field of ecology as a core concept within ecosystems (Holling 1973); in recent years, it has been widely used to express the ability of a system to absorb perturbation or disturbances. As concerns natural disaster, resilience is the capacity to resist and recover from disaster losses (Kleina et al. 2003; Zhou et al. 2010). One of the more widespread definition of resilience includes three components: i) response to disturbance; ii) capacity to self-organize; and iii) capacity to learn and adapt.

Since the social and environmental consequences of disasters (including pollution events) are strongly increasing, and cause high costs to society, there is the need to involve stakeholders and population in strategies to manage the impacts (Djalante 2012; Lei et al. 2014). This is especially required in heavily populated areas, or in those sectors in the proximity of big cities and metropolitan areas, where the intertwining among natural and anthropogenic factors may be very strong (Serre and Barroca 2013). This is also true in karst, due to the extreme intrinsic fragility of the environment, and, as a consequence, of the likely negative effects that human actions may produce to the natural landscape, the natural resources therein contained, and the karst ecosystems as well.

In karst, the concept of disturbance produced by man to the natural environment can be analyzed by means of the Karst Disturbance Index (KDI) proposed by van Beynen and Townsend (2005), and later revised by North et al. (2009). KDI takes into account a number of indicators, subdivided into five different categories (Geomorphology, Atmosphere, Hydrology, Biota, Cultural), to evaluate the disturbance to the karst environment deriving by anthropogenic actions. As shown by applying the index to different karst settings in the world (Calò and Parise 2006; De Waele 2009; North et al. 2009; Day et al. 2011), it appears that knowledge of the main features of the karst environments, encompassing many different fields and disciplines of interest, is fundamental for a proper understanding of the changes occurring, and for linking such changes to specific actions by man or to variability of other factors (i.e., climate changes at the global or local scale).

Repeated episodes of mismanagement in karst, with events of pollution, both at the surface and underground (Parise and Pascali 2003; Delle Rose et al. 2007), or with severe consequences in particular situations, as post-war scenarios (Calò and Parise 2009), highlight the fragility of this setting (Fig. 17.5) and, once again, the necessity of a

Fig. 17.5. Examples of cave destruction and pollution: a) overall view of a open quarry, showing destruction of the very large cave entrance to the right, and the cave opening at the quarry level (to the left), now almost completely filled with rock wastes; b) a cave access masked by rock heaps within a quarry in southern Italy; c) solid wastes in the Monte Pelosello cave, Apulia, Italy (photo courtesy F. Lomastro); d) pollution in the Bussento spring, campania, Italy (photo courtesy N. Damiano).

direct involvement of the population in order to create an environmental awareness, addressed to safeguard the delicate karst setting and the natural resources it contains.

The concept of resilience in karst should therefore be further object of analysis, involving all the three components above mentioned, and in particular performing an effort in adapting it to the peculiarity of karst environments.

17.4 CONCLUSIONS

Safeguarding the unique groundwater ecosystems that are present in karst from natural and man-made hazards (Parise & Gunn, 2007; Gutierrez, 2010; Parise, 2010; De Waele et al. 2011), including climate change at the global and local scale, and mitigating the related risks are the main priority actions to be followed in order to reach a sustainable management in this fragile setting.

A very high pressure is acting on karst environment, and its natural resources, starting from the extremely important water resources. In order to face such pressure, approaches dedicated to manage the delicate balance existing in karst, and in particular the trans-boundary karst water resources, are necessary. These latter, together with the precious habitats and species living therein, need to be dealt with through specific, dedicated actions of safeguard, aimed at not destroying the high-value biodiversity in karst (Brancelj and Culver 2005; Culver and Pipan 2009), and at mitigating as much as possible the occurrence of geohazards.

One of the best example of trans-boundary karst is represented by the lakes Ohrid and Prespa, located in the Balkan Peninsula, and shared by three countries: Macedonia, Albania, and Greece (Popovska and Bonacci 2007). The two lakes, which location is greatly controlled by tectonics, with a number of caves in the surrounding areas (Trocino et al. 2010), are a remarkable site for biodiversity and cultural heritage. Lake Ohrid, declared a World Heritage Site by UNESCO in 1980, is also fed by the waters coming from Lake Prespa, that has no surface outflow (Amataj et al. 2007). Hydrologically, the system of the two lakes should be ideally managed as a trans-boundary water resource. However, so far no coordinated water resources management among the three countries has been established.

As also shown by the example above, karst cannot be dealt with methodologies and studies typically used in other environments. Its peculiarity requires a specific approach, that necessarily has to take into account the interrelationships between the surface and the underground worlds, as well as the time and space variability of many of the factors playing a role in karst ecosystems. Karst aquifers are a huge natural resource in many areas worldwide, and biodiversity in karst represents a remarkable and valuable character. Therefore, the human actions should be addressed toward definition of standard procedures and guidelines to safeguard the natural landscape and resources, and protecting them from mismanagement actions.

17.5 REFERENCES

Amataj, S., Anovski, T., Benischke, R., Eftimi, R., Gourcy, L. L., Kola, L., Leontiadis, I., Micevski, E., Stamos, A. and J. Zoto 2007. Tracer methods used to verify the hypothesis of Cvijic about the underground connection between Prespa and Ohrid lake. *Environmental Geology* 51 (5): 749-753.

Andriani, G. F. and N. Walsh 2003. Fabric, porosity and water permeability of calcarenites from Apulia (SE Italy) used as building and ornamental stone. Bull. Eng. Geol. Environ. 62: 77-84.

Assad, F. A. and H. Jordan 1994. Karst terranes and environmental aspects. *Environ. Geol.* 23: 228-237.

Bakalowicz, M. 2005. Karst groundwater: a challenge for new resources. *Hydrogeol. Journal* 13: 148-160.

Bonacci, O., Ljubenkov, I. and T. Roje-Bonacci 2006. Karst flash floods: an example from the Dinaric karst Croatia. *Natural Hazards Earth System Science* 6: 195–203.

Brancelj, A. and D. C. Culver 2005. Epikarstic communities. In: Culver, D. C. and W. B. White (eds.), *Encyclopedia of caves*. Elsevier Academic Press: 223-229.

Brinkmann, R., Parise, M. and D. Dye 2008. Sinkhole distribution in a rapidly developing urban environment: Hillsborough County, Tampa Bay area, Florida. *Engineering Geology* 99: 169-184.

Bruno, E., Calcaterra, D. and M. Parise 2008. Development and morphometry of sinkholes in coastal plains of Apulia, southern Italy. Preliminary sinkhole susceptibility assessment. *Engineering Geology* 99: 198-209.

Calò, F. and M. Parise 2006. Evaluating the human disturbance to karst environments in southern Italy. *Acta Carsologica* 35(2): 47-56.

Calò, F. and M. Parise 2009. Waste management and problems of groundwater pollution in karst environments in the context of a post-conflict scenario: the case of Mostar (Bosnia Herzegovina). *Habitat International* 33: 63-72.

Culver, D. C. and T. Pipan 2009. *The Biology of Caves and other Subterranean Habitats.* Oxford University Press, Oxford.

Day, M., Halfen, A. and S. Chenoweth 2011. The cockpit country, Jamaica: boundary issues in assessing disturbance and using a karst disturbance index in protected areas planning. In: Van Beynen, P. E. (ed), *Karst Management.* Springer, Dordretch: 399-414.

Delle Rose, M. and M. Parise 2010. Water management in the karst of Apulia, southern Italy. Proc. Int. Interdisc. Scientific Conf. *"Sustainability of the karst environment. Dinaric karst and other karst regions"*, Plitvice Lakes (Croatia). IHP-UNESCO, Series on Groundwater 2: 33-40.

Delle Rose, M., Andriani, G. F. and M. Parise 2007. Evaluating the impact of quarrying on karst aquifers of Salento (southern Italy). In: Parise, M. and J. Gunn (eds.) *Natural and Anthropogenic Hazards in Karst Areas: Recognition, Analysis and Mitigation.* Geological Society of London, 279: 153–171.

De Waele, J. 2009. Evaluating disturbance on Mediterranean karst areas: the example of Sardinia (Italy). *Environmental Geology* 58(2): 239-255.

De Waele, J. and M. Parise 2013. Discussion on the article "Coastal and inland karst morphologies driven by sea level stands: a GIS based method for their evaluation" by Canora F., Fidelibus D., Spilotro G. *Earth Surface Processes and Landforms* 38 (8): 902-907.

De Waele, J., Gutierrez, F., Parise, M. and L. Plan 2011. Geomorphology and natural hazards in karst areas: a review. *Geomorphology* 134(1-2): 1-8.

Djalante, R. 2012. Adaptive governance and resilience: the role of multi-stakeholder platforms in disaster risk reduction. *Natural Hazards Earth System Sciences* 12: 2923-2942.

Ekmekci, M. 1993. Impact of quarries on karst groundwater systems. In: Gunay, G., Johnson, A. I. and W. Back (eds.), Hydrogeological processes in karst terranes. IAHS publ 207: 3-6.

Farfan, H., Dias, C., Parise, M. and C. Aldana 2010. Scenarios of groundwater pollution in a karst watershed: a case study in the Pinar del Rio province at Cuba. In: Carrasco, F., La Moreaux, J. W., Duran Valsero, J. J. and B. Andreo (eds.) *Advances in research in karst media.* Springer: 287-292.

Ford, D. C. and P. Williams 2007. *Karst hydrogeology and geomorphology.* John Wiley and Sons, Chichester, UK.

Formicola, W., Gueguen, E., Martimucci, V., Parise, M. and G. Ragone 2010. Caves below quarries and quarries above caves: problems, hazard and research. A case study from southern Italy. *Geological Society of America Abstracts with Program* 42 (5).

Goldscheider, N. 2005. Karst groundwater vulnerability mapping: application of a new method in the Swabian Alb, Germany. *Hydrogeology Journal* 13: 555–564.

Goldscheider, N. and D. Drew (eds.) 2007. *Methods in Karst Hydrogeology.* International Contributions to Hydrogeology 26. Int Ass Hydrogeol, Taylor & Francis, London.

Gunn, J. 1993. The geomorphological impacts of limestone quarrying. *Catena* 25: 187-198.

Gunn, J. 2004. Quarrying of limestones. In: Gunn, J. (ed.) *Encyclopedia of cave and karst science*. Routledge, London: 608-611.

Gunn, J. 2007. Contributory area definition for groundwater source protection and hazard mitigation in carbonate aquifers. In: Parise, M. and J. Gunn (eds.) *Natural and Anthropogenic Hazards in Karst Areas: Recognition, Analysis and Mitigation.* Geological Society of London 279: 97-109.

Gutiérrez, F. 2010. Hazards associated with karst. In: Alcántara, I. and A. Goudie (eds.), *Geomorphological Hazards and Disaster Prevention.* Cambridge University Press, Cambridge: 161–175.

Gutiérrez, F., Parise, M., De Waele, J. and H. Jourde 2014. A review on natural and human-induced hazards and impacts in karst. *Earth Science Reviews*, submitted.

Hermosilla, R. G. 2012. The Guatemala City sinkhole collapses. *Carbonates and Evapor.* 27 (2): 103–107.

Holling, C. S. 1973. Resilience and stability of ecological systems. *Ann. Rev. Ecol. Syst.* 4: 1-23.

Iovine, G., Parise, M. and A. Trocino 2010. Breakdown mechanisms in gypsum caves of southern Italy, and the related effects at the surface. *Zeitschrift Geomorphology* 54 (suppl 2): 153-178.

Kleina, R. J. T., Nicholls, R. J. and F. Thomalla 2003. Resilience to natural hazards: How useful is this concept? *Environmental Hazards* 5: 35-45.

Klimchouk, A. and V. Andrejchuk 2002. Karst breakdown mechanisms from observations in the gypsum caves of the Western Ukraine: implications for subsidence hazard assessment. *International Journal of Speleology* 31(1/4): 55–88.

LaMoreaux, P. E. 1997. Legal aspects of karst and insurability. In: Gunay, G., Johnson, A. I. and W. Back (eds.), *Hydrogeological processes in karst terranes.* Proc. Antalya Symposium and Field Seminar, October 1990, IAHS publ. 207: 11-17.

LaMoreaux, P. E., Powell, W. J. and H. E. LeGrand 1997. Environmental and legal aspects of karst areas. *Environ. Geol.* 29: 23-36.

Lei, Y., Wang, J., Yue, Y., Zhou, H. and W. Yin 2014. Rethinking the relationships of vulnerability, resilience, and adaptation from a disaster risk perspective. *Natural Hazards* 70: 609-627.

Margiotta, S., Negri, S., Parise, M. and R. Valloni 2012. Mapping the susceptibility to sinkholes in coastal areas, based on stratigraphy, geomorphology and geophysics. *Natural Hazards* 62 (2): 657-676.

Milanovic, P. 2002. The environmental impacts of human activities and engineering constructions in karst regions. *Episodes* 25: 13-21.

Nicod, J. 1972. *Pays et paysages du calcaire.* Presses Universitaires de France, Paris.

North, L. A., van Beynen, P. E. and M. Parise 2009. Interregional comparison of karst disturbance: west-central Florida and southeast Italy. *Journal of Environmental Management* 90: 1770-1781.

Palmer, A. N. 2007. *Cave geology*. Cave Books.

Parise, M. 2003, Flood history in the karst environment of Castellana-Grotte (Apulia, southern Italy). *Natural Hazards Earth System Science* 3(6): 593-604.

Parise, M. 2008. Rock failures in karst. In: Cheng, Z., Zhang, J., Li, Z., Wu, F. and K. Ho (eds.) *Landslides and Engineered Slopes*. Proc. 10th Int. Symp. on Landslides, Xi'an (China), 1: 275-280.

Parise, M. 2010a. Hazards in karst. In: Bonacci, O., (ed.), *Proc. Int. Sc. Conf. "Sustainability of the Karst Environment, Dinaric Karst and Other Karst Regions"*, Plitvice Lakes, Croatia, 23–26 September 2009, IHP-UNESCO, Series on Groundwater 2: 155–162.

Parise, M. 2010b. The impacts of quarrying in the Apulian karst. In: Carrasco, F., La Moreaux, J. W., Duran Valsero, J. J. and B. Andreo (eds.) *Advances in research in karst media.* Springer: 441-447.

Parise, M. 2012. A present risk from past activities: sinkhole occurrence above underground quarries. *Carbonates and Evaporites* 27 (2): 109-118.

Parise, M. and J. Gunn (eds.) 2007. *Natural and anthropogenic hazards in karst areas: Recognition, Analysis and Mitigation.* Geological Society of London, sp. publ. 279.

Parise, M. and V. Pascali 2003. Surface and subsurface environmental degradation in the karst of Apulia (southern Italy). *Environmental Geology* 44: 247-256.

Popovska, C. and O. Bonacci 2007. Basic data on the hydrology of Lakes Ohrid and Prespa. *Hydrological Processes* 21: 658-664.

Quinlan, J. F. 1986. Legal aspects of sinkhole development and flooding in karst terranes – 1. Review and synthesis. *Environmental Geology Water Science* 8 (1/2): 41-61.

Ryka, W. and Z. Werner 1985. Ten centuries of salt mining in Poland. *Episodes* 8 (3): 169–172.

Santo, A., Del Prete, S., Di Crescenzo, G. and M. Rotella 2007. Karst processes and slope instability: some investigations in the carbonate Apennine of Campania (southern Italy). In: Parise, M. and J. Gunn (eds.) *Natural and Anthropogenic Hazards in Karst Areas: Recognition, Analysis and Mitigation.* Geological Society of London, 279: 59–72.

Serre D. and D. Barroca 2013. Natural hazard resilient cities. *Nat. Hazards Earth System Sc.* 13: 2675-2678.

Trocino, A., Parise, M. and A. Rizzi 2010. Ricerche speleologiche in Albania: primi dati sulle cavità nei pressi del lago di Prespa. *Proc. XII Reg. Meeting Speleol. "Spelaion 07"*, Altamura, 7–9/12/ 2007: 246-259.

van Beynen, P. E. and K. Townsend 2005. A disturbance index for karst environments. *Environmental Management* 36: 101-116.

Waltham, T. and Z. Lu 2007. Natural and anthropogenic rock collapse over open caves. In: Parise, M. and J. Gunn (eds.), *Natural and anthropogenic hazards in karst areas: recognition, analysis and mitigation*. Geological Society of London, sp publ 279: 13-21.

Waltham, T., Bell, F. and M. Culshaw 2005. *Sinkholes and subsidence: karst and cavernous rocks in engineering and construction.* Springer, Berlin

White, W. B. 1988. *Geomorphology and Hydrology of Karst Terrains*. Oxford University Press, New York.

White, W. B. 2002. Karst hydrology: recent developments and open questions. *Engin. Geology* 65: 85–105.

White, E. L. and W. B. White 1984. Flood hazards in karst terrains: lessons from the Hurricane Agnes storm. In: Burger, A. and L. Dubertret (eds.) *Hydrogeology of Karst Terrains*, 1: 261-264.

Williams, P. W. 2008. The role of the epikarst in karst and cave hydrogeology: a review. *International Journal of Speleology* 37: 1-10.

Zhou, H., Wang, J., Wan, J. and H. Jia 2010. Resilience to natural hazards: a geographic perspective. *Natural Hazards* 53: 21-41.

Zwahlen, F. 2004. *Vulnerability and Risk Mapping for the Protection of Carbonate (Karstic) Aquifers.* Final report COST action 620. European Commission, Brüssel.

CHAPTER 18

CO-BENEFITS OF INNOVATIVE WATER SAVING SOLUTIONS IN AN URBAN WATER SYSTEM

Irina Ribarova[1], Peyo Stanchev[1], Galina Dimova[1], Dionysis Assimacopoulos[2]

[1]University of Architecture, Civil engineering and Geodezy, 1404 Sofia, Bulgaria

[2]National Technical University of Athens, Athens GR-15780, Greece

ABSTRACT

The paper shows how the sustainable use of water resources was considered in an urban water system. Along with the usual quantitative impact on the fresh water resources, the importance of other impact categories, associated with air, water and soil pollution as well as resource depletion is demonstrated. For this purpose, two water saving alternatives are compared. The first alternative is supply-side oriented with a main target of reducing physical water losses in the water distribution network. The second alternative is focused on demand-side water savings. The alternatives include a different energy saving element. The results show that: 1) Eco-efficiency assessment, standardized with ISO 14045, is a powerful tool for thorough comparison of alternatives; 2) Water saving solutions bring significant environmental co-benefits, which should not be neglected in the decision making process and 3) The simple indicator "total volume of the abstracted water" is sufficient to compare different solutions within the boundaries of one system in terms of water quantity, but other indicators should be developed when different systems are the subject of comparison.

18.1 INTRODUCTION

Sustainable water use is a complex and interdisciplinary topic, whose study demands the application of approaches that go beyond narrow disciplinary tools. A relatively recent trend prompts to "move away from a focus on theories that ignore or underestimate co-benefits" (Gleick et al. 2011). The term "co-benefits" appeared in scientific literature in the nineties, but attracted wider interest only around ten years later (Miyatsuka et al. 2010). Since then it has been gaining increasing popularity (Hamilton et al. 2010; Strand et al, 2010).

This paper shows how the intended positive side effects, the so called "co-benefits", of water saving solutions in the urban water system of Sofia, the capital of Bulgaria have been assessed.

It is well known that "the largest, least expensive and most environmentally sound source of water to meet future needs is the water currently being wasted in every sector of our economy" (Gleick et al. 2003). In the recent EU water policy related documents, "Addressing the challenge of water scarcity and droughts in the European Union" and "Blueprint to Safeguard Europe's Water Resources", the Commission has identified several actions that could greatly improve water efficiency (COM, 2007; COM, 2012). Among them are the cost-effective water-saving technologies.

To consider sustainable water use, Brooks and co-authors have suggested "soft path approach to water management" (Brooks et al. 2009). It is a management strategy that "frees up water by curbing water waste" (Brooks et al. 2009). It considers the application of super-efficient technologies for reducing human uses of water without jeopardizing quality of life and ensuring satisfaction of the needs of the ecosystem.

In line with Brooks' approach, this work considers two solutions for reduction of freshwater withdrawal, which relate respectively to demand-side and supply-side of an urban water system. The guiding principles for selection of the alternatives are: i) the capability to free water; ii) the innovative performance; iii) the potential of bringing co-benefits such as reducing the negative environmental impact.

The standardized recently eco-efficiency assessment, which is a systematic way of analyzing and comparing of alternatives, has been applied (ISO14045, 2012). It is a tool, which relates environmental performance to the product value, thus it addresses two of the three pillars of the sustainability concept. The environmental assessment is based on life cycle assessment (LCA), standardized with ISO 14040 and ISO 14044. The only drawback of the current LCA practice is the lack of a generally accepted way of assessing the impact of a product system on the freshwater resources from a quantitative point of view. Only recently, new methodologies, emphasizing freshwater use inventory schemes, have started to be developed (Boulay et al. 2011a; Boulay et al. 2011b). The potential environmental impacts of freshwater use considering various cause–effect relationships have been assessed as well (Bayart J.B, et al.2010; Milà i Canals et al. 2009; Pfister et al. 2009).

Considering the lack of a standardized LCA regarding freshwater use, the simplest indicator for assessment of the quantitative impact of the water use, "total volume of the abstracted water", was applied in this study. Its advantages and disadvantages are discussed in the paper. Other eleven indicators, based on LCA midpoint impact categories, are used to assess the co-benefits of the solutions.

18.2 TERMINOLOGY

18.2.1 Water Use and Water Consumption

The difference between freshwater use and freshwater consumption is crucial when environmental impact is assessed (Koehler et al. 2012). Figure 18.1 summarizes the recent classification of these terms (Gleick et al. 2011; Koehler et al. 2012; Milà i Canals et al. 2009; Pfister et al. 2009).

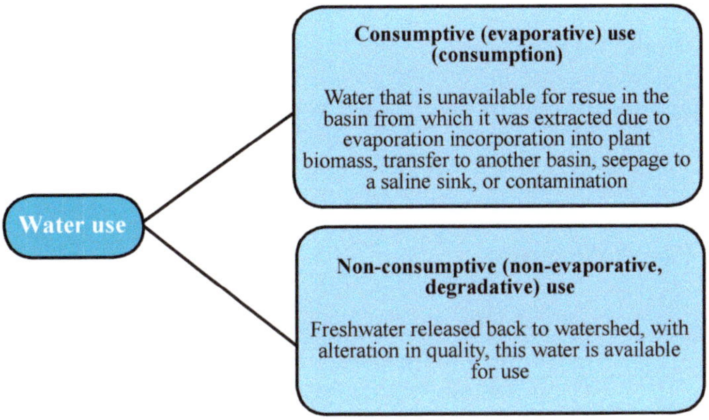

Fig. 18.1. Classification of the water use.

18.2.2 Water Exploitation Index

Different definitions and indicators of the water stress could be found in the literature. Here Water Exploitation Index (WEI) of the European Environmental Agency is adopted (EEA, 2003).

$$WEI = \frac{Total\ abstracted\ water\ (annual)}{Total\ available\ resources\ (annual)} \quad (18.1)$$

According to this methodology, water stress exists if WEI >0.2 and there is severe water stress if WEI >0.4.

18.2.3 Co-Benefits

Although there is no agreed definition, many authors consider "co-benefits" as intended positive side effects of a policy. It differs from unintended positive side effects, called "ancillary benefits" (Miyatsuka et al. 2010).

18.2.4 Innovation

Based on his etymological study, Frankelius defines that innovation means something "1) new with high-level of originality, 2) in whatever area 3) that also breaks in to (or obtains a foothold in) society, often via the market, and 4) means something revolutionary for people. An innovative process has not been fulfilled until customers or others for whom it may be of benefit have acknowledged and accepted a new thing" (Frankelius, 2009).

18.2.5 Eco-Efficiency

According to ISO14045, the eco-efficiency is an "aspect of sustainability relating the environmental performance of a product system to its product system value" (ISO14045, 2012). In this definition three terms are further clarified in the standard: i) Environmental performance: it is "measurable results related to environmental aspects"; ii) Product system is "collection of unit processes with elementary and product flows, performing one or more defined functions, and which models the life cycle of a product"; iii) Product system value is "worth or desirability ascribed to a product system" (ISO14045, 2012).

18.3 METHODS

18.3.1 Eco-Efficiency Assessment

The most commonly used mathematical expression for quantification of the eco-efficiency, adopted also in this study, is:

$$\textit{Eco-Efficiency} = \frac{\textit{Product system value}}{\textit{Environmental performance}} \tag{18.2}$$

18.3.2 Assessment of the Economic Performance

The economic performance (product system value) is estimated by equation (18.3) (Ribarova et al. 2014):

$$\textit{Product system value} = \frac{1}{\textit{costs}} \tag{18.3}$$

The denominator includes all costs through the urban water cycle (water abstraction, distribution, use, waste water collection, and final treatment), belonging to the main actors in the system – water operator and users. They are divided into three groups:

1) Fixed costs – these are the costs, which do not depend on the amount of the abstracted water. These might be salaries, depreciation, amortization, maintenance, taxes, etc.

2) Operational costs – these are the costs, which depend on the amount of the abstracted and/or used water. For example quantity of the reagents for water purification in the water

purification plant depends on the amount of the treated water, the energy for heating water in the households depends on the hot water use, etc. So, generally here are included the costs for the system's input material flows: water abstraction from the nature; energy, associated with water use, purification and wastewater treatment; reagents, associated with water purification and wastewater treatment; transport for reagent and sludge disposal.

3) Investment costs – these are the costs for implementation of the innovative technologies (price of the equipment, etc.). If an upgrade of existing technology is foreseen, only the price of the upgrade is considered. For example, replacing old dishwashing machine with new water and energy efficient one, only the difference in prices between the old and the new one is considered.

All costs are calculated at annual basis.

18.3.3 Assessment of the Environmental Performance

The assessment of the environmental performance is done according to LCA approach, standardized with ISO 14040 and ISO 14044. The two mandatory steps of the LCIA are included: 1) Selection of impact categories and classification; 2) Characterization. For this purpose, the material flows are disaggregated to components considering their upstream supply chain, which are further identified with corresponding impact categories (Ribarova et al. 2014).

Eleven midpoint environmental categories are selected as the most representative for the urban water systems. For them, the environmental indicators are calculated to perform the environmental assessment of the system.

The environmental impact factors per unit of produced good in the background processes are obtained from ELCD and Ecoinvent 2.2 databank with SimaPro 7 software using CML 2001, TRACI and ReCiPe.

The environmental performance in regard to quantitative freshwater impact is assessed using an indicator "total volume of the abstracted water".

18.4 THE CASE STUDY OF SOFIA URBAN SYSTEM

18.4.1 Justification of the Study

Sofia is situated in the Upper Iskar basin, which is under water stress, as indicated by the WEI varying around 0.4 (Ribarova, 2009). Water stress is attributed to natural and manmade causes: seasonal river run off, alternation of periodic droughts and floods, conflicts among water users, increased urbanisation and population growth of the capital, old pipelines in water networks, high percentage of drinking water losses in the water supply network, inefficient management of the water resources, etc. This provides the stimulus for improvement of the municipal water use.

Sofia's urban water system serves a population of about two million, of whom some 1.3 million are permanent residents. The water is taken from Iskar reservoir, situated at Iskar River. The water supply system consists of around 4000 km distribution network and 112 water reservoirs, placed 20 to 100 m above the altitude of the city, which determines gravity operation. The current disadvantage of this scheme is that there are many sites at the network with pressure above the recommended optimal one, especially at night time when the water use and flow rate are low. This results in higher physical water losses through leakages as well as in more system failures. The water losses amount at about 60%.

18.4.2 Supply-side Innovative Solution

One of the possible solutions for overcoming the high physical water losses in the system, thus reducing the amount of fresh water withdrawal, is pressure management. Traditionally it is realised by pressure reduction valves (PRV). In the recent years an innovative solution has been developed – pressure management with pressure reduction turbines (PRT). Together with solving the high pressure problem, described above, PRTs utilize the hydro potential energy and produce electricity, thus bringing co-benefit by improving the energy efficiency of the system. Because of these reasons, they were selected as a solution to be investigated in this study.

18.4.3 Demand-side Innovative Solution

It has been reported that users are the most significant urban water system component in terms of energy and water savings (Flower et al. 2007). Reducing the water use in households and replacing the water appliances with more water and energy efficient ones looks like promising solution bringing number of environmental co-benefits and eventually financial benefits due to decrease in bills. Therefore, our study assessed the potential for water saving and for increse of the eco-efficiency of the domestic water use through replacement of the sanitary appliances.

18.4.4 Assumptions

18.4.4.1 Clustering of the users

The case study includes different water users. To facilitate the assessment, the users with the same patterns of the water use are grouped in clusters. There are two main types of urban water users - Domestic and Non-Domestic Customers. Domestic users are further divided into two sub-clusters in relation to the type of water heating, respectively to: 1) domestic users with central water heating; 2) domestic users with local water heating.

While all domestic water users, nevertheless the way water is heated, have relatively similar pattern of the water use and wastewater discharge, there is no such pattern for the non-domestic users. Their water use and pollution load depends on the production needs and varies significantly from one user to another. In order to overcome this challenge, the approach of substituting non-domestic users with equivalent citizens in regard to their water use and BOD pollution was applied according to the Urban Wastewater Directive (Directive 91/271). In relation to the BOD pollution level, non-domestic water users are divided into 3 sub-clusters: 1) non-domestic 1st category; 2) non-domestic 2nd category; non-domestic 3rd category. However, electricity consumption and other pollution loads for non-domestic users were not considered, because of the impossibility to assess them due to the variety of the industries and products.

18.4.4.2 Setting of a baseline scenario

The baseline scenario is the one, which represents the current state of the system, before the implementation of the innovative solutions. Year 2011 was selected as a base, because our research started in 2012, thus the last year with available full data set was 2011. More important, in 2011 there was a national census, results of which were used to determine some crucial parameters of the system.

In our work we created virtual baseline scenario. In 2011 in many district metering areas (DMAs) the pressure management was already applied, by means of pressure reduction valves. The real situation is that now these valves could be changed with pressure reduction turbines. In order to simulate the effect on water saving before and after implementation of pressure management solution using real data provided from the water operator, the virtual baseline scenario was built under the assumption that the pressure reduction valves had not been installed yet. Thus real measurements for the water withdrawal before and after installation of the pressure reduction valves were used in our calculations. The average observed effect of pressure management on water losses reduction was about 20%.

18.4.4.3 Setting of a supply-side scenario

There are about 180 Pressure management areas in Sofia's water distribution network. Not all of the PRVs are appropriate for replacement with turbines. The criteria, which were used to select the PRVs to be changed with PRTs are:

1. Mean average daily flow rate bigger than 26 l/s, which guarantees minimum flow, necessary for the turbine operation;

2. Net pressure drop bigger than 20 meters for average daily flow rate to guarantee the operational head of the turbine;

3. Power of the turbines bigger than 6 kW, because for lower power the investment is not efficient.

With these screening criteria, the total number of the appropriate PRVs become 44. Only contribution of these PRVs on the reduction of the water losses was accounted in the study.

18.4.4.4 Setting of a demand-side scenario

As there were no available data about the water demand and energy consumption in the households, an additional study was carried out. It included 5 representative districts, for which data from the national census of 2011 was bought from the national statistical institute. For the same districts, the water operator provided record of the water meter accounts in households and revenue water meters in the buildings. The analysis of these data showed that:

1. The water consumption depends on the way the water is heated. Two main types of households are distinguished: 1) households with local water heating and 2) households with district water heating.
2. Almost all households have washing machines;
3. About 25% of the households have dishwashers. These households are called “relatively rich” in the study.

Based on the records of the water accounts of the individual water meters and data provided from NSI the average water demand for each type of households was calculated for the year 2011.

In terms of water appliances efficiency, households are assigned into two categories. The first one is called “relatively rich” households (around 25%). According to the information, gained by the census, they have already implemented water efficiency measures, such as dishwashers or more energy efficient washing machines. It is assumed that the average energy class for dishwashers and washing machines in “relatively rich” households is “A”. The second category is called in the study “relatively poor” (75% of the households). Most of them do not have dishwashers, as declared by them in the national census’ questionnaires. It is assumed that their water appliances are less water efficient. Their washing machines are old, consuming more energy and water. Thus an average energy class “C” is assumed for them.

Based on the above, the scenario is set up as follows:

The group of “Relatively poor” households: It is assumed that 70% of them will replace their old water appliances, inefficient toilets and shower heads with more efficient ones and 30% of them will replace their washing machines with water and energy efficient ones. Expected improvement of the average energy class is from "C" to "A";

The group of "Relatively Rich" households: It is assumed that all of them already have water efficient faucets, low flush toilets and water saving shower heads, but 30% of them will change their washing machines and dishwashers with more water and energy efficient ones. The expected improvement of the average energy class is from "A" to "A+".

18.5 RESULTS

18.5.1 Conceptual Model of Sofia Urban Water System

The conceptual model, which was developed to study the environmental performance of Sofia's urban water system, is presented in Fig. 18.2. Urban water systems have two major external material inflows – energy and reagents. The determination of the system boundary requires clarification of the length of these chains, e.g. should the assessment include the production of these two groups of consumables (with their respective values and environmental impact). In this study two subsystems were considered – foreground system and background system, impacts of which were summed up in the final assessment of the environmental impact of the urban water system (Fig. 18.2). The foreground system follows the main water flow – the abstraction from the nature (water resource), water supply network, users, sewerage system and discharge back to the nature. The background system ends at the urban water system and starts at the production start of the corresponding flow. For example, the energy is a flow, entering as an input for many processes in the foreground system, but the whole energy production cycle is followed in the background system.

Figure 18.2 shows further the 11 environmental impact categories, which were selected in our study. They are divided into two subgroups for this visualization: 1) environmental impact due to resource depletion and 2) environmental impact due to pollution (air, soil, water).

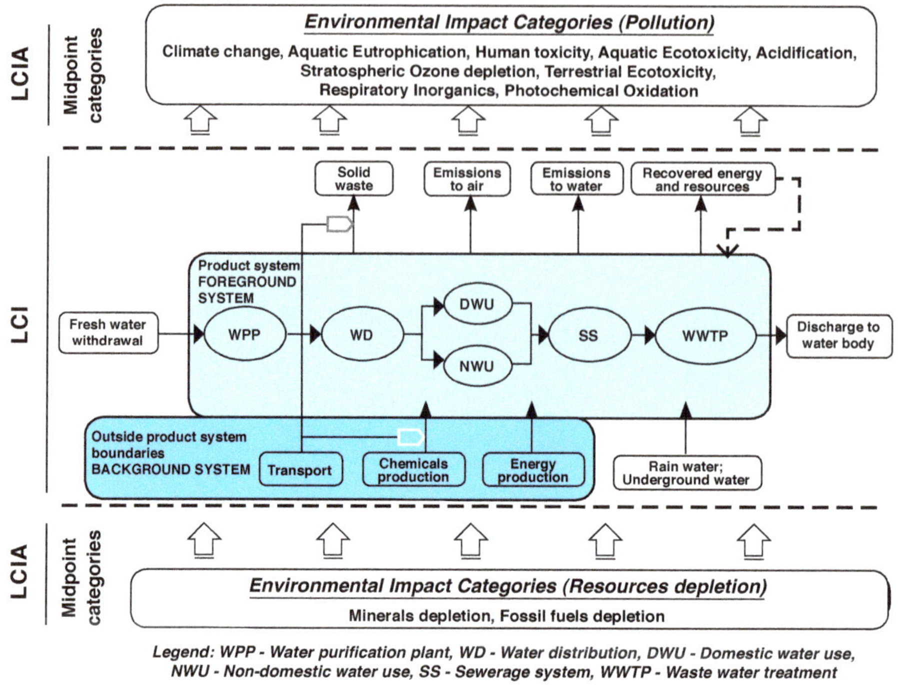

Fig. 18.2. Conceptual model of Sofia's urban system.

18.5.2 Eco-Efficiency of the Solutions in Regard to Their Main Target

In order to overcome the lack of agreed LCA indicator of the quantitative aspect of the water use in urban water systems, the environmental performance of the system is estimated simply based on the indicator "Total amount of the abstracted water", considering that the only difference between baseline and other two scenarios will be exactly this parameter. The rest - number of citizens, households, industry, etc., remains unchanged, according to the way of setting the scenarios. It is obvious that reduction of this indicator means mitigation of a negative environmental impact given the fact that less water will be taken from the nature.

In Fig. 18.3 the calculated values for the environmental performance and for the eco-efficiency of the two scenarios are related to the corresponding baseline's values.

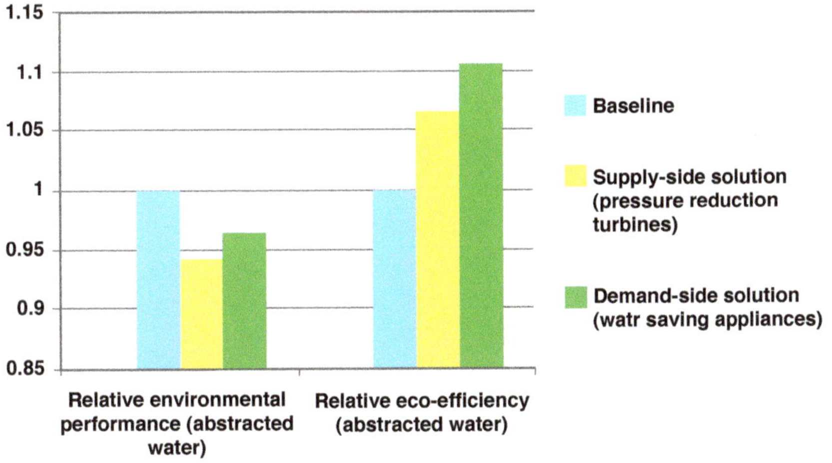

Fig. 18.3. Environmental performance and eco-efficiency for the indicator "Total abstracted water".

Figure 18.3 shows that both scenarios improve their environmental performance (the quantity of the abstracted water decreases in comparison with the baseline scenario). This had been expected, considering that the solutions were chosen to reduce the water use. However, in the time of setting the scenarios, it was not clear which solution will have better effect. Fig. 18.3 shows that supply side solution has better environmental performance in terms of water savings.

The second unclear issue before the calculations was the eco-efficiency performance of the two solutions regarding the quantitative aspect of the freshwater resources' impact. Figure 18.3 shows that that here the trend is opposite – the demand-side solution has better eco-efficiency performance. This means that it shows better relation of the environmental impact and the financial costs for achieving it.

18.5.3 Eco-Efficiency of the Solutions in Regard to Their Co-Benefits

Co-benefits of the solutions were measured by 11 indicators in selected midpoint environmental categories. For visualization of the results, spider diagrams were used. Baseline scenario values were taken for 1 representing the current state of the system. Each of the indicators is presented in one arrow of the spider with its relative value referred to the value in the baseline scenario.

Figure 18.4 shows that for all 11 indicators the environmental impacts were reduced in both scenarios (the scenario's loops are inside the baseline's loop), i.e. both solutions bring environmental co-benefits.

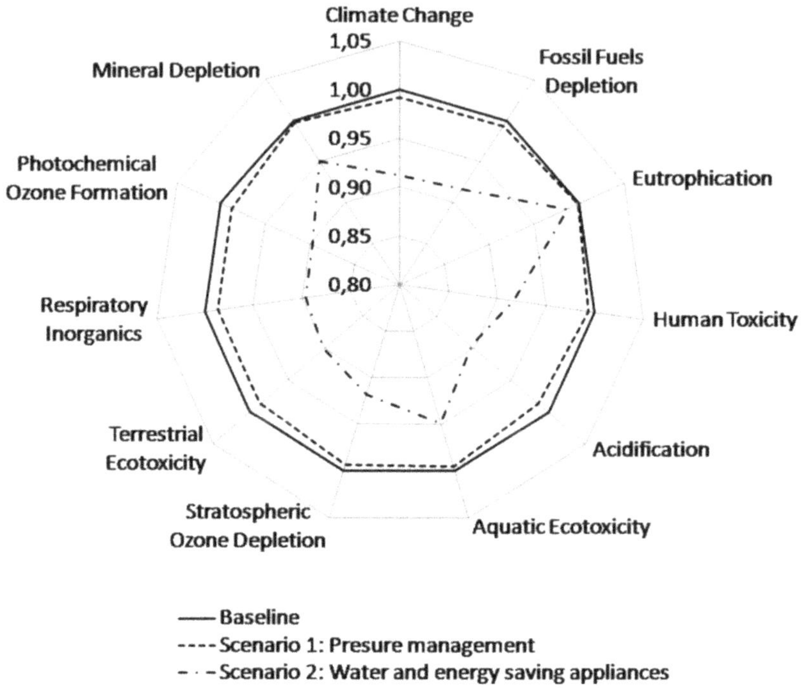

Fig. 18.4. Co-benefits expressed as improved environmental impact of the solutions.

Scenario 1, the supply side solution by pressure reduction turbines mitigates less the negative environmental impact than Scenario 2.
Figure18.5 shows the estimated eco-efficiency indicators, for which the environmental impact was related to the costs according to equation (18.2).

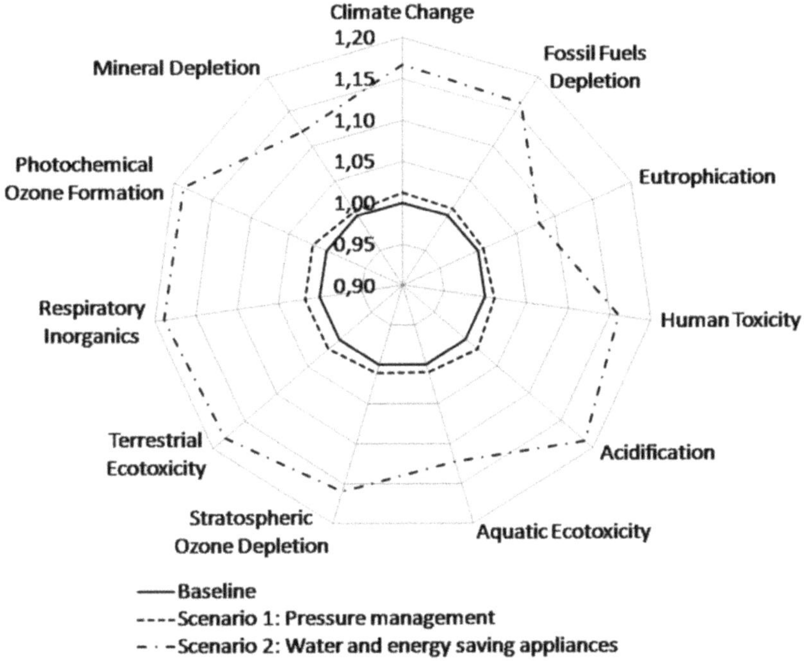

Fig. 18.5. Co-benefits, expressed as improved eco-efficiency of the solutions.

Both solutions improve the eco-efficiency performance of the system (scenarios' loops outside baseline's loop). Similar to the environmental performance, demand side solution shows much better performance.

18.6 DISCUSSION

18.6.1 LCA and Water Use in Urban Water Systems

The quantitative impact of the urban water systems on the freshwater resources is not easy to be assessed with the currently available LCA approaches. The main difficulty is associated with the fact that urban water use does not belong entirely to one of the two types of water use, shown in Fig. 18.1. The share of consumptive and degradative water use depends on the settlement specificities: presence of industry and its type; availability of private gardens in detached houses; presence of city gardens and the water sources for their irrigation, etc.

In the case of Sofia urban water system, the water operator provided data that around 65% of the abstracted water is used by domestic users, rest 35% is used by non-domestic users. With regard to domestic users there are enough data to do a rough estimation of the proportion of the two types of water use, according to Fig. 18.1. With regard to non-domestic users, however, even rough assumption is impossible, because of their heterogeneity and lack of recorded data for the pattern of their water use.

Another non-solved issue is associated with the way of addressing the degradative water use in LCA. Milà i Canals and co-authors suggest that non-evaporative (degradative) water used, which is subsequently returned to the water source does not lead to relevant environmental impacts from a resource perspective, thus should be disregarded in the LCIA phase (Milà i Canals et al, 2009). Although they left their opinion open for practitioners, no opposite opinion was found in the reviewed literature.
In the case of Sofia, the water abstraction and discharge sites are situated at a distance of about 60 km. This means that river flow in this reach is times less than its original flow. So, although non-evaporative in regard to basin scale, water use in Sofia has environmental impact on a particular reach, which should not be disregarded.

18.6.2 The Indicator "Total Abstracted Water"

To compare environmental and eco-efficiency performance of the alternatives with main target on water saving, a simple indicator "total abstracted water" was applied. There might be arguments that it is too simple, but we are convinced that in this particular case it is the most appropriate, due to the facts that: 1) it is reliable, being measurable; 2) it gives direct information of the quantity of the abstracted water and the conclusions are obvious – less water abstracted, better environmental performance; 3) there is no need to relate it to number of citizens, served or any other parameter, because it is a comparative study of the same engineering system and each parameter is related to baseline scenario. So, even if we adopt indicator "water used per citizen or household", given the assumption that there will not be changes of the number of citizens the ratio between baseline and scenario of interest will be the same as compared to indicator "total abstracted water".

The main drawback of this simplistic indicator is that it is not appropriate for comparing different systems, which serve different number of citizens and have different ratio of domestic to non-domestic users. In these cases there is a need for selection of other indicators.

Another drawback might seem to be that it does not count on the availability of the resources in the water basin, e.g. does not distinguish between water abundant and water scarcity regions. But, again, even if some characterization coefficients, counting on regional water availability are applied, the final picture will be the same; because everything is relative in our approach (the ratio will be mathematically simplified).

Last, but not the least, this indicator does not distinguish evaporative from non-evaporative use. But, if needed, provided that data are available, it will be easy to calculate these two proportions.

18.6.3 The Main Target and the Co-Benefits of the Solutions

Two solutions aimed at water saving with co-benefits were compared in regard to: 1) their environmental performance and 2) their eco-efficiency performance.

The environmental performance shows an interesting pattern – the solution with better environmental performance in regard to the main target, shows worse environmental performance in regard to co-benefits (Fig. 18.2 and Fig. 18.3). So, if a decision should be taken based on this analysis only, it would not be an easy task.

It is useful to look deeper into the results of the study in order to explain them. An analysis was done on the contribution to the final results of the background and foreground system as well as of the different stages in the urban water chain. This analysis will be given in detail in a forthcoming paper, but the main conclusions are that the most significant role for the environmental performance of the urban water system play energy flow and the water use stage. It is in compliance with findings of other authors. (Flower et al. 2007 and Gleick et al. 2003). So, our second solution is more co-beneficial, because: 1) it concerns the most significant stage of the urban chain – the users; 2) it is associated with energy savings, which in our calculations are higher than the energy, generated by the PRTs.

However, the assessment of the eco-efficiency emphatically shows that second solution (water saving appliances) should be the preferable ones, if the task is to select only one of them (Fig. 18.2 and Fig. 18.4). The eco-efficiency assessment considers both environmental performance and system value, in our case expressed with monetary terms. Said with other words, for less money, higher environmental impact will be achieved. So, eco-efficiency gives much more information to the decision makers.

18.6.4 Sustainability Aspects in Our Work

Although there are some recent publications, discussing the need for measuring the sustainability with more dimensions, still the widest applied concept is that it has three pillars: social, economic and environment (Seghezzo, 2009). The eco-efficiency tool provides quantification of two of them – economic and environmental performance of the system. Our study showed that there are different results when the selected solutions for water saving are compared only in regard to their environmental impact on fresh water withdrawal and when the costs for achieving the target are included in the comparison (Fig. 18.2, Fig. 18.3 and Fig. 18.4). Eco-efficiency tool is not the only one, which provides such information. In all water infrastructural investment projects cost-benefit analysis are carried out to select the alternative, which is less expensive and equally technically good. What eco-efficiency tool provides more, is the inclusion of environmental performance in the entire life cycle, thus bringing forward the co-benefits of the alternatives in the assessment (Fig. 18.3).

The third pillar of the sustainability, the social one, is not measured by the eco-efficiency tool. But the comparison of the studied alternatives in this aspect does not seem difficult. Alternative 1 (supply- side solution: pressure reducing turbines) does not have direct impact on people's well-being. The second alternative (demand-side: water and energy saving appliances) obviously will increase the living comfort of the citizens.

18.7 CONCLUSIONS

Most of the LCA studies on the quantitative impact on the freshwater resources deal with evaporative water use, most probably because of its significance given that it "disappears" from the basin and is not available for reuse. There are not many LCA studies on urban water systems and there definitely is a lack of generally accepted methodology. Our study shows the importance of studies on the prevailing degradative water use in urban water systems, because urban water use brings environmental co-benefits, which should not be neglected.

We demonstrate that the simple indicator “total volume of the abstracted water” is applicable and sufficient when different alternatives within one urban water system are compared in regard to quantitative impact on freshwater resources. But we conclude that when different systems are the subject of comparison, a methodology should be developed to quantify consumptive and degradative types of water use in the heterogeneous urban water systems and to develop comparable indicators, for example “quantity of water per unit of user”.

This study is one of the first, which applies the recently published eco-efficiency standard ISO 14045 to an urban water system. The results prove that eco-efficiency is a powerful tool for comparison of alternatives, considering two of the three pillars of sustainability.

The conclusions of our work have a very useful practical meaning. Locally, decision makers, researchers and practitioners direct their efforts to find solutions for reducing physical water losses in the water supply system, which are too big (50-60%) compared to the average rate for the EU cities. When only water saving potential is assessed, even simple mathematical calculations could easily show that supply side solutions give better results. However, our study shows that with the assumptions regarding the number of the households which will apply the measures, the demand-side solution is the much better one. It shows better eco-efficiency of the indicator “total volume of the abstracted water” and more significant co-benefits for all 11 indicators for air, water, soil pollution and natural resource depletion.

18.8 ACKNOWLEDGEMENTS

This work was supported financially by the European Commission, 7th Framework program, EcoWater project, Grand agreement N 282882. The authors thank Sofiyska voda AD for the fruitful discussions and data providing.

18.9 REFERENCES

Bayart, J-B., Bulle, C., Deschênes, L., Margni, M., Pfister, S., Vince, F., and Koehler, A. 2010. A framework for assessing off-stream freshwater use in LCA. *Int J Life Cycle Assess* 15(5):439–453.

Boulay, A-M., Bouchard, C., Bulle, C., Deschênes, L., and Margni, M. 2011a. Categorizing water for LCA inventory. *Int J Life Cycle Assess* 16 (7):639–651.

Boulay, A-M., Bulle, C., Bayart, J-B., Deschênes, L., and Margni, M. 2011b. Regional characterization of freshwater use in LCA: modeling direct impacts on human health. *Environ Sci Technol* 45(20): 8948–8957.

Brooks, D.B., Brandes, O.M., and Gurman, S., eds. 2009.Making the most of the water we have: the soft path approach to water management. London: Earthscan.

Communication from the Commission to the European Parliament and the Council, 2007 (414 final). Addressing the challenge of water scarcity and droughts in the European Union, http://eur-lex.europa.eu/LexUriServ/LexUriServ.do?uri=COM:2007:0414:FIN:EN:PDF.

Communication from the Commission to the European Parliament, the Council, the European economic and social committee and the committee of the regions, 2012 (673 final). A Blueprint to Safeguard Europe's Water Resources, European Environment Agency, 2012. Towards efficient use of water resources in Europe, EEA Report No 1/2012, ISSN 1725-9177.

Council Directive concerning urban wastewater treatment, Directive 91/271, 21 May 1991, Official journal of the EU Communities, L 135/40.

EEA. 2003. Dobris report - Environmental Assessment Report, *European Environmental Agency*, 10.

Flower, DJM., Mitchell, VG., and Codner, GP., 2007. Urban water systems: Drivers of climate change?, http://search.informit.com.au/documentSummary;dn=889378592864493;res=IELENG.

Frankelius, P. 2009. Questioning two myths in innovation literature, *Journal of High Technology Management Research*, Vol. 20 (1): 40–51.

Gleick, P.H., Haasz, D., Henges-Jeck, C., Srinivasan, V., Wolff, G., Cushing, K.K., and Mann, A., 2003.Waste not, want not: the potential for urban water conservation in California. Oakland, CA: Pacific Institute.

Gleick, P., Christian-Smith, J., Cooley, H. 2011. Water-use efficiency and productivity: rethinking the basin approach, *Water International*. Vol. 36 (7): 784–798.

ISO 14045: 2012. Environmental management - Eco-efficiency assessment of product systems - Principles, requirements and guidelines", CEN.

ISO 14040: 2006. Environmental management - Life cycle assessment - Principles and framework. Second edition. International Organization for Standardization, Genève.

ISO 14044: 2006. Environmental management - Life cycle assessment - Requirements and guidelines. International Organization for Standardization, Genève.

Hamilton, K. and S. Sameer Akbar, 2010. Assessing the Environmental Co-Benefits of Climate Change Actions, 2010 Environmental strategy, The World bank group

Koehler, A. and D. Thylmann. 2012. Introduction to Water Assessment in GaBi Software, Version 1.0 November 2012, PE International AG, Germany.

Milà i Canals, L., Chenoweth, J., Chapagain, A., Orr, S., Antón, A., and Clift, R. 2009. Assessing freshwater use impacts in LCA: part I—inventory modelling and characterisation factors for the main impact pathways. *Int J Life Cycle Assess* 14(1):28–42.

Miyatsuka, A. and E. Zusman. 2010. Fact Sheet No.1 What are Co-benefits? Asian co-benefits partnership.

Pfister, S., Koehler, A., and Hellweg, S. 2009. Assessing the environmental impacts of freshwater consumption in LCA. *Environ Sci Technol* 43(11):4098–4104.

Ribarova, I. 2009. Sharing water under conditions of scarcity and stress, in "Water Stress Mitigation: The Aquastress Case Studies". In Assimacoppoulos, D. (Ed). Aiga publishing, 84-99.

Ribarova, I., Stanchev, P., Dimova, G., and Assimacopoulos, D. 2014. A first iteration of an eco-efficiency assessment of Sofia's urban water system, *Procedia engineering*, 70: 1411-1420.

Seghezzo, L. 2009. The five dimensions of sustainability, *Environmental Politics*, vol.18, N4: 539-556.

Strand, J. and M. Toman. 2010. "Green Stimulus", Economic Recovery, and Long-Term Sustainable Development, Policy research working paper 5163, The World Bank Development Research Group Environment and Energy Team.

CHAPTER 19

ASSESSMENT OF THE IMPACT OF CLIMATE CHANGE TO ALLUVIAL AQUIFER: AN UNCERTAINTY ANALYSIS

Emanuele Romano[1], Stefania Camici[2], Luca Brocca[2], Tommaso Moramarco[2], Federico Pica[1], Elisabetta Preziosi[1]

[1] CNR - Water Research Institute, Area della Ricerca di Roma 1-Montelibretti, via Salaria km 29,300, PB 10 – 00015 Monterotondo Stazione, Rome, Italy

[2] CNR - Research Institute for Geo-Hydrological Protection, via della Madonna Alta 126, 06128, Perugia, Italy

ABSTRACT

In this paper, an uncertainty analysis of the processing chain from General Circulation Models scenarios to impact on an alluvial aquifer is presented. To this goal, we examined the forecast climate change impact on the Umbra Valley aquifer (central Italy), in which the human pressures in terms of groundwater usage are significant. In order to understand which part of the modelling processing chain mainly contributes to the final uncertainty we considered the combined effects of: 1) three different General Circulation Models; 2) two different downscaling procedures; 3) the stochastic generation of daily T and P time series; 4) the parameterization of the Thornthwaite-Mather soil water balance model, adopted to estimate the effective infiltration to the aquifer; 5) the uncertainty arising from the procedure adopted to calibrate/validate a groundwater flow model developed for the study area using the MODFLOW2005 code. The final results show that the uncertainty due to the groundwater flow model calibration/validation is comparable to that arising from the whole processing chain from the GCM choice to the effective infiltration estimates. Moreover, while the uncertainty on the effective infiltration estimate appears to be centered around the mean of the scenarios, the uncertainty on the piezometric heads presents an asymmetry clearly related to the non-linearity of the aquifer system response to external drivers: this issue should be carefully taken into account when the impact of climate change on groundwater is assessed.

19.1 INTRODUCTION

Despite the critical importance of groundwater resources in many areas of the planet, researches dealing with the possible impacts of climate change on groundwater recharge, and definitely on groundwater availability, appeared only recently in the international literature (see the reviews of Green et al. 2011; Kumar, 2012; Taylor et al. 2013), based on ground monitoring and/or on local or global scale modelling (Allen et al. 2004; Scibek and Allen, 2006; Crosbie et al. 2010; Dams et al. 2010; Oude Essink et al. 2010; Rozell and Wong, 2010; Green et al 2011). A growing body of literature is now providing evidence on the combined effects of changes in precipitation and temperature on the groundwater recharge in many part of the world (Green et al 2011; Ali et al 2012) especially in those areas in which negative trends of the precipitation and an increased frequency of drought are currently observed, including the Mediterranean area (Pachauri and Reisinger, 2007; Goderniaux et al 2009; IPCC, 2012; Romano and Preziosi, 2013). The exploitation of groundwater may mitigate droughts, because many aquifers have a large storage capacity and are potentially less sensitive to climate change than surface water bodies, which often rely on groundwater discharge to maintain their baseflow (Dragoni and Sukhija, 2008). As a consequence, it is of considerable interest to assess the effects of the possible precipitation decline combined with the increasing temperature on groundwater availability.

To this goal, a chain of models is often used (Green et al. 2011; Kumar, 2012): scenarios of recharge to the aquifer are generated through infiltration models based on scenarios of temperature and precipitation issuing from General Circulation Models scenarios, possibly after statistical/dynamical downscaling to local observations (Brouyere et al 2004; Goderniaux et al 2009; Guyennon et al. 2013). Groundwater flow models, typically calibrated and validated on the past hydrometeorological time series and/or observed piezometric heads, are forced by GCM-dowscaled scenarios to forecast possible future trends of groundwater heads and aquifer discharge.

However, the assessment of climate change impacts on hydrogeological and hydrological systems is affected by the uncertainty in the modelling chain, including: a) the variability of the climate system; b) the model structure and parameterizations at different spatial and temporal scales; c) the downscaling techniques and bias correction methods; d) the choice of future climate scenarios (Senatore et al 2011); e) the choice and parameterizations of the infiltration model; f) the goodness of the groundwater flow and modeling calibration.

In addition to this, one should also consider the so called indirect impacts of climate change, resulting from human intervention in water systems (Green et al. 2011), that are expected to have the largest short-term effects on groundwater resources (Holman, 2006; Zhou et al. 2010; Taylor et al. 2013). Typical examples of indirect impacts on groundwater are the increased abstractions particularly for irrigation. For these reasons, the evolution of groundwater heads may not reflect the effects of climate change alone but the sum of climate effects and human abstractions. Within this framework, the assessment of the final uncertainty on the piezometric heads is crucial for water managers which should be aware of the extent of such uncertainty. Planning for future water exploitation and distribution in the next decades could head in the wrong direction, should this uncertainty be underestimated. On the other hand, to understand which part of the modelling chain from GCM to groundwater flow modeling mainly contribute to the final uncertainty, is crucial to address the efforts of the scientific community for a better comprehension of the whole hydrologic cycle in a context of climate change. Concerning this point, there is not a general consensus.

Green et al. (2011) underline the need for a better estimation of the responses of hydrological systems to projected climate changes while individuating the greatest source of uncertainty in the estimation of projected climate, due to the large variability among GCMs. The same groundwater flow model may provide divergent results in terms of discharge variation/ heads decline depending on the GCM choice, downscaling technique adopted, goodness of the fit of the groundwater flow model. Holman et al. (2009) found that most of the uncertainty in evaluating the impact of climate change on groundwater resources arise from the choice of downscaling methods. Also Stoll et al. (2011) pointed out "*the weaknesses of the downscaling methods which directly influence the predicted hydrological fluxes, and thus also the predicted groundwater levels*"

In this paper we examine the forecast of climate change impact on the Umbra Valley aquifer, a large alluvial aquifer in central Italy (about 330 km^2), in which the human pressures are significant both in terms of groundwater usage (drinking water and irrigation) and input of chemicals.

In order to understand which part of the modelling processing chain from GCM to groundwater modelling mainly contributes to the final uncertainty, we consider the combined effect of: 1) three different General Circulation Models (Section 19.3); 2) two different downscaling procedures (Section 19.4.1); 3) the stochastic generation of hourly T and P time series (Section 19.4.2); 4) the parameterization of the soil water balance model; 5) the uncertainty arising from the procedure adopted to calibrate/validate a groundwater flow model developed for the study area using the MODFLOW2005 code (Harbough, 2005). The whole processing chain analyzed in this study is shown in Fig. 19.1. The final uncertainty is assessed in terms of head variation in the groundwater model.

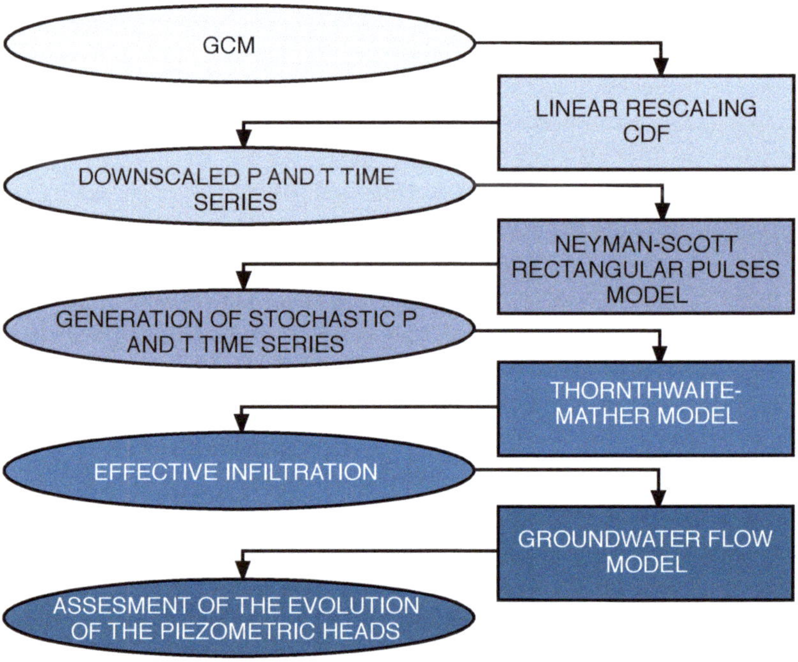

Fig. 19.1. Processing chain analyzed in this study.

In the uncertainty analysis presented in this paper, the indirect impacts of climate change on groundwater are neglected. In other words, we assume that the exploitation system (demand, infrastructure, etc.) is constant in time and independent from the actual climatic conditions.

19.2 THE CASE STUDY

The Umbra Valley aquifer (approximately 330 km^2) is located in the Umbria Region (Central Italy), in the upper Tiber river basin, between the Martani western ridges, the Subasio Mount and Foligno-Spoleto eastern ridges. It is an alluvial aquifer consisting of Pleistocene-Olocene alluvial deposits, that lie on a marly-arenaceous substrate of Miocene age, bordered to the South, West and North-East by lacustrine and fluvio-lacustrine Villafranchian deposits (Fig. 19.2). It is a mono-layer phreatic system in the northern and southern part while in its central part a thick layer of clay deposits divides a deep confined aquifer from a shallow phreatic aquifer. The whole aquifer is mainly fed by zenithal recharge; local inflow from the boundary ridges, as well as losses from the rivers crossing the plain, contributes to the total recharge.

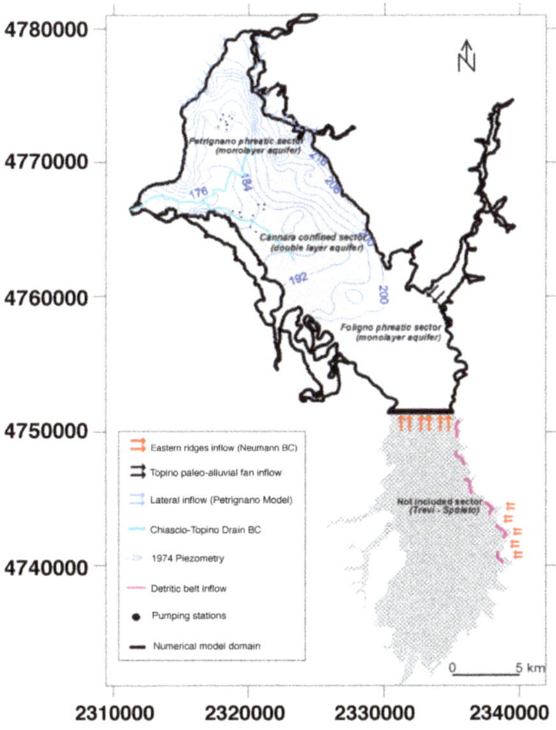

Fig. 19.2. Hydrogeological scheme of the Umbra valley aquifer. The piezometric map refers to the undisturbed condition, prior to the exploitation through the well fields of Petrignano d'Assisi and Cannara.

The whole aquifer is exploited for human consumption through two well fields (Fig. 19.2), one (Petrignano d'Assisi) operating since the 1970s and exploiting the phreatic sector located in the North of the valley and the second (Cannara) operating since the 1980s and exploiting the confined sector located in the central part. Few other wells, located to the south with respect to the Cannara pumping station (the Cantone and S. Pietro 1 and 2 wells) are used for human consumption. Groundwater is also abstracted through private wells for irrigation purposes. For a brief description of the conceptual model of the aquifer and of the flow model refer to Section 19.6.

In the study area, a dense (about 83 stations) hydrometeorological monitoring network has provided daily time series at the Tiber basin scale for the period 1957-2012 as well as hourly time series of rainfall and temperature for the period 1989–2012. A negative trend in the 1957-2012 total annual precipitation has been detected, with values passing from 1050 to 740 mm (30% reduction). Mean and standard deviation of precipitation have been estimated about 895 and 190 mm, respectively, for the period 1961-1990. The decreasing trend began roughly in the early 1990s and it is superimposed to a cyclic behaviour with a period of approximately 15 years, related to the *NAO* oscillations (Romano and Preziosi, 2013). Such variations strongly impact both on the surface hydrological regime (Romano et al. 2013) and on the groundwater regime (Preziosi and Romano, 2013).

19.3 GENERAL CIRCULATION MODELS

Three GCMs have been chosen for this study from the CMIP3 database (http://pcmdi9.llnl.gov/esgf-web-fe/, Table 19.1): the Parallel Climate Model (PCM), the Community Climate System Model, version 3.0 (CCSM3) and the ECHAM5 model. In the following analysis, daily rainfall and temperature data for two different scenarios and periods are used: (i) the 20C3M scenario (as defined in the Special Report Emission Scenarios, SRES) for the baseline period, 1961 to 1990 and (ii) the SRES A2 scenario for the future period 2010-2099. The A2 scenario has been chosen among the four basic storylines of SRES, as a reasonable "worst case" scenario, identifying the future world as heterogeneous and more concerned for economic growth than environmental aspects (Nakicenovic et al. 2000). However, it should be noted that more recent studies (Dettinger, 2012) have defined the A2 scenarios as representative of middle-of-the-road-change projections.

Table 19.1. Main characteristics of the investigated GCM models. μ and σ refer to the mean and standard deviation of precipitation.

Model	Research Centre	Nation	Grid size	1961-1990 annual precipitation for the Umbra Valley area μ (σ) [mm]	2010-2070 annual precipitation for the Umbra Valley area μ (σ) [mm]	1961-1990 observed P data for the Umbra Valley area μ (σ) [mm]
PCM	National Centre Atmospheric Research	USA	2.8° x 2.8°	690 (110)	670 (115)	895 (190)
CCSM3	National Centre Atmospheric Research	USA	1.4° x 1.4°	560 (118)	554 (105)	
ECHAM5	Max Planck Institute for Meteorology	Germany	1.87 x 1.87	435 (108)	403 (102)	

The main characteristics of the selected GCMs are listed in Table 19.1 along with the mean and standard deviation of annual cumulated precipitation extracted from pixels covering the Umbra Valley study area for the present (1961-1990) and future period (2010-2070). The three models share a general tendency to underestimate the observed annual precipitation, both in term of mean and standard deviation, suggesting the need of applying a downscaling procedure prior to use GCM data. Table 19.2 shows observed temperature data for the period 1961-1990 and the averaged mean variations, ΔT, provided by the GCMs for future periods.

Table 19.2. Observed temperature data for the period 1961-1990 and variations, ΔT, provided by the GCMs for future periods. μ and σ refer to the mean and standard deviation temperature.

Observed T data for the Umbra Valley area (1961-1990) μ (σ)	Model	ΔT 2010-2039 period	ΔT 2040-2069 period	ΔT 2070-2099 period
14 °C (7.45)	**PCM**	+0.80 °C	+1.50 °C	+2.50 °C
	CCSM3	+1.58 °C	+2.80 °C	+4.47°C
	ECHAM5	+1.20 °C	+2.35 °C	+3.74 °C

19.4 DOWNSCALING PROCEDURE AND STOCHASTIC GENERATION OF PRECIPITATION AND TEMPERATURE TIME SERIES

The temperature and precipitation time series from the three considered GCMs (PCM, CCSM3 and ECHAM5) have been downscaled using two different downscaling procedures for precipitation (thus obtaining six different time series of P), and one single procedure for temperature (Section 19.4.1). Both daily precipitation and temperature have been downscaled with respect to one single station, whose data have been provided by the Regional Agency for Environmental Protection (ARPA Umbria): the station of Bastia for precipitation (1960-2008) and the station of Petrignano d'Assisi for temperature (1960-2009). We shall refer to these downscaled time series as *GCM+DSC*.

Then, for each time series the Neyman-Scott Rectangular Pulses (NSRP, Cowpertwait et al. 1996) model is used for stochastic rainfall generation, while the Fractionally Differenced ARIMA (FARIMA, Montanari et al. 1997) model is used for temperature simulations. Through the stochastic models, for each downscaled time series, 100 hourly scenarios, statistically equivalent, have been generated (Section 19.4.2). We shall refer to such time series as *GCM+DSC+STOC*.

These scenarios have been analysed in terms of cumulative density function of the mean annual temperature and cumulative annual precipitation, in order to infer the uncertainty arising from the processing chain *GCM+DSC+STOC* (Section 19.4.3).

19.4.1 Downscaling Procedure

To obtain data for the future period 2010-2070 compatible with the climate of the study area, two statistical downscaling procedures, the linear rescaling (LINRESC) method and the quantile mapping (CDF) method, are applied to rainfall, P, whereas the delta change method is applied to temperature, T.

19.4.1.1 Linear rescaling method (LINRESC)

The daily variables provided by GCMs for the baseline period are modified in order to fit their annual mean and standard deviation to the ones of the daily observed data, OBS^{basper}, for the same baseline period. Then, the same changing factors are applied to the future GCM predictions as:

$$GCM_{rescal}^{fut} = \left(GCM^{fut} - \overline{GCM^{basper}}\right) \bullet \frac{\sigma(OBS^{basper})}{\sigma(GCM^{basper})} + \overline{OBS^{basper}} \qquad (19.1)$$

where σ is the standard deviation operator and OBS^{basper} are the observed data for the baseline period.

19.4.1.2 Cumulative Distribution Function method

The Cumulative Distribution Function (CDF) method (also known in literature as Quantile Mapping method, Wood et al. 2004; Ines and Hansen 2006; Dequeé et a., 2007; Li et al. 2010; Piani et al. 2010; Themeßl et al. 2011; Gudmundsson et al. 2012; Guyennon et al. 2013; Camici et al. 2014) is a relatively simple approach that has been successfully used in hydrologic and climate impact studies. This method allows correcting the GCM data mapping distribution onto that of the observed data. For the time series provided by GCMs for the baseline period, the method can be written as:

$$P\_GCM_{k,QM}^{basper} = F^{-1}_{P\_OBS_K^{basper}}\left(F_{P\_GCM_K^{basper}}\left(P\_GCM_k^{basper}\right)\right) \quad k = 1,...,12 \qquad (19.2)$$

where $F^{-1}_{P\_OBS_k^{basper}}$ is the inverse empirical CDF (or quantile function) of the observations for the month k, $F_{P\_GCM_k^{basper}}$ is the empirical CDF of the GCM data, both computed for the baseline period. Then, for a given percentile in the month k, the ratio R_k between the empirical CDFs of the GCM and observed value for the baseline period can be written as (Li et al. 2010):

$$R_k\left[F_{P\_GCM_k^{fut}}\left(P\_GCM_k^{fut}\right)\right] = \frac{F^{-1}_{P\_OBS_k^{basper}}\left(F_{P\_GCM_k^{fut}}\left(P\_GCM_k^{fut}\right)\right)}{F^{-1}_{P\_GCM_k^{basper}}\left(F_{P\_GCM_k^{fut}}\left(P\_GCM_k^{fut}\right)\right)} \qquad (19.3)$$

where $F^{-1}_{P\_GCM_k^{fut}}$ is the empirical CDF of the GCM data in the future period.

Thus, assuming the hypothesis of stationarity of the ratio R_k also for the future period, the corresponding time series can be computed as:

$$P\_GCM_{k,QM}^{fut} = P\_GCM_k^{fut} \bullet R_k \qquad k = 1,...,12 \qquad (19.4)$$

19.4.1.3 The Delta Change method

The well-known Delta Change (DC, Hay et al. 2000; Camici et al. 2014) method is based on the calculation of a monthly change factor for temperature, $\delta_{T,\kappa}$, as follows:

$$\delta_{T,k} = \overline{T\_GCM_k^{fut}} - \overline{T\_GCM_k^{basper}} \qquad k = 1,...,12 \qquad (19.5)$$

where $T\_GCM_k^{basper}$ and $T\_GCM_k^{fut}$ represent the daily temperature time series provided by GCMs for the baseline (basper) and the future period (fut), respectively; the overbar is the monthly average of time series for the *k* month. Then, the daily time series of temperature for the future period, $T\_OBS_{k,DC}^{fut}$, are estimated by adding the observed time series to the respective change factor, $\delta_{T,k}$:

$$T\_OBS_{k,DC}^{fut} = T\_OBS_k^{basper} + \delta_{T,k} \qquad (19.6)$$

where $T\_OBS_k^{basper}$ refer to the observed daily data of temperature recorded within the study area in the baseline period. The main drawback of this procedure is that the future and baseline scenarios differ only in terms of their respective means while all other statistical properties of the data, such as the variance or skewness, remain unchanged.

Summarizing, the Delta Change method allows obtaining, for the future period, daily climate time series by changing only the monthly mean of the observed data; the Linear Rescaling provides future daily time series in which both the mean and standard deviation are modified according to the actual observations, whereas the Cumulative Distribution Function method approach modifies all the statistical properties of the GCM daily time series such as the variance and skewness, besides the mean.

19.4.2 Stochastic Generation of Temperature and Precipitation Time Series

To obtain long time series of rainfall and temperature with hourly resolution, stochastic generation models are applied. Specifically, the Neyman-Scott Rectangular Pulse (NSRP, Cowpertwait et al. 1996) model is applied for rainfall generation, while the ARIMA model with fractional differentiation (FARIMA, Montanari et al. 1997) is used for temperature simulation.

Briefly, the NSRP model is characterized by a flexible structure in which the model parameters broadly relate to underlying physical features observed in rainfall fields. The model has a total of five parameters, estimated through sampling statistics computed from the observed data. The main feature of the model is its ability to preserve statistical properties of a rainfall time series over a range of time scales. Full details of the NSRP may be found in Cowpertwait et al. (1996). Based on NSRP model, 100 non-stationary hourly rainfall time series of length 60 years have been generated. For details refer to Camici et al. (2014).

The FARIMA model is able to fit the autocorrelation function characterized by a slow decay suggesting the presence of long-term persistence. This dependence was detected in many time series of hydrological data and, very often, in the air temperature series. The procedure for the implementation of the FARIMA model is not straightforward, particularly in the identification phase for the preliminary evaluation of model parameters. The method employed in this study is the one suggested by Montanari et al. (1997). Similarly, 100 temperature time series of length 60 years were generated.

19.4.3 Statistical Analysis of the Temperature and Precipi-tation Scenarios

The scenarios of hourly temperature and precipitation generated through the stochastic procedure described in the Section 19.4.2 (time series *GCM+DSC+STOC*) have been aggregated at annual scale for the period 2010-2070. For each time series, the annual (mean for temperature and cumulative for precipitation) have been computed and the resulting probability distribution has been analysed in terms of cumulative density function. The analysis allowed us to infer the uncertainty arising from the downscaling procedure for each variable.

In the Tables 19.3 and 19.4, the values of the 5th, 50th and 95th percentile, for temperature and rainfall respectively, are shown for each downscaled scenario. The variation of the mean values with respect to the mean of the period 1957-2012 is also reported.

Table 19.3. 5th, 50th and 95th percentile of the mean annual temperature (expressed in °C) for each climate scenario and related variation expressed as difference with respect to the mean annual temperature of the period 1957-2012.

TEMPERATURE							
	PCM		CCSM3		ECHAM5		MEAN 1957-2012
	T (°C)	ΔT (°C)	T (°C)	ΔT (°C)	T (°C)	ΔT (°C)	
5th	15.4	+1.5	16.7	+2.8	16.3	+2.4	
50th	15.9	+2	16.9	+3	16.5	+2.6	13.9
95th	16.4	+2.5	17.5	+3.6	16.6	+2.7	

Table 19.4. 5th, 50th and 95th percentile of the mean annual precipitation (expressed in mm) for each climate scenario and related variation expressed as percentage with respect to the mean annual precipitation of the period 1957-2012.

PRECIPITATION							
	PCM		CCSM3		ECHAM5		MEAN 1957-2012
	CDF	LINRESC	CDF	LINRESC	CDF	LINRESC	
5th	816	877	851	862	798	750	
% ΔP	-3.2 %	+4.0 %	+0.9 %	+2.3 %	-5.3 %	-11.0 %	
50th	852	909	891	907	826	787	843
% ΔP	+1.1 %	+7.8 %	+5.7 %	+7.6 %	-2.0 %	-6.6 %	
95th	893	946	934	938	857	825	
% ΔP	+5.9 %	+12.2 %	+10.8%	+11.3 %	+1.7 %	-2.1 %	

The results presented in Tables 19.3 and 19.4 can be summarized as follows:

1. The increasing trend for temperature projected by the three GCMs, is confirmed also after the stochastic generation procedure.

2. The variability among stochastic generated temperature time series (estimated through the difference between the 5th and the 95th : 1.0 °C, 0.8 °C and 0.3 °C for PCM, CCSM3 and ECHAM5, respectively) appears to be comparable to the variability among the GCM scenarios. Indeed by comparing the 50th percentile, the difference in the mean annual temperature is equal to 1.0 °C between PCM and CCSM3 and 0.4 °C between CCSM3 and ECHAM5. As a consequence, from the temperature scenarios, the uncertainty arising from the downscaling procedure is comparable to that one arising from the GCMs.

3. Concerning the precipitation, the climate signal projected by the three GCM is not unique:

   - CCSM3 indicate an increasing trend, regardless the adopted DSC method and for all the stochastic scenarios;

   - ECHAM5 forecast a decreasing trend for precipitation, no matter the stochastic generation if the LINRESC downscaling procedure is adopted and for the 85% of the time series if the CDF is adopted;

   - the PCM CDF time series indicates a tendency to increasing precipitation, although some time series stochastically generated show a decreasing trend, while the PCM model downscaled through the LINRESC procedure indicates a clear tendency to increase.

4. The uncertainty arising from the downscaling procedure appears to be quite high: if we consider the 50th percentile, the difference, ΔP, with respect to the annual precipitation averaged over the GCM (862 mm) ranges between -6.60% (ECHAM5 LINRSC) and +7.8% (PCM LINRSC). This uncertainty is

comparable to the variability arising from the stochastic generation: the difference between the 5th and the 95th is in the order of 5% for each downscaled scenario (the percentage is computed with respect to the correspondent 50th percentile). Thus also for precipitation the uncertainty arising from the downscaling procedure is comparable to that one arising from the GCMs.

19.5 EFFECTIVE INFILTRATION SCENARIOS

In order to evaluate the recharge to the aquifer, the Thornthwaite-Mather model (1951) has been applied using all the *GCM+DSC+STOC* time series of precipitation and temperature generated through the procedure described in Section 19.4.

The Thornthwaite-Mather model adds a new element of non-linearity to the processing chain presented in Section 19.2 (Fig. 19.1): a) the relation between temperature and potential evapotranspiration is non-linear; b) the generation of runoff takes place when the soil is saturated, i.e. when the field capacity is reached due to net precipitation (precipitation minus potential evapotranspiration). As a consequence a n% of decreasing precipitation and a m% of increasing temperature does not result in a net variation of (m-n)% of effective infiltration.

In order to consider for each downscaled scenario all the possible combination of temperature and precipitation, the 5th, 50th and 95th of the stochastically generated time series *GCM+DSC+STOC* have been taken into account. The field capacity considered to compute the effective infiltration is the same for all the scenarios and has been set equal to 200 mm.

The mean annual effective infiltration scenarios I_{eff} are shown in Table 19.5, adopting the following notation:

- **Column 1** (SCENARIO): P5_T5_PCM_CDF refers to the effective infiltration computed considering the 5th percentile of the precipitation and temperature time series obtained from the PCM model and downscaled through the CDF method. Similarly for the other notation.

- **Column 2** (MEAN I_{eff} 1957-2012): mean annual effective infiltration computed considering the observed temperature and precipitation.
- **Column 3** ($\overline{GCM}$): mean annual effective infiltration computed considering the mean of the three GCM scenarios.

Table. 19.5. Summary of the mean annual effective infiltration scenarios.

<table>
<tr><th colspan="9">MEAN ANNUAL EFFECTIVE INFILTRATION</th></tr>
<tr><th>1</th><th>2</th><th>3</th><th colspan="2">4</th><th colspan="2">5</th><th colspan="2">6</th></tr>
<tr><th>SCENARIO</th><th>MEAN Ieff 1957-2012</th><th>MEAN GCM</th><th colspan="2">GCM</th><th colspan="2">GCM+DSC</th><th colspan="2">GCM+DSC+STOC</th></tr>
<tr><td></td><td>mm</td><td>mm</td><td>mm</td><td>%</td><td>mm</td><td>%</td><td>mm</td><td>%</td></tr>
<tr><td>P5_T95_PCM_CDF</td><td rowspan="12">335</td><td rowspan="12">326</td><td rowspan="4">325</td><td rowspan="4">-0.3%</td><td rowspan="2">308</td><td rowspan="2">-5.5%</td><td>284</td><td>-12.9%</td></tr>
<tr><td>P95_T5_PCM_CDF</td><td>337</td><td>3.4%</td></tr>
<tr><td>P5_T95_PCM_LINRESC</td><td rowspan="2">341</td><td rowspan="2">4.6%</td><td>314</td><td>-3.7%</td></tr>
<tr><td>P95_T5_PCM_LINRESC</td><td>377</td><td>15.6%</td></tr>
<tr><td>P5_T95_CCSM3_CDF</td><td rowspan="4">337</td><td rowspan="4">3.4%</td><td rowspan="2">326</td><td rowspan="2">0.0%</td><td>297</td><td>-8.9%</td></tr>
<tr><td>P95_T5_CCSM3_CDF</td><td>376</td><td>15.3%</td></tr>
<tr><td>P5_T95_CCSM3_LINRESC</td><td rowspan="2">347</td><td rowspan="2">6.4%</td><td>301</td><td>-7.7%</td></tr>
<tr><td>P95_T5_CCSM3_LINRESC</td><td>367</td><td>12.6%</td></tr>
<tr><td>P5_T95_ECHAM_CDF</td><td rowspan="4">316</td><td rowspan="4">-3.1%</td><td rowspan="2">325</td><td rowspan="2">-0.3%</td><td>313</td><td>-4.0%</td></tr>
<tr><td>P95_T5_ECHAM_CDF</td><td>339</td><td>4.0%</td></tr>
<tr><td>P5_T95_ECHAM_LINRESC</td><td rowspan="2">306</td><td rowspan="2">-6.1%</td><td>281</td><td>-13.8%</td></tr>
<tr><td>P95_T5_ECHAM_LINRESC</td><td>342</td><td>4.9%</td></tr>
</table>

- **Column 4** (*GCM*): mean annual effective infiltration computed considering the mean scenario of each single GCM (PCM, CCSM3 and ECHAM5). The percentage variations are given with respect to the $\overline{GCM}$.
- **Column 5:** mean annual effective infiltration averaged over each single GCM downscaled using a given method (CDF or

LINRSC). The percentage variations are given with respect to the $\overline{GCM}$.

- **Columns 6:** mean annual effective infiltration computed through the complete chain processing *GCM+DSC+STOC* generation. The two values reported at column 6 refer to the extreme time series for a given GCM and downscaling method: for instance, the time series P5_T95_PCM_CDF refers to the PCM general circulation model, downscaled with the CDF method, taking into account the 5th percentile of the stochastically generated time series of precipitation and the 95th percentile of the stochastically generated time series of temperature. Similarly for the other notations. The percentage variations are given with respect to the $\overline{GCM}$.

Results presented in Table19.5 show that:

1. The trend of the effective infiltration depends on the GCM: PCM and ECHAM predict on average a decreasing recharge (from 335 mm to 325 mm and 316, respectively, corresponding to -3.0% and -5.7%); CCSM3 forecasts an almost stationary trend (337 mm, equal to +0.5%).

2. The uncertainty introduced by the GCMs is at maximum in the order of ±3% (column 4).

3. The uncertainty introduced by the downscaling procedure is higher (approximately ±6%). Thus, it appears that considering the processing chain *GCM+DSC*, the uncertainty in the effective infiltration estimate equally arises from the choice of the GCM and from the adopted downscaling procedure.

4. The uncertainty in the effective infiltration estimates introduced by the stochastic generation procedure ranges between +15.6% and - 13.8%, thus more than twice the uncertainty arisen from the first part of the processing chain (*GCM+DSC*). It can be concluded that the uncertainty on the forecast of the effective infiltration doubles for each step of the implemented process, passing from approximately 3% (*GCM*) to 6% (*GCM+DSC*), to ≈15% (*GCM+DSC+STOC*).

In Fig. 19.3, an overview of the estimated uncertainty of the annual effective infiltration from the $\overline{GCM}$ (averaged over all scenarios) to the *GCM+DSC+STOC* is shown.

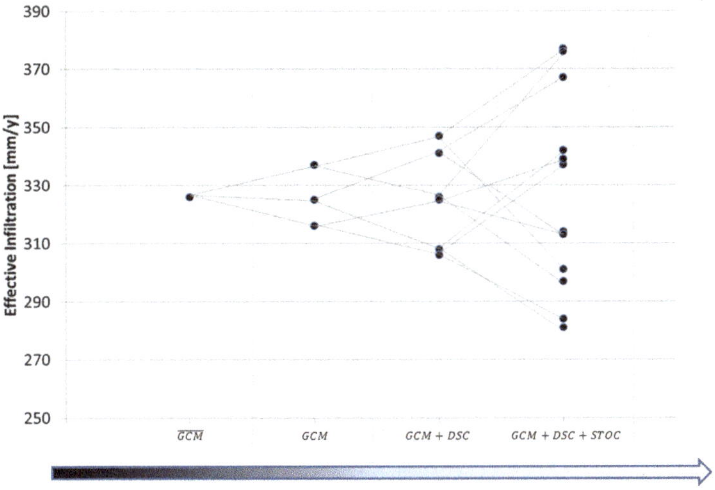

Fig. 19.3. **Effective Infiltration: propagation of the uncertainty through the processing chain from *GCM* to *GCM+DSC+STOC*. The values of I_{eff} are expressed in mm/y.**

19.6 AQUIFER MODELING

In order to assess the possible impact to the Umbra Valley aquifer of the climate change scenarios, a numerical model has been developed and calibrated using hydro-meteorological past observations, both under undisturbed conditions (1974) and under exploitation conditions (1998-2010) using the finite-difference code MODFLOW2005 (Harbough, 2005).

The development of the model and the calibration procedure for past climate conditions is summarized in the following (Romano and Preziosi, 2010; Pica et al. 2013):

1. The bottom of the aquifer, reconstructed through well-log and geophyical surveys, is located about -90 m and -140 m with respect to the ground surface in the central part of the aquifer, in correspondence to the Petrignano d'Assisi and Cannara well field

areas, respectively (Idrotecno – R.p.a., 1974). It rises to approximately -20 m toward the boundaries of the aquifer.

2. The aquifer is mainly fed by the zenithal effective infiltration. As presented in Section 19.5, the mean annual effective infiltration estimated by means of the Thornthwaite-Mather model on the ground of the 1957-2012 observed data is equal to 335 mm.
3. Based on the available piezometric maps, lateral inflow has been hypothesized both on the Petrignano and on the Cannara areas (Fig. 19.4). The total amount of such inflows have been inferred through a calibration procedure.
4. The Chiascio river crossing the northern sector of the plain, acts as a drain in its lower course, where it is in direct hydraulic contact with the aquifer. Discharge measurements performed in 2008 estimated an inflow of approximately 0.2 m^3/s.
5. The main rivers crossing the Cannara and Foligno sector of the valley are the Topino, the Ose and the Clitunno rivers (Fig. 19.4). Discharge measurements carried out on the Topino river suggest a flow from the aquifer to the surface water bodies seasonally ranging from 0 to 1 m^3/s in the central sector of the Valley.

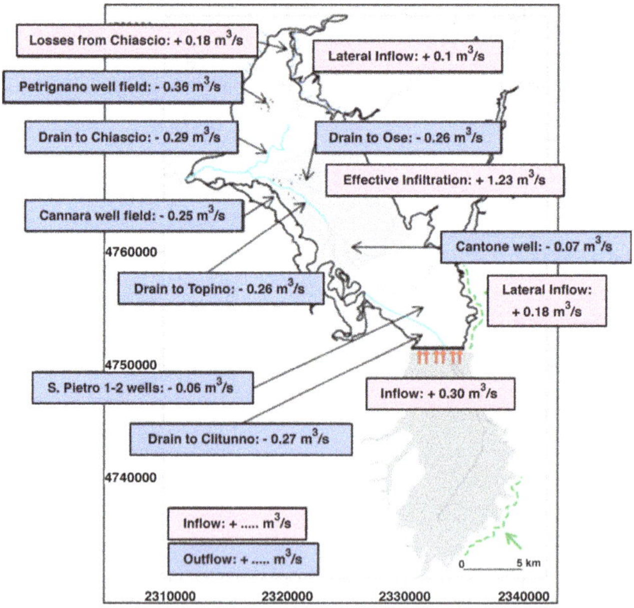

Fig. 19.4. **Total budget of the Umbra Valley aquifer under exploitation conditions. Red and blue boxes refer to inflow and outflow terms, respectively.**

6. The Umbra Valley aquifer is exploited through the well fields of Petrignano d'Assisi (360 l/s) and Cannara (250 l/s), and through the wells of S. Pietro 1 and 2 (65 l/s) and Cantone (70 l/s). Additionally there is a number of private wells for irrigation purposes, whose total abstraction was estimated 300 l/s during the growth season (May to August/September). The effective exploitation varies seasonally or annually in relation to the demand which may strongly increase especially during droughts. However, the average values indicated above are representative of the current mean exploitation conditions of the aquifer.

7. Under undisturbed conditions (before the pumping station was built in the 1970's) the groundwater flow in the Petrignano sector was approximately directed from North to South, following the Chiascio River course. Since its exploitation, a large depression cone became evident (Fig. 19.5); the available piezometric time series show, starting from 2001, an enlargement of the depression cone that possibly determines a variation of the capture zone and a modification of the river-aquifer interaction regime.

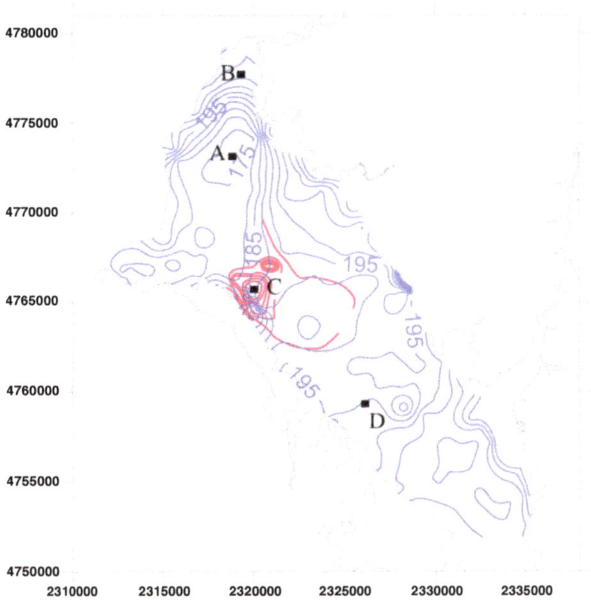

Fig. 19.5. Piezometric map of 2007 (source: Regione Umbria, Regional Geological and Seismic Service). Red lines represent the piezometric heads of the confined aquifer in the Cannara sector.

8. Concerning the Cannara sector, a continuous silty-clay layer (70-80 m thick nearby Cannara) constitutes an aquitard separating a deep confined aquifer from a phreatic shallow aquifer, the first one being artesian under undisturbed conditions. The exploitation at the Cannara well field since the 1980's, has caused a significant depletion of the confined aquifer heads, inverting the vertical gradient from upward to downward.

9. To model the different hydraulic behavior of the North (Petrignano), South (Foligno) and central (Cannara) sectors (phreatic the first two and confined the third one), a 3D layer model has been developed. In the Petrignano and Foligno sector the vertical hydraulic conductivity has been set constant for the three layers in order to reproduce a monolayer model for those areas, while in the Cannara sector a lower conductivity (10^{-7} m/s) has been assigned to the second layer, to simulate the aquitard dividing the shallow phreatic aquifer from the deep confined one. Steady-state conditions are considered.

10. The hydraulic conductivity has been calibrated in steady state conditions through the numerical code MODAC (Guo and Zhang, 2000). The adopted target was the piezometric field under undisturbed conditions, thus before 1970. A comprehensive description of the adopted calibration procedure can be found in Romano and Preziosi (2010). In Fig. 19.6 (dark triangles) a comparison between the observed and simulated piezometric heads is presented.

11. The performed calibration has been finally validated with respect to the observed piezometric heads of 2007 (Fig. 19.6, grey circles). The residues are approximately zero-mean (+ 0.15 m), with a standard deviation equal to 2.6 m. In the analysis of the uncertainty propagation through the processing chain from GCMs to aquifer impact assessment, we will consider the value ±2.6 m as the uncertainty related to groundwater modeling.

The results obtained by the calibration-validation procedure were considered satisfactory in terms of residues, taking into account the high heterogeneity of the study aquifer (particularly concerning the conductivity field) and the uncertainty on the aquifer budget.

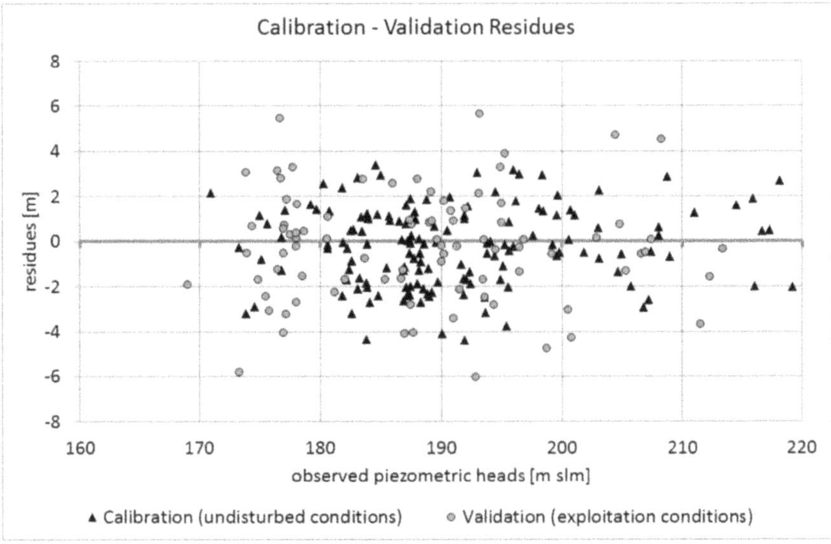

Fig. 19.6. **Results of the calibration under undisturbed conditions (dark triangles) and validation under exploitation conditions (grey circles).**

19.7 SCENARIOS OF IMPACT ON THE AQUIFER.

In order to assess the growing uncertainty through the processing chain from GCM to the impact assessment on the Umbra Valley aquifer, the piezometric heads have been computed in steady state conditions, by the numerical flow model presented in Section 19.6, assigning as the recharge input the values of effective infiltration presented in Table 19.5.

To compare the results, four points have been chosen, whose location is shown in Fig. 19.5: two in the Petrignano d'Assisi sector (point A located in the middle of the Petrignano well field and point B located approximately 2 km upstream) and two in the Cannara sector (point C located in the middle of the Cannara well field and point D located approximately 3 km upstream, in the south direction). Concerning the Cannara sector, the monitoring targets have been set in the third layer (the confined one).

The obtained results are shown in Fig. 19.7 for the Petrignano sector and in Fig. 19.8 for the Cannara sector.

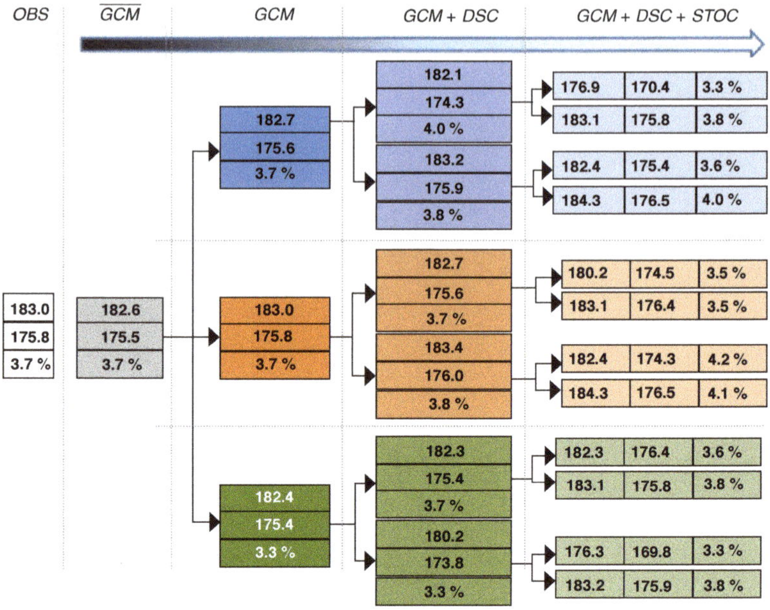

Fig. 19.7 **Uncertainty propagation of the piezometric head computed in the Petrignano d'Assisi sector.**

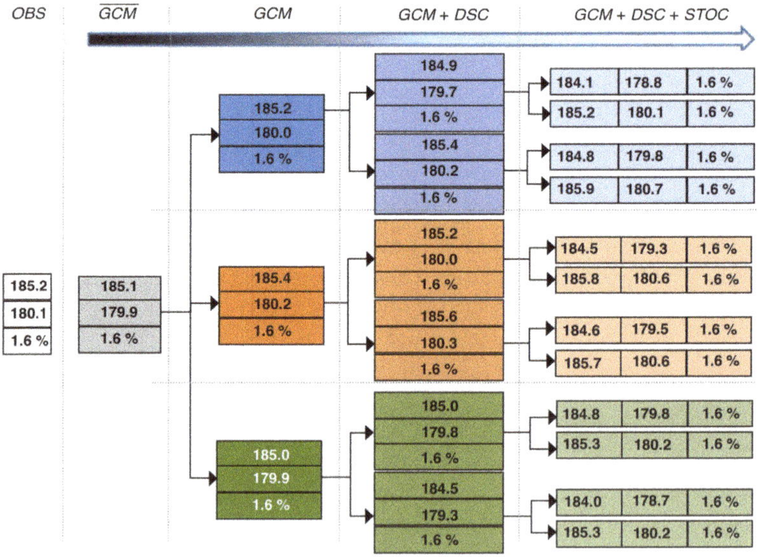

Fig. 19.8. As in Fig. 19.7, but for the Cannara sector.

For each simulation the first and the second value refer to the piezometric head computed at the point B and A of Fig. 19.5. The third value refers to the computed hydraulic gradients. In the last column, piezometric head values are referred to the more (upper row) and less (lower row) severe scenario obtained by the combination of time series of stochastic precipitation and temperature corresponding, respectively, to 5th and 95th percentile (more severe) and viceversa (less severe).

For each simulation the first value refers to the piezometric head computed at the points B and D (located upstream in the Petrignano and Cannara sector, respectively), the second value to the piezometric head computed at the point A and C (in the middle of the respective well fields), and the third value refers to the resulting hydraulic gradient. In the last column of Fig.19.8 (GCM+DSC+STOC) it is shown, for each GCM and downscaling technique, the piezometric head computed considering, on the one hand, the more severe scenario (upper) identified by the combination of the stochastic rainfall and temperature time series corresponding to 5th and 95th percentile, respectively, of generated time series and, on the other hand, the less severe scenario (lower) obtained by the stochastic time series corresponding to the 95th and 5th percentile for rainfall and temperature, respectively.

As one can argue from Fig. 19.7 and 19.8, the different scenarios of effective infiltration strongly impact on the simulated piezometric heads both in the Petrignano and in the Cannara sector. Focusing the attention on the well field areas, the difference between the wettest and the driest scenario is 6.7 m for Petrignano and 2.0 m for Cannara. Such difference between the two sectors can be inferred analyzing the hydraulic gradients: in the Cannara sector the gradient does not change with the recharge, while in the Petrignano sector it does. This is due to the different characteristics of the aquifers: in the Petrignano area the aquifer is a phreatic one, directly recharged by the effective infiltration; in the Cannara sector the well field exploits the confined layer: as a consequence the piezometric heads and, eventually, the hydraulic gradients, are less affected by the change of the I_{eff} scenarios.

19.8 UNCERTAINTY ANALYSIS.

As a final step, the uncertainty estimated as the final result of the flow model validation procedure (± 2.6 m) has been added to the results obtained in terms of piezometric heads from the different climatic scenarios (Section 19.7). The final result of our analysis is shown in Fig.

19.9, where the piezometric head computed in the middle of the Petrignano d'Assisi field is shown. The numbers reported in Fig. 19.9 show for each processing step the percentage variations of the aquifer thickness (piezometric head minus the bottom level) with respect to the aquifer thickness computed considering the effective infiltration. For each processing step the highest and lowest piezometric heads are taken into account.

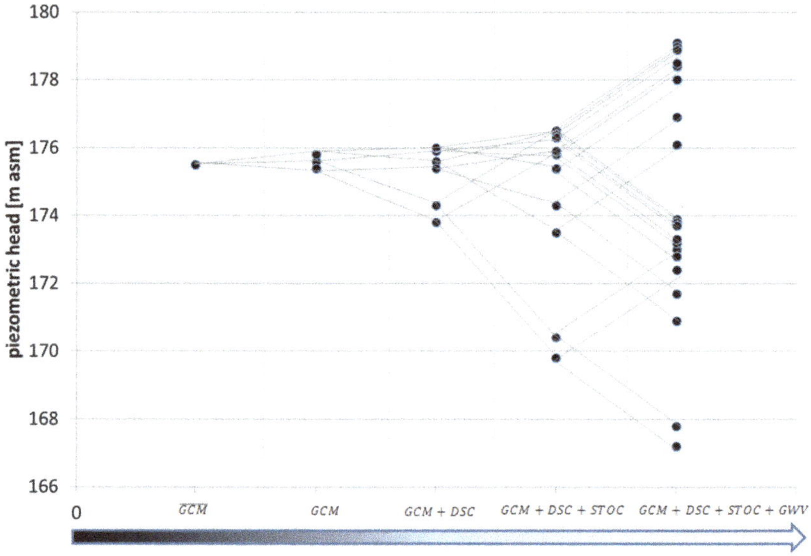

Fig. 19.9. Uncertainty propagation of the piezometric head computed in the middle of the Petrignano d'Assisi well field (point A in Fig. 19.5). GWV stands for the uncertainty of ±2.6 m resulting from the calibration/validation procedure of the groundwater flow model and added to the results obtained at the end of the processing chain *GCM+DSC+STOC*.

A significant difference in the processing chain can be pointed out by comparing the uncertainty propagation of the effective infiltration (Fig. 19.3) and the uncertainty propagation of the piezometric head (Fig. 19.9): the first one appears to be centred around the mean of the scenarios, while in the second one a clear negative tail can be inferred. In other words, although the forcing uncertainty is approximately distributed in a symmetric way, the impact on the aquifer is not linear, especially for the Petrignano sector. This is due to the non-linear behaviour of the phreatic aquifer. Two are the elements of non-linearity: 1) the transmissivity of the aquifer, that depends on the thickness of the aquifer itself hence on the piezometric heads except in the confined aquifer, and 2) the exchanges between the surface and ground water, that depend on the relative position of the hydrometric level in the river and the water table.

Such non-linearity results in a different propagation of the uncertainty. In fact, if one considers the positive uncertainty, the percentage variation of the aquifer thickness is +1.2% taking into account the processing chain *GCM+DSC+STOC* and increases to +4.5% adding the uncertainty from the calibration/validation of the groundwater flow modelling. As a consequence, for positive value of the uncertainty the errors from the flow model appear to be higher than the errors from the effective infiltration estimates. On the contrary, if one considers the negative uncertainty, the percentage variation of the aquifer thickness is -7.1% taking into account the processing chain *GCM+DSC+STOC* and increases to -10.3% adding the uncertainty from the calibration/validation of the groundwater flow modelling. As a consequence, for negative value of the uncertainty the errors from the flow model appear to be lower than the errors from the effective infiltration estimates.

These results confirm the importance of correctly modelling the non-linear processes involved in the groundwater flow, as suggested also by previous studies. For example, Allen et al. (2004) performed a sensitivity analysis for the Grand Forks aquifer (Canada) taking into account the impact on groundwater due to the variations of the recharge and of the river stage (boundary conditions for the aquifer). They found that the variations in the surface water - groundwater interaction (non-linear process) impact on piezometric heads more strongly than the recharge. Similar conclusions are drawn by Gosling et al. (2011).

19.9 CONCLUSIONS

The assessment of climate change impacts on groundwater resources is a key point, especially for management purposes. The modelling chain, from the GCM choice to the piezometric heads computation, is subject to a certain degree of uncertainty, which propagates at each step. The evaluation of the weakest link of the whole chain, is of paramount importance in order to assess the sensitivity of the modelling process and address further research.

The uncertainty estimation related to the whole processing chain from General Circulation Models, providing reference climate regimes for the future, to flow numerical models, that calculate the consequent variation in terms of aquifer heads and/or water budget, is essential in order to correctly plan the water resources management.

The present work aimed at assessing which part of such processing chain mainly contributes to the total uncertainty by analysing the possible climate change impact on a large alluvial aquifer in central Italy. To this goal the following processing chain was implemented: 1) three different General Circulation Models (the Parallel Climate Model, the Community Climate System Model, version 3.0 and the ECHAM5 model have been considered; 2) two different downscaling procedures (the Linear Rescaling Method and the Cumulative Distribution Function method) have been applied to the time series of precipitation and temperature issuing from the three GCM; 3) for each downscaled time series the Neyman-Scott Rectangular Pulses model has been used for stochastic rainfall generation, and the Fractionally Differenced ARIMA (FARIMA) model has been used for temperature simulations, obtaining 100 scenarios, statistically equivalent; 4) the Thornthwaite-Mather method, whose input is constituted by the P and T time series obtained from the previous step, has been applied to obtain scenarios of effective infiltration; 5) a flow model of the Umbra Valley aquifer has been developed, calibrated and validated through the MODFLOW2005 code. Finally, the increasing uncertainty has been evaluated for each processing step in terms of heads variation in the groundwater model, leading to the following results:

1. The uncertainty on the projections of the effective infiltration is in the order of ±3% if one considers only the ensemble of the GCM, without considering the downscaling steps.

2. The downscaling step (both the methods of downscaling and the stochastic generation) significantly increases the uncertainty from approximately 3% (*GCM*) to 6% (*GCM+DSC*), to ≈15% (*GCM+DSC+STOC*).

3. Although the uncertainty of the forcing (effective infiltration) is symmetrically distributed around the mean, the impact on the aquifer is not linear, showing a clear divergence toward the low values of the piezometric heads (especially in the areas where the aquifer is

phreatic). This result confirms the importance of correctly modelling the non-linear processes involved in the groundwater flow, as suggested also by previous studies.

4. The uncertainty in terms of piezometric heads arising only from the groundwater model calibration/validation (thus considering the effective infiltration unaffected by uncertainty) is higher than the uncertainty from the effective infiltration estimates if one considers scenarios characterized by an increase of the recharge. The opposite can be found if one considers scenarios characterized by a decrease of the recharge.

19.10 ACKNOWLEDGMENT

This work was partly funded by the POR-FESR Project "Drought and Climate Changes", Umbria Region.

19.11 REFERENCES

Allen, D. M., Mackie, D. C. and Wei, M. (2004), Groundwater and climate change: a sensitivity analysis for the Grand Forks aquifer, southern British Columbia, Canada, *Hydrogeology Journal*, 12, 270 – 290

Allen, D. M., Cannon, A. J., Toews, M. W. and Scibek, J. (2010), Variability in simulated recharge using different GCMs, *Water Resources Research*, Vol. 46. doi:10.1029/2009WR008932

Brouyère S., Carabin G., Dassargues A. (2004), Climate change impacts on groundwater resources: modelled deficits in a chalky aquifer, Geer basin, Belgium, *Hydrogeology Journal* (2004) 12:123–134 DOI 10.1007/s10040-003-0293-1

Bates, B.C., Kundzewicz, Z.W., Wu, S., Palutikof, J.P., Eds. (2008), Climate Change and Water. Technical Paper of the Intergovernmental Panel on Climate Change. IPCC *Secretariat. Geneva*, 210 pp

Camici, S. Brocca, L., Melone, F., Moramarco, T, (2014), Impact of climate change on flood frequency using different climate models and downscaling approaches, *Journal of Hydrologic Engineering* 19(8):04014002.

Cowpertwait, P. S. P., O'Connell, P. E. O., Metcalfe, A. V., and Mawdsley, J. A. (1996), Stochastic point process modelling of rainfall, I. Single-site fitting validation. *J. Hydrol.*, 175, 17-46.

Crosbie, R. S., McCallum, James L., Walker, Glen R. and Chiew, Francis H. S. (2010), Modelling climate-change impacts on groundwater recharge in the Murray-Darling Basin, Australia, *Hydrogeology Journal*, 18, 1639 – 1656

Dams, J., Salvadore, E. and Batelaan, O. (2010), Predicting Impact of Climate Change on Groundwater Dependent Ecosystems, 2nd International Interdisciplinary Conference on Predictions for Hydrology, Ecology and Water Resources Management: Changes and Hazards caused by Direct Human Interventions and Climate Change, 20-23 September 2010, Prague, Czech Republic.

Dettinger, M. (2012). "Projections and downscaling of 21st century temperatures, precipitation, radiative fluxes and winds for the Southwestern US, with focus on Lake Tahoe." Clim. Change, DOI 10.1007/s10584-012-0501-x.

Dragoni, W. and Sukhija, B. S. (2008), Climate change and groundwater: a short review, *Geological Society*, London, Special Publications

Goderniaux, P., Brouyere, S., Fowler, H.J., Blenkinsop, S., Therrien, R., Orban, P., Dassargues, A., 2009. Large scale surface–subsurface hydrological model to assess climate change impacts on groundwater reserves. *J. Hydrol.* 373, 122–138.

Gosling, S.M., Taylor, R.G., Arnell, N.W., Todd, M.C., (2010), A comparative analysis of projected impacts of climate change on river runoff from global and catchment scale hydrological model, *Hydrol. Earth Syst. Sci.*, 15, 279-294

Green, T.R., Taniguchi, M., Kooi, H., Gurdak, J.J., Allen, D.M., Hiscock, K.M., Treidel, H., Aureli, A., (2011), Beneath the surface of global change: impacts of climate change on groundwater. *J. Hydrol*. 405, 532–560

Gudmundsson, L., Bremnes, J. B., Haugen, J. E. Engen Skaugen, T. (2012), "Technical Note: Downscaling RCM precipitation to the station scale using quantile mapping – a comparison of methods." *Hydrol. Earth Syst. Sci. Discuss*., 9, 6185–6201, doi:10.5194/hessd-9-6185-2012.

Guyennon N., E. Romano, I. Portoghese, F. Salerno, S. Calmanti, A. B. Petrangeli, G. Tartari, D. Copetti (2013). Benefits from using combined dynamical-statistical downscaling approaches – lessons from a case study in the Mediterranean region. *Hydrol. Earth Syst. Sci.*, 17, 705-720. doi:10.5194/hess-17-705-2013

Guo, X., and C.M. Zhang. (2000), Hydraulic gradient comparison method to estimate aquifer hydraulic parameters under steady-state conditions. *Ground Water 38*(6), 815–826

Hay, L. E., Wilby, R. L., Leavesley, G. H. (2000), A comparison of delta change and downscaled GCM scenarios for three mountainous basins in the United States, *Journ, of the Americ. Water Resourc. Ass.*, 36, 2.

Harbough, A.W. 2005. The U.S. Geological Survey Modular Ground-Water Model—the Ground-WaterFlow Process, U.S.Geological Survey Techniques of *Water Resources Investigation Book 6*, pp. 253

Holman I.P. (2006). Climate change impacts on groundwater recharge-uncertainty, shortcomings, and the way forward? Hydrogeology Journal, 14, 637-647

Holman I.P., Tascone D., Hess T.M. (2009) A comparison of stochastic and deterministic downscaling methods for modelling potential groundwater recharge under climate change in East Anglia, UK: implications for groundwater resource management, *Hydrogeology Journal*, 17, 1629-1641

IPCC, (2012), Managing the Risks of Extreme Events and Disasters to Advance Climate Change Adaptation. A Special Report of Working Groups I and II of the Intergovernmental Panel on Climate Change [Field, C.B., V. Barros, T.F. Stocker, D. Qin, D.J. Dokken, K.L. Ebi, M.D. Mastrandrea, K.J. Mach, G.-K. Plattner, S.K. Allen, M. Tignor, and P.M. Midgley (eds.)]. Cambridge University Press, Cambridge, UK, and New York, NY, USA, 582 pp.

Kumar C. P. 2012. Climate Change and Its Impact on Groundwater Resources. RESEARCH INVENTY: International *Journal of Engineering and Science*. ISSN: 2278-4721, Vol. 1, Issue 5 (October 2012), PP 43-60. www.researchinventy.com 43

Li, H., Sheffield, J., and Wood, E. F. (2010), "Bias correction of monthly precipitation and temperature fields from Intergovernmental Panel on Climate Change AR4 models using equidistant 15 quantile matching." *J. Geophys. Res.*, 115, D10101, doi:10.1029/2009JD012882, 2010.

Nakicenovic, N. et al (2000), "Special Report on Emissions Scenarios: A Special Report of Working Group III of the Intergovernmental Panel on Climate Change." Cambridge University Press, Cambridge, U.K., 599 pp.

Montanari, A., Rosso, R., and Taqqu, M.S. (1997), "Fractionally differenced ARIMA models applied to hydrologic time series: identification, estimation and simulation." *Water Resour. Res.*, 33, 1035-1044.

Pachauri R, Reisinger A (eds) (2007) Climate change 2007: synthesis report. Contribution of working groups I, II and III to the 4th assessment report of the intergovernmental panel on climate change. Core Writing Team, *IPCC*, Geneva, p 104

Piani, C., Haerter, J., and Coppola, E. (2010), "Statistical bias correction for daily precipitation in regional climate models over Europe." *Theor. Appl. Climatol.*, 99, 187–192, doi:10.1007/s00704-009-0134-9.

Oude Essink, G. H. P., van Baaren, E. S. and de Louw, P. G. B. (2010), Effects of climate change oncoastal groundwater systems: A modeling study in the Netherlands, *Water Resources Research*, 46

Preziosi E., E. Romano (2013), Are large karstic springs good indicators for Climate Change effects on groundwater? *Geophysical Research Abstracts,* 15, EGU2013-9367

Romano E., E. Preziosi. (2010). The sustainable pumping rate concept: lessons from a case study in Central Italy. *Ground Water*, 48, 217-226. DOI: 10.1111/j.1745-6584.2009.00628.x

Romano E, E. Preziosi, A.B. Petrangeli (2011). Spatial and Time Analysis of Rainfall in the Tiber River Basin (Central Italy) in relation to Discharge Measurements (1920-2010). *Procedia Environmental Sciences*, 258–263. doi:10.1016/j.proenv.2011.07.045

Romano E, E. Preziosi (2013). Precipitation pattern analysis in the Tiber River basin (central Italy) using standardized indices. *International Journal of Climatology.* 33 (7) , 1781-1792 doi: 10.1002/joc.3549

Romano, E., Camici, S., Brocca, L., Moramarco, T., Pica, F., Preziosi, E. (2013), On the variables to be considered in assessing the impact of climate change to alluvial aquifers: a case study in central Italy, *Procedia engineering Elsevier.*

Rozell, Daniel J. and Wong, Teng-fong (2010), Effects of climate change on groundwater resources atShelter Island, New York State, USA, *Hydrogeology Journal*, 18, 1657-1665

Scibek, J. and Allen, D. M. (2006), Modeled impacts of predicted climate change on recharge and groundwater levels, *Water Resources Research*, 42, W11405, 18p

Senatore A, Mendicino G, Smiatek G, Kunstmann H 2011. Regional climate change projections and hydrological impact analysis for a Mediterranean basin in Southern Italy. doi:10.1016/j.jhydrol.2010.12.035

Stoll, S., Hendricks Franssen, H. J., Butts, M., and Kinzelbach, W.: Analysis of the impact of climate change on groundwater related hydrological fluxes: a multi-model approach including different downscaling methods, *Hydrol. Earth Syst. Sci.*, 15, 21-38, doi:10.5194/hess-15-21-2011, 2011

Taylor R.G. et al. 2013. Ground water and climate change. *Nature Clim. Change*, http://dx.doi.org/10.1038/nclimate1744 (2012)

Themeßl, M. J., Gobiet, A., and Leuprecht, A. (2011). "Empirical-statistical downscaling and error correction of daily precipitation from regional climate models." *Int. J. Climatol.*, 31, 1530–1544, doi: 10.1002/joc.2168.

Zhou, Y., Zwahlen, F., Wang, Y., and Li, Y. (2010). Impact of climate change on irrigation requirements in terms of groundwater resources, *Hydrogeology Journal*, 18, 1571-1582

CHAPTER 20

POSSIBLE RESPONSE OF TWO WATER SYSTEMS IN CENTRAL ITALY TO CLIMATIC CHANGES

W. Dragoni[1], C. Giontella[1], M. Melillo[2], C. Cambi[1], L. Di Matteo[1], D. Valigi[1]

1)Dept. of Physics and Geology, University of Perugia, via Alessandro Pascoli snc, 06123 Perugia, Italia

2) CNR-IRPI, Sede di Perugia, via Madonna Alta 126, 06128 Perugia, Italia

ABSTRACT

Aiming to building realistic possible scenarios of the impacts of the ongoing climate change on the hydrologic systems of Central Italy, a rainfall-runoff model was used to simulate two strategic water systems. The model, made by several tanks in series and adapted to the specific investigated systems, was applied to the Trasimeno Lake to simulate its levels and to Bagnara spring to simulate its discharge; the model works at a monthly scale. In both cases the model was properly calibrated and validated over different sets of lake levels and spring discharge data. The calibrated models have been then used to simulate the systems behaviour under different climatic scenarios; these scenarios were set in accordance with the indications obtained by the analysis of historical rainfall and temperature series recorded in the systems surroundings. For the Trasimeno Lake the model indicates that in about two decades the lake would seasonally get dry if the yearly rainfall would decrease by -20% with respect to 1984-2006.

As for Bagnara spring, simulation shows that a rainfall variation of about -11% with respect to 1989-2009, leads the discharge to decrease of about 17% assuming the temperature to stay constant, and of about 20% considering a positive temperature trend. These results are plausible "scenarios", not forecasts: in spite of this, they demonstrate that the two systems are very sensible even to slight variation of rainfall. Given that in the area, for the next decades, most of the scenarios point towards a decrease of future rainfall, the stakeholders should consider these results when updating the management strategies.

20.1 INTRODUCTION

This paper synthesizes some results of a research about the impact of the ongoing climate change on two hydrological water systems in the Mediterranean area, in Central Italy.

According to the last IPCC report (IPCC, 2013) between 1880 and 2012 the average global atmospheric temperature has risen by about 0.65 to 1.06 °C. In the Mediterranean basin, a mean warming of about 0.01 °C/year has been estimated during the interval 1951-2000 (Gualdi et al. 2013). This estimate is similar to those put forward in many previous works (IPCC, 1996; Dragoni, 1998; Maugeri and Nanni, 1998; Brunetti et al. 2000a; 2000b; 2002).

All general Circulation Models (GCMs) and most of the empirical extrapolations and considerations indicate that such tendency should continue. In particular, according to the IPCC (2013) an increase of global mean surface temperatures for 2081-2100 relative to 1986-2005 should be in the extreme range of 0.3°C to 4.8°C, depending on the GCM used and the future greenhouse gases (GHG) emissions.

An increase of atmospheric temperature implies, at global scale, an increase of evaporation, which corresponds to an increase of average rainfall. Indeed, according to the last IPCC report (IPCC, 2013) in the Northern Hemisphere "*mid-latitude land areas do show a likely overall increase in precipitation (medium confidence prior to 1951, but high confidence afterwards)*".

In Southern Europe and the Mediterranean, precipitation and water resources availability are decreasing (Dragoni and Sukhija, 2008; Gao and Giorgi, 2008; Gualdi et al. 2013; IPCC, 2013; Romano and Preziosi, 2013; Schneider et al. 2013). Furthermore analyses of climatic series by means of climatic indexes such as SPI and RCCI (McKee et al. 1993; Giorgi, 2006) indicate the Mediterranean as a region the climate of which is especially responsive to global change (Giorgi, 2006) and that the frequency of extreme droughts are increasing in Central and Southern Italy (Bartolini et al. 2008; Nanni et al. 2008; Fiorillo and Guadagno, 2012; Di Matteo et al. 2013). It can be told that past and present tendencies are in general known; nevertheless, there is still a certain degree of uncertainty on the current tendencies and the future evolution of climate at local scale, when considering specific geographic areas, and taking into account precipitation trends and distribution and impact on hydrologic resources (Strzepek and Yates, 1997; Todisco and Vergni,

2008). In particular some scenarios from GCMs and Regional Circulation Models (RCMs) indicate that rainfall in some areas of the Mediterranean could increase (Gualdi et al. 2013).

From the economic and social point of view one of the key points is to understand how climate changes affect and will affect water resources in different geographic areas and different geological contexts. The available information show that some important springs located in the Umbria-Marche Apennines (Bagnara, Pescara di Arquata del Tronto) are negatively influenced by the ongoing climate change and show statistically significant negative discharge trends (Cambi and Dragoni, 2000; Dragoni et al. 2003). Negative effects can also be observed on lake systems, on which climate changes induce decreases in levels and outlet discharges, amplified by constantly increasing withdrawals (Dragoni et al. 2006; Di Matteo et al. 2010b).

One more key point in this context is creating instruments able to predict how specific water systems would react to different possible climatic scenarios and allowing to study the better management schemes. Properly built and calibrated rainfall-runoff models of specific strategic systems can be conveniently used with this aim (Cambi and Dragoni, 2000; Dragoni et al. 2013; Romano et al. 2014). In this work a lumped rainfall-runoff model was used to investigate the reactions of two strategic hydrogeological systems in Central Italy to different possible climatic scenarios.

20.2 CONSIDERED SYSTEMS

Two different hydrogeological systems located in Central Italy, Trasimeno Lake and Bagnara spring, have been analyzed in detail and modelled, in order to investigate their reaction to possible future climatic scenarios. Such scenarios have been built on the basis of the statistical analysis of long-term climatic series in the neighborhood systems.

20.2.1 Trasimeno Lake

20.2.1.1 General information

The Trasimeno Lake, located in the Umbria Region (Central Italy) is one of the largest Italian lake by surface area, covering about 122 km^2; the mean elevation of its catchment is 290 m a.s.l. Lake Trasimeno has no natural emissaries, and for the last few decades it has had a maximum depth of less than six metres. The area of the catchment draining into the lake is 262 km^2. In addition to rainfall directly on the lake and surface runoff from the catchment, the lake is fed by the groundwater surrounding it; according to the piezometric map of the aquifer it appears that, in practice, the surface basin corresponds to the hydrogeological basin. Lake Trasimeno is extremely important from both economical and environmental point of view since it constitutes an important water resource and a unique cultural and tourist centre with highly significant historical landscape values and unique or rare vegetable and animal species. Lake Trasimeno is subjected to strong anthropic pressure. This corresponds basically to farming activities, causing fertilizers pollution, and water intake for irrigation. In summer the touristic activity, absolutely important for the economy of the area, causes an increase of population, with a consequent increase of the polluting loads. In recent years the effects of the anthropic pressure, normally kept under reasonably good control, have become a secondary consideration in relation to the consequences of a droughty climatic phase which has led to a substantial drop in the level of the lake which, in 2004, reached a maximum depth of only four meters. In addition to causing hygiene problems and navigation difficulties, this decrease of the water level contributed to the deterioration of the quality of the water, with negative impacts on the ecosystem and on local economic activities and tourism, which during the crisis of 2002-2006 have been reduced considerably (Fig. 20.1).

Fig. 20.1. Lake Trasimeno, Summer 2004. The low level of the lake caused heavy environmental problems and made navigation almost impossible (photo by W. Dragoni).

20.2.1.2 Geological and hydrogeological framework

The lake is sited in the inner portion of the Northern Apennines, a NE verging arc-shaped thrust belt, originated by the Oligocene-Early Miocene eastward convergence between the Alpine orogen and the Adriatic promontory (Gasperini et al. 2010). Extensional deformations related to the opening of Northern Tyrrhenian sea has interested western Umbria and in the Early-Middle Pliocene; the extensional tectonics has originated two major NW/SE-trending extensional basins, the Val di Chiana and Valtiberina grabens, between which Trasimeno Lake is located (Gasperini et al. 2010).

Before 1959 Trasimeno Lake catchment was about 309 km^2 wide, the lake surface being, in average, 122 km^2 wide. Between 1959 and 1961 the basin was artificially expanded by connecting the catchments of four creeks (Moiano, Maranzano, Tresa and Rio Maggiore) some of which had been diverted outside the lake in the late Middle age (Dragoni et al. 2012) to solve the problem of high levels and floods: the catchments was so enlarged to 383.4 km^2. All of the main Northern Apennines units are present in the Trasimeno Lake basin. Ligurian units crop out in few places on the southwestern of the lake (e.g. Castiglione del Lago, Sanfatucchio). The Ligurian units overlay the Tuscan turbidites (Macigno Fm., Oligocene-Early Miocene in age), which, in turn, lay over the easternmost Umbria-Marche turbidites (Marnoso-Arenacea Fm., Early/Middle Miocene in age).

The Tuscan and Umbria-Marche units occupy most of the catchment area and are, in the western part, overlaid by marine and fluvio-lacustrine plio-pleistocene sediments. Olocenic alluvial deposits are found around the Lake and along the main streams.

The most recent work on Trasimeno catchment geology is dated 2010 (Gasperini et al. 2010). Nevertheless, on the hydrogeological point of view and for the aims of this work, the simplified geological sketch shown in Fig. 20.2 seems to of proper use.

The permeability of the Oligocene-Miocene turbidite formations belonging to the Tuscan and Umbria-Marche units can be considered, in overall, from low to very low. On the contrary the permeability of plio-pleistocene sediments outcropping on the western part of the catchment ranges from low to medium-high. These sediments host therefore local aquifers having as a base the turbidites. Due to the presence of the

Fig. 20.2. **Geological map of Trasimeno Lake catchment with location of rainfall and temperature stations (Modified after Dragoni, 1982).**

1) Tuscan turbidites (Oligocene-Early Miocene) 2) Marine pliocenic sediments 3) fluvio-lacustrine plio-pleistocene sediments 4) Olocene alluvial deposit 5) Catchments artificially linked in the early 60s 6) Natural lake catchment 7) Rainfall stations 8) Temperature stations 8) Hydrometer.

impervious substratum represented by the turbidite formations, these aquifers drain toward the lake and the hydrogeological basin can be virtually considered the same as the surface basin.

20.2.1.3 Climatic characteristics

In order to define the average climatic characteristics of Trasimeno Lake catchment on a monthly scale, rainfall data recorded in 17 rainfall stations located within the basin and in its neighborhood and data from a temperature station close to the lake have been analyzed. The location of temperature and rainfall stations is shown in Fig. 20.2.

20.2.1.4 Mean monthly temperature over the catchment

Monthly temperature data recorded in Monte del Lago during the interval 1966-2010 station have been used to define the mean annual temperature over the lake catchment. This station has been running almost continuously since 1926. A few data missing during the periods 1990-1995, 1996-2000 and 2000-2007 have been reconstructed by means of multiple regression, by using respectively the data recorded in the Perugia, Castiglione del Lago and Isola Polvese stations, after having verified the good degree of correlations of data coming from these stations with those recorded in Monte del Lago. The mean annual temperature of the catchment, also defined on the basis of these data, is 14.1 °C.

20.2.1.5 Mean monthly rainfall over the catchment

To define monthly rainfall over the lake catchment, the data recorded in the rainfall stations indicated in Fig. 20.2 have been used. Since the topography of the lake catchment is characterized by limited elevation variations, it has been possible to estimate the monthly rainfall by means of Thiessen polygons. The polygons have been built by using different sets of rainfall stations, depending on the availability of data during different intervals of the 1966-2010 period. The data analysis also allowed to define mean annual rainfall both over the catchment and over the lake surface, which are respectively 742 and 704 mm/year.

20.2.1.6 Climatic trend

The estimated monthly rainfall over the catchment has been analyzed in order to define whether there are significant climatic trends in this area. The statistical analysis, run for the 1966-2010 period by using the Mann-Kendall test, indicated a negative trend for the January-March trimester of -1.39 mm/trimester, with a statistical significance of 90%. The mean annual temperature has a significant positive linear trend with a gradient of 0.038 °C/year.

20.2.1.7 Lake model

A mathematical model was used to simulate the lake levels changes at monthly scale. The model simulates volume and level variations by calculating the water budget starting from rainfall, temperature, water withdrawals and outlet discharge data, and aims to reproduce the observed lake levels variations by minimizing the differences between observed and simulated levels by using the mean squared error criterion. Lake levels are calculated as the elevation of the water surface over the maximum lake deepness. The model, implemented by modifying and updating previous models (De Felice et al. 1993; Dragoni and Valigi, 1995; Cambi et al. 2003; Di Matteo et al. 2005; Di Matteo et al. 2010b; Dragoni et al. 2012), is lumped and is made by three tanks in series, respectively representing the lake catchment (TANK1), the aquifer draining to the lake (TANK2) and the lake itself (LAKE) (Fig. 20.3)

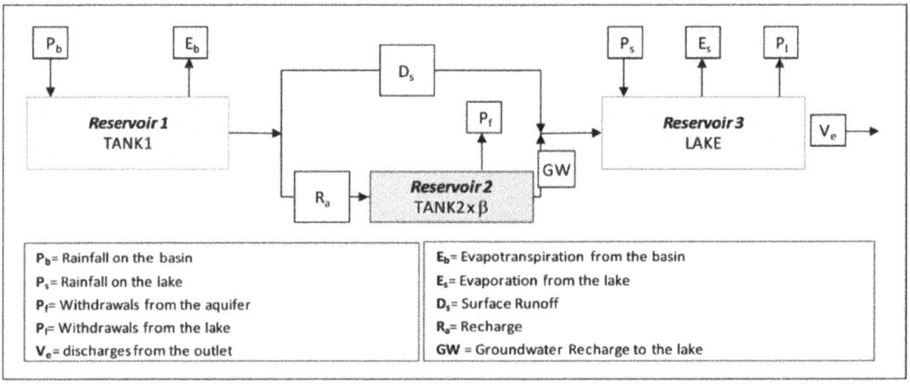

Fig. 20.3. Lake model conceptual scheme.

TANK 1 receives rainfall (P_b) as an input, and has got three outputs: actual evapotranspiration (E_b), surface runoff (D_s) and groundwater recharge (R_a). Evapotranspiration is estimated by using a modified Thornthwaite-Mather method (Thornthwaite and Mather, 1957) and the surface runoff is calculated by means of the equation proposed by Vandewiele (1992). The water left when subtracting E_b, and D_s from monthly rainfall is the groundwater recharge (R_a) feeding TANK2, which represents the aquifer. TANK2, in which withdrawals from the aquifer (P_f) are an output, implicitly uses Darcy equation, delaying the arrival of groundwater with respect to surface water, to create a second output (GW), which feeds the lake. The amount of the delay is regulated by a β coefficient to be calibrated. GW, generated by TANK2, the surface runoff generated directly by TANK1 (D_s) and the rainfall over the lake surface (P_s) are inputs to LAKE, which represents the lake itself. LAKE generates evaporation (E_s) on the basis of an empirical equation proposed by Dragoni and Valigi (1994). The evaporation generated this way is corrected by means of two coefficients to be calibrated. Two more outputs from LAKE are represented by withdrawals from the lake (P_l) and outflow from the outlet (V_e).

The model was calibrated and validated, with fairly good results on the basis of the monthly lake levels recorded between Aug 1984 and Jan 2006.

The calibrated coefficients were verified by means of a validation process run over two different periods (Aug 1966-Aug 1984 and Jan 2006 - May 2010). Table 20.1 shows Nash-Sutcliffe Efficiency (NSE) values (Nash and Sutcliffe, 1970; Moriasi et al. 2007) for calibration and validation processes.

Table 20.1. Nash-Sutcliffe Efficiency (NSE) for calibration and validation process for the Trasimeno Lake model.

Period	Process	NSE
1984-2006	calibration	0.98
1966-1984	validation	0.90
2006-2010	validation	0.97

Figure 20.4 shows the simulation for both calibration and validation periods.

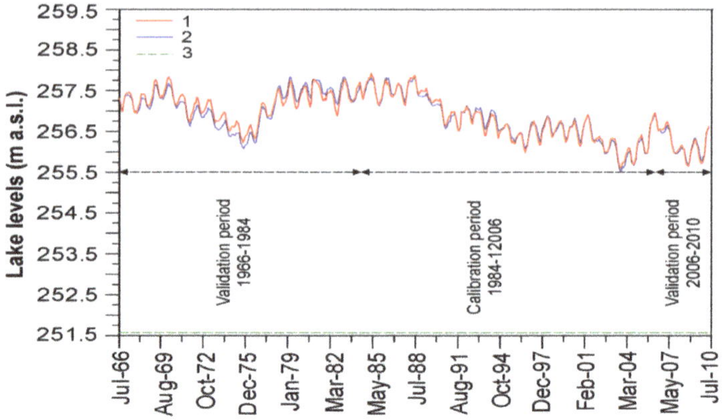

Fig. 20.4. Model simulation for calibration and validation periods (modified after Dragoni et al. 2012).

It is interesting to notice that lake levels have generally decreased through the analyzed period and that strong lake crises have occurred with increasing frequency in the latest years, which is in accordance with the increasing frequency of severe draughts observed in other areas of Southern and Central Italy (Fiorillo and Guadagno, 2012; Di Matteo et al. 2013).

20.2.1.8 Future scenarios

In order to quantify the possible lake levels changes in the next years, the calibrated model was run assuming the rainfall to be lower than that effectively recorded in the 1984-2006 period. Figure 20.5 shows what lake levels would have been if rainfall, in 1984-2006, was 5%, 10%, 15% and 20% lower than that recorded and had the same monthly distribution.

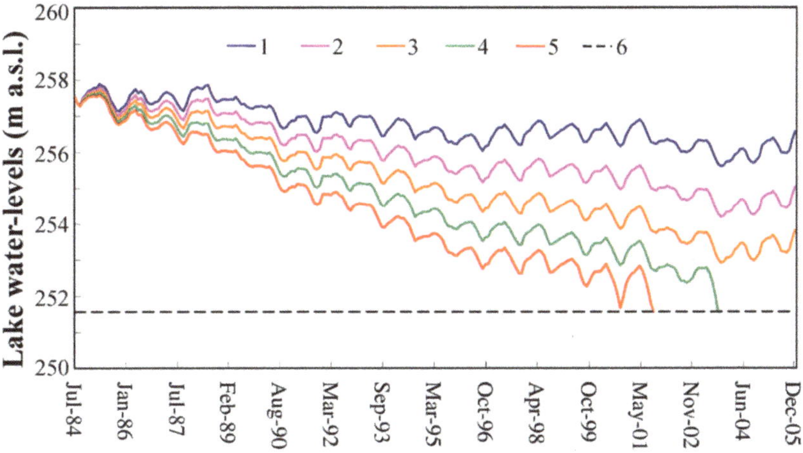

Fig. 20.5. **Lake levels simulations with different rainfall decrease percentage with respect to 1984-2006. 1) Observed levels; 2) -5%; 3) -10%; 4) -15%; 5) -20%, 6) Lake bottom (modified after Dragoni et al. 2012).**

The results indicate that, in the hypothesis of a rainfall reduction higher than 10%, and staying the same other factors, such as evapotranspiration, evaporation and water withdrawals, there would have been a dramatic decrease of lake levels, and the lake would seasonally get dry. The lake would reach this situation very fast, in no more than two or three decades. In the hypothesis of rainfall reductions of 10% and 5% the lake would reach new average levels of about 253.5 and 254.7 m a.s.l., corresponding to surfaces of 90 and 110 km^2 respectively. Despite the approximations of such simulations it is evident that even low rainfall decrease would have dramatic effects on the lake and its ecosystem.

It has to be noticed that, according to the model results, the influence of evaporation on lake levels is much stronger than that of human withdrawals; this suggests that reducing withdrawals would not be sufficient to minimize the impact of climate changes on the lake.

20.2.2 Bagnara Spring

20.2.2.1 General information

Bagnara Spring is one of the main water resources feeding the aqueduct network of Umbria region, being part of the system Bagnara-San

Giovenale-Aretusa, which supplies from 220 l/s to 365 l/s to the aqueduct systems, depending on the season (Regione Umbria, 2007).

The mean annual discharge of Bagnara spring, calculated in the 1984-2010 period, is about 110 l/s.

20.2.2.2 Geological and hydrogeological characteristics

Bagnara spring is located on the west slope of an asymmetric anticline (Monte Pennino), at the contact between the marl formation of Scaglia Variegata and the limestone formation of Scaglia s.l. (Fig. 20.6). The Monte Pennino is an east-vergent asymmetric anticline, with a main thrust structure and some secondary thrusts on the eastern slope, and back thrusts structures on the western slope.

The rocks outcropping in the area belong to the Umbria-Marche geological sequence, consisting of Jurassic-Miocene carbonates and marls in which fractured and/or stratified limestones host aquifers separated from each other by marl aquicludes. In this area the Umbria Marche reduced sequence outcrops in which, unlike in the complete sequence, the Jurassic platform limestone of Calcare Massiccio, outcropping in the Monte Pennino anticline core, is separated from the cretaceous pelagic limestone of Maiolica by transition Jurassic formations mainly made by micritic limestones. On the groundwater circulation point of view, in the Umbria Marche sequence a key role is played by the Marne a Fucoidi formation, which is made by very low permeability, marls separating the older permeable formations (from Calcare Massiccio to Maiolica) from the younger ones (Scaglia s.l.). In the Monte Pennino area, due to the presence of back thrusts structures, which break the continuity of the Marne a Fucoidi, there is a hydraulic connection between the older and the younger permeable formations, so that the anticline core feeds the spring. In Fig. 20.6 the geological sketch of the potential recharge area (about 7.5 km^2 wide) is represented, which is bounded by low permeability marl formations. It has to be noticed that the water coming from the anticline core partly feeds the spring, and partly is connected to a deeper flow directed westward. This scheme is consistent with that proposed for other Apennines anticlines of the same kind as Monte Pennino, in which the estimated aquifer recharge is significantly higher than the discharge coming out from the spring (Cambi and Dragoni, 2000; Cambi et al. 2010; Di Matteo et al. 2013).

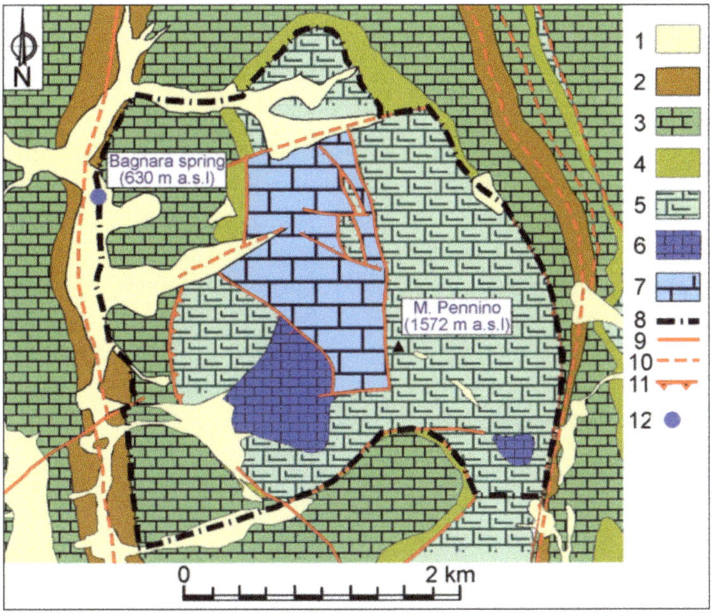

Fig. 20.6. **Hydrogeological scheme of Bagnara spring recharge area. Rock formations similar in age and permeability are unified: 1) Alluvial deposits 2) Scaglia variegata and cinerea (low permeability), 3) Scaglia s.l. (medium-high permeability), 4) Marne a Fucoidi (low permeability), 5) Maiolica (high permeability), 6) Jurassic formations (medium-high permeability), 7) Calcare Massiccio (high permeability), 8) recharge area boundary, 9) fault, 10) hypnotized fault 11) thrust 12) Bagnara spring (modified after Cambi and Dragoni, 2000).**

20.2.2.3 Climatic characteristics

In order to define the average climatic characteristics of the Bagnara spring catchment on a monthly scale, temperature and rainfall data recorded in climatic stations sited in its neighborhood were used.

20.2.2.4 Mean monthly temperature

Since there are no temperature stations located within the Bagnara spring catchment, data recorded in the Nocera Umbra stations by the hydrographic Service of Umbria Region were used to define the mean monthly temperature of the area. This station, located at an elevation of 535 meters a.s.l., is operating since 1988, is the closest to the recharge area and, unlike other stations in this area, is rather continuous, with less than 7% data missing since 1988. Since in this area there are no temperature recordings at high elevations, in order to define the monthly temperature at the mean elevation of the catchment (1150 m a.s.l.), a negative gradient of 0.6 °C/100 m was applied to the data measured in Nocera Umbra. This gradient is consistent with that used by other authors in the same area (Arcaleni et al. 1996; Ardizzone et al. 1999). The mean monthly temperature defined this way was used to calibrate and validate the Bagnara spring model, which will be described in the following paragraphs. The analyzed data also allowed the defining of mean annual temperature over the recharge area which was estimated equal to 8.0 °C.

20.2.2.5 Mean monthly rainfall

As for temperature, no rainfall stations are located within the catchment area. To define the mean monthly rainfall over the catchment, data from rainfall stations located in the spring neighborhood were used.

The data recorded after 1954 in seven of the analyzed stations, all having a percentage of missing data lower than 15%, were selected to define the elevation/rainfall relationship. Table 20.2 shows the stations used to define this relationship and their geographic coordinates.

Table 20.2. Location of rainfall stations and distance from spring.

Rainfall stations	Elevation (m a.s.l.)	Coordinates WGS84		Distance from spring (km)
		Lon	Lat	
Bagnara	620	19.3118292	42.6632402	0
Bolognola	1070	19.6496262	42.5064203	33
Gelagna Alta	711	19.4481583	42.6171125	13
Pioraco	441	19.4524838	42.7163239	13
Serravalle del Chienti	647	19.4017412	42.6172102	10
Sorti	716	19.4062284	42.661243	8
Ville Santa Lucia	664	19.3251176	42.7384656	9

The detected relationship allowed to estimate an annual rainfall of 1680 mm at the average elevation of the catchment; this value was compared with the annual rainfall recorded in Bagnara, which is the station closest to the spring. The ratio between the two values allowed to define a coefficient equal to 1.38, by which the monthly rainfall recorded in Bagnara was multiplied to obtain the monthly rainfall over the catchment.

20.2.2.6 Climatic trend

The estimated monthly rainfall over the catchment has been analyzed in order to define whether there are significant climatic trends in this area. The statistical analysis, run for the 1954-2010 period by using the Mann-Kendall test, indicated a statistically significant decrease in precipitation. Table 20.3 shows the seasonal and annual trends and their significance.

Table 20.3. Trend and Significance Level (SL, Mann-Kendall test) rainfall over the catchment (1954-2010 period).

Period	Trend	SL (%)
Jan-Mar	-2.39 mm/trimester	96
Apr-Jun	-1.27 mm/trimester	83
Jul-Sep	-1.89 mm/trimester	89
Oct-Dec	-1.62 mm/trimester	67
Annual	-7.15 mm/year	99

This trend is mainly to be related to the statistically significant negative trend detected in the January-March trimester, with a gradient of -2.39 mm/trimester.

A statistical analysis was also run for the monthly temperature estimated over the basin, for the same period as rainfall (1954-2010). The results of the analysis, run by using the Mann-Kendall test, indicate the presence of a statistically significant positive trend in annual, six-month and seasonal temperature. The only season with no trend is winter (January-March).

20.2.2.7 Spring model

A rainfall runoff model was used to simulate the monthly discharge of Bagnara spring and to estimate the possible discharge variations over the 2011-2040 period.

A lumped model was used, which is a modification of rainfall-runoff models previously implemented and used to simulate the discharge of some catchment in Central Italy (De Felice and Dragoni, 1993; Dragoni and Valigi, 1995; Cambi et al. 2003; Di Matteo et al. 2010a).

The water budget is calculated monthly and the calibration is run by means of the minimum squares error criterion comparing the measured and simulated monthly spring discharges.

The input data necessary to run the model are the latitude, the monthly rainfall over the catchment (P_b), the mean monthly temperature (T_m) and the measured monthly discharges (Q_u). Since the average catchment elevation is 1150 a.s.l., the model allows to generate the effect of snow by using the SNOW PACK routine in which precipitation (S_w) is treated as snow; the precipitation fallen during a specific month is entirely available to generate evapotranspiration, runoff and recharge within that month only when the mean temperature is higher than -0.5°C. For lower temperatures, only 15% of precipitation is made available within that month, whereas the 85% left is shifted to following months and made available as melted snow (M) just in the first month with temperature higher than -0.5°C.

The model is conceptually made of two tanks in series (TANK1 and TANK2); TANK1 simulates the soil and generates real evapotranspiration (E_b), surface runoff (D_s) and aquifer recharge (R_a). TANK2 conceptually follows Darcy's equation and generates both spring discharge (Q_u) and the deep regional flow (Q_l) (Fig. 20.7).

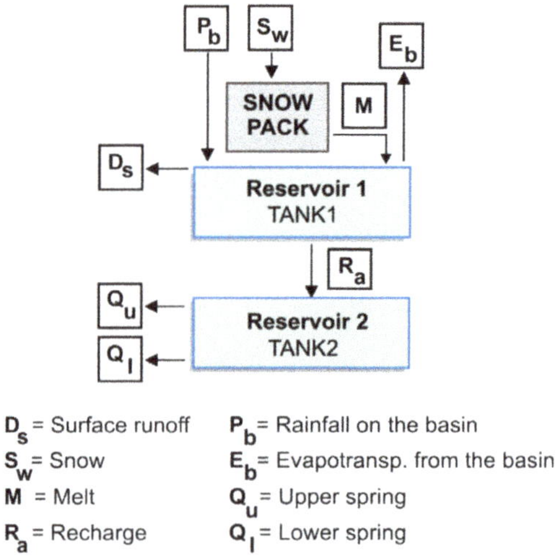

Fig. 20.7. Spring model conceptual scheme.

TANK1 generates potential evapotranspiration by using the equation proposed by Dragoni and Valigi (1994), modified by using a wave function with four calibrated coefficients. E_b is calculated, starting from potential evapotranspiration, by means of the Thornthwaite-Mather criterion (Thornthwaite and Mather, 1957).

Surface runoff (D_s) is estimated on the basis of the Vandewiele equation (Vandewiele et al. 1992) starting from the monthly rainfall and is considered as a system output.

Groundwater recharge (R_a) is calculated by TANK1 as the difference between P_b and E_b net of the surface runoff; this amount conceptually recharges the aquifer represented by TANK2, which feeds two outputs located at different elevation, respectively representing Bagnara spring and the deep regional flow. The amount of water flowing from TANK2 in a specific month depends on the water available in the TANK2 and on the recharge generated during that month by TANK1.

Q_u is generated by TANK2 by means of a coefficient (β_0) to be calibrated, which regulates the aquifer drainage (De Felice et al. 1993). Once Q_u has been generated Q_l is calculated by means of the equation proposed by Chapman (1999).

The best set of coefficient is chosen by comparing the actual and simulated discharges and by minimizing the standard error.

The model was calibrated, with fairly good results on the basis of the discharge data recorded between Aug 1998 and Aug 2009 (mean discharge = 99 l/s) and verified by means of a validation process run over the Jan 1989 - Dec 1996 period (mean discharge =108 l/s). It is important to notice that data used to calibrate the model are referred to the upper spring discharge, which is effectively measured, being the deep flow discharge calculated by the model as a consequence of the water available in TANK 2 and the discharge of the upper spring.

When the amount of water stored in TANK2 within a month is not high enough to feed both outputs the tank only feed the lower output, representing the regional flow, whereas the discharge of the higher output, representing the spring, gets to nil.

Table 20.4 shows the results of calibration and validation.

Table 20.4. Calibration and validation process results for the Bagnara spring model; in italic the values calibrated by the model.

Parameter	**Calibration (1998 – 2009)**	**Validation (1989 - 1996)**
Mean monthly temperature °C	8.9	8.4
Rainfall (P_b) over the recharge area (mm/year)	1537	1569
Evapotranspiration (E_b) (mm/year)	*634*	*618*
Surface runoff (D_s) (mm/year)	*330*	*343*
Recharge (R_a) (mm/year)	*573*	*608*
Water surplus (mm/year)	*892*	*932*
Simulated upper spring discharge (Q_u)(l/s)	*100*	*96*
Simulated lower spring discharge (Q_u) (l/s)	*39*	*39*

Table 20.5 shows Nash-Sutcliffe Efficiency (NSE) values for calibration and validation processes.

Table 20.5. Calibration and validation process results for the Bagnara spring model.

Period	**Process**	**NSE**
1998-2009	calibration	0.83
1989-1996	validation	0.73

It can be noticed that, according to the model, both in calibration and validation period, the deep flow discharge is, in average, less than 30% of the total discharge coming from the system and about 40% of the upper spring discharge. Nevertheless the ratio between the deep flow and upper spring discharge, changes month by month depending on the amount of water available in TANK 2; the less the water available the higher the ratio between deep flow and upper spring discharge.

Figure 20.8 shows the discharges simulation for both calibration and validation periods. It is interesting to notice that in the calibration period (1998-2009) minimum monthly discharge values approached zero more

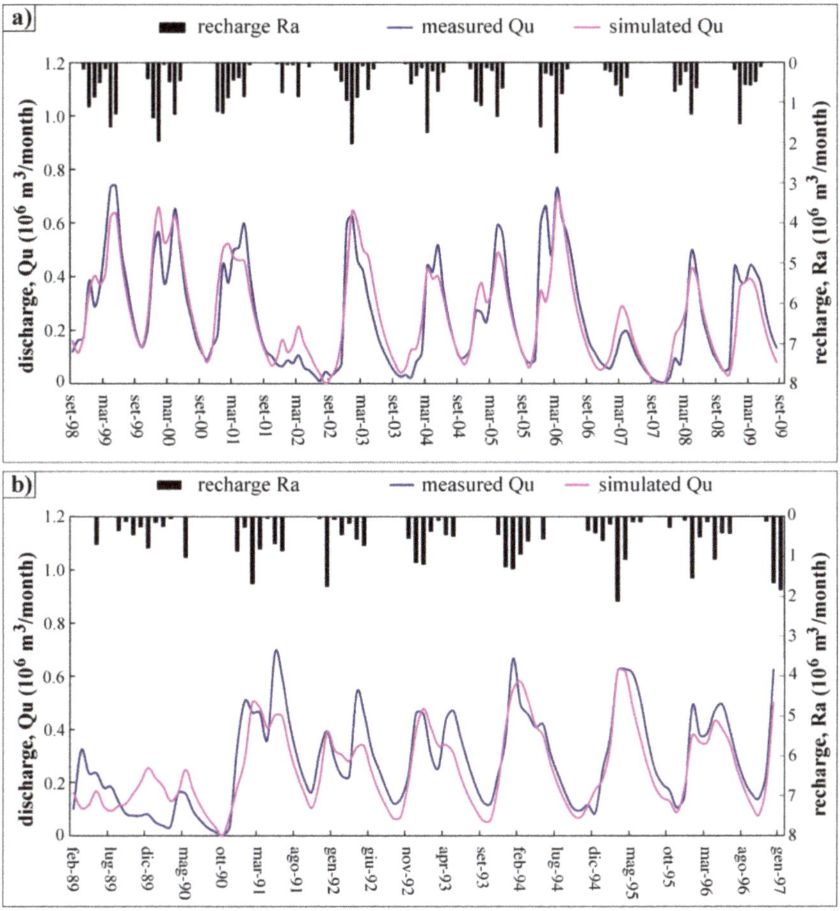

Fig. 20.8. Bagnara spring discharge simulations for the calibration (a) and validation period (b).

times that in the validation period (1989-1996). This behavior can be attributed to the increase of extreme draughts events detected in the latest decades (Fiorillo and Guadagno, 2012; Di Matteo et al. 2013).

The model results indicate that the surface runoff is about 20% of rainfall, and 37% of water surplus. These results agree with the lithological characteristics of the recharge area, which is made by permeable limestones (Celico, 1986; Civita, 2005).

20.2.2.8 Future scenarios

In order to quantify the possible discharge reduction of Bagnara spring in the next 30 years, the calibrated model was run under different climatic scenarios. The climatic scenarios were set assuming that rainfall and temperature monthly trends detected in the last 60 years would continue over the next 30. The simulations were run both assuming the temperature to remain the same over the next thirty years and assuming the monthly trends of the 1954-2010 interval to continue. This allowed the verifying the influence of temperature variations on the monthly discharge. It is important to notice that in the assumed climatic scenarios rainfall decreases but its distribution remains the same as in the 1954-2010.
Table 20.6 shows the simulation results.

Table 20.6. Simulations of the Bagnara spring discharge for the next thirty years.

Parameter	Period (1989-2009)	Simulation for the next 30 years	
		No T trend, P trend (1954-2010)	T trend, P trend (1954-2010)
Mean monthly temperature (°C)	8.5	8.5	9.1
Rainfall (P_b) over the recharge area (mm/year)	1553	1386	1386
Simulated upper spring discharge (Q_u)(l/s)	98	81	78
Simulated deep flow discharge (Qd) (l/s)	58	52	51

Figure 20.9 shows the simulated spring hydrograph for the next thirty years assuming both temperature and rainfall to change.

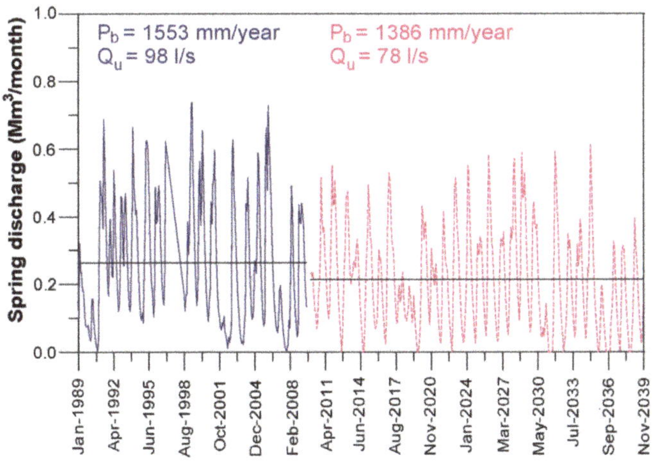

Fig. 20.9. Simulation of present (1989-2009) and future (next thirty years) spring hydrographs.

Results show that a rainfall variation of about -11% with respect to 1989-2009, leads the discharge to decrease of about 17% assuming the temperature to stay constant, and of about 20% considering a positive temperature trend. These results indicate that the investigated system does not react linearly to a change in rainfall, even though rainfall distribution remains the same. This non-linearity is for sure to be attributed, at least partly, to the fact that the systems have two outputs (i.e. exits) at different levels, and to the fact that the distribution of water between the two outputs changes by varying the available water. It is also interesting to notice that in the simulated scenario the springs dries up with a significantly higher frequency with respect to the previous interval, which is in accordance with the increment in droughts periods detected in the latest decades.

20.3 DISCUSSION AND CONCLUSIONS

In this work two relatively small strategic water systems were analyzed and modeled in order to better understand their behavior and their possible reaction to climatic variations. Once the models had been calibrated and validated they were used to simulate the systems behaviour assuming the climatic conditions to change, and particularly assuming rainfall to decrease and temperature to stay constant or to increase, in accordance with the trends detected in the latest decades in the areas where the systems are located. The choice of setting future climatic scenarios according to these criterions is due to a few reasons. One is that in the considered area there is a high spatial variability of rainfall, so that even the high resolution RCMs do not represent properly the rainfall on small recharge areas as that of Bagnara (about 7.5 km^2). Besides this, GCMs and RCMs give, for Central Italy, precipitation scenarios which are sensibly different, and thus also the results obtained using downscaling methods would differ. Indeed, according to Russell et al. (2011), Teutschbein et al. (2011), Chen et al. (2011) and Camici et al. (2013), climate scenarios built up from downscaling methods have to be used with caution due to their high uncertainty. In this framework, the local available data sets indicate a long going decreasing trend of rainfall: as our aim is to get an idea of what would happen to the water resources if these trends continue, we just extrapolated them to the next 30 years. The extrapolation is in some way justified by the fact that in the considered area there is a significant inverse relationship between average yearly temperature and yearly rainfall, indicating that if in the future the temperature increases, on the average the rainfall should decrease (Dragoni, 1998; IPCC, 2013). As for Trasimeno Lake, possible future scenarios were assimilated to those built assuming the rainfall to decrease up to -20% with respect to the 1984-2006 interval. Such decrease corresponds to the worst situation also according to GCMs. Model results indicate that, even assuming temperature to remain constant and rainfall decrease to be as small as -5%, the water stored in the lake and the average lake deepness would decrease significantly. According to the model indications, the lake would become seasonally dry for a rainfall decrease of about 15% of that of the last decades. Model results indicate that the preeminent effect on lake levels is due to evaporation from lake surface and to evapotranspiration from the catchment, being other outputs such as withdrawals for irrigation one order of magnitude smaller. This suggests that, in the hypothesis of a strong rainfall decrease, lake management would become particularly difficult, since even a reduction in withdrawals could not be sufficient to mitigate lake levels decrease.

As for Bagnara spring, future scenarios were built on the basis of the rainfall and temperature trends detected in the time series recorded in the spring surroundings. The positive linear trend detected in temperature and the negative linear trend detected in rainfall was extrapolated to the next thirty years to create scenarios under which running the previously calibrated model. Although there are no evidences of the fact that the detected trend will continue in the future, given the uncertainty of GCMs projections, this seemed to be a proper procedure to get indications on the possible future behavior of the spring, considering also that extrapolating the trends leads to scenarios compatible with the indications of GCMs. The key point is that both GCMs and the detected instrumental trend indicate that rainfall is decreasing and temperature is rising so that a negative reaction of water system is to be expected and proper management schemes have to be prepared. With this perspective what seems to be important is that the model is able to quantify in which way the Bagnara spring system behavior would change in the hypothesis of a decrease in recharge, making it evident that the ratio between upper spring and deep flow discharge would decrease when decreasing effective infiltration, so that the upper spring would be more affected by a rainfall drop than the deep, unexploited, regional flow. The indications of the lumped model presented here agree with those formerly obtained by Cambi and Dragoni (2000) by means of a finite difference spring model.

Due to the inaccuracies about the input data, the approximations of the modelling and, above all, the uncertainties about the future climate, these results are no more than plausible scenarios. Nonetheless they demonstrate that the two systems are very sensible even to slight variation of rainfall. Given that in the area, as well as all over the Italian peninsula, for the next decades most of the scenarios point towards a decrease of future rainfall, the stakeholders should consider these results when updating the management strategies (cf. just for example, Dragoni et al. 2013; Baiocchi et al. 2014).

20.4 ACKNOWLEDGMENTS

The sections regarding the Trasimeno Lake was in part supported by Fondazione Cassa di Risparmio di Perugia, (progetto 2010.011.0442); the sections about Bagnara Spring summarize part of the results obtained within the framework of project PRIN-2008YYZKEE and project SECLI 2011 (Regione Umbria). The authors wish to thank all these Institutions for their support.

20.5 REFERENCES

Arcaleni, A., Ardizzone, F., Mariani, M. and C. Pasquali. 1996. *Progetto per la definizione delle aree di salvaguardia della sorgente di Bagnara*. Rapporto Umbra Acque (not published).

Ardizzone, F., Dragoni, W., Marchetti, G., Mariani, M. and G. Vacca. 1999. Definizione delle aree di salvaguardia della sorgente di Bagnara di Nocera Umbra. Quaderni di Geologia Applicata 1:1305-1313.

Baiocchi A., Dragoni W., Lotti F., Piscopo V. 2014. Sustainable yield of fractured rock aquifers: the case of crystalline rocks of Serre Massif (Calabria, southern Italy). Chapter 5 of "Fractured Rock Hydrogeology", edited by John M. Sharp, CRC Press, Series: IAH - Selected Papers on Hydrogeology, pp. 402. ISBN 9781138001596.

Bartolini, B., Morabito, M., Torrigiani, T., Petralli, M., Cecchi, L., Orlandini, S., Baldi, M., Grifoni, D., Dalu, G., Pasqui, M. and G. Maracchi. 2008. Climatologia delle temperature ed eventi estremi estivi a scala nazionale e regionale. In: *Clima e Cambiamenti Climatici - Le attività di Ricerca del CNR*, Carli, B., Cavarretta, G., Colacino, M. and S. Fuzzi (eds). Consiglio Nazionale delle Ricerche: Roma, 233-236.

Brunetti, M., Cecchini, S, Maugeri, M. and T. Nanni. 2000a. Solar and terrestrial signals in precipitations and temperature in Italy from 1865 to 1996. Advances in Geosciences, W. Schroeder (ed.). IAGA: 124-133.

Brunetti, M., Maugeri, M. and T. Nanni. 2000b. Variations of temperature and precipitation in Italy from 1866 to 1995. Theor Appl Climatol 65:165-174.

Brunetti, M., Maugeri, M. and T. Nanni. 2002. Atmospheric circulation and precipitation in Italy for the last 50 years. Int J Climatol 22:1455-1471.

Cambi, C. and W. Dragoni. 2000. Groundwater, recharge variability and climatic changes: Considerations arising from the modelling of a spring in the Umbria-Marche Apennines. Hydrogéologie 4:11-25.

Cambi, C., Dragoni, W. and D. Valigi. 2003. Water management in low permeability catchments and in times of climatic change: the case of the Nestore River (western central Italy). Phys Chem Earth 28/4-5:201-208.

Cambi, C., Valigi, D. and L. Di Matteo. 2010. Hydrogeological study of data-scarce limestone massifs: the case of Gualdo Tadino and Monte Cucco structures (central Apennines, Italy). B Geofis Teor Appl 51:4:345-360.

Camici, S., Brocca, L., Melone, F. and T. Moramarco. 2013. Impact of Climate Change on Flood Frequency Using Different Climate Models and Downscaling Approaches. J. Hydrol. Eng., 10.1061/(ASCE)HE. 1943-5584.0000959 (in press).

Celico, P. 1986. *Prospezioni Idrogeologiche.* Vol. 1. Liguori editore, pp 735.

Chapman, T.G. 1999. A comparison of algorithms for stream flow recession and baseflow separation. *Hydrol Proc* 13:701–714.

Chen, J., François, P. B. and R. Leconte. 2011. Uncertainty of downscaling method in quantifying the impact of climate change on hydrology. *Journal of Hydrology* 401:190–202.

Civita, M. 2005. *Idrogeologia Applicata e Ambientale.* Casa editrice Ambrosiana, pp 794.

De Felice, A. M., Dragoni, W. and G. Giglio. 1993. Comparison of basin hydrological characteristics using only one lumped parameter: preliminary note. 1° Conference on Methods of Hydrologic Basin Comparison, Oxford, UK, 28 september - 2 October 1992. Robinson M. (ed.), published by Institute of Hydrology, Wallingford, UK, Report 120:112-122.

Di Matteo, L., Gnucci, L. and F. Lotti. 2005. Problemi ambientali ed idrogeologici dei principali laghi dell'Italia Centrale. Convegno: Giornata mondiale dell'acqua: lagune, laghi e invasi artificiali italiani. Accademia Nazionale dei Lincei, Roma, 22 Marzo 2005. In: Atti dei Convegni Lincei, 222:247-259.

Di Matteo, L., Dragoni, W., Giontella, C. and M. Melillo. 2010a. Impact of climatic change on the management of complex systems: the case of the Bolsena Lake and its aquifer, Central Italy. In: *Global Groundwater Resources and Management (Chapter 5).* Paliwal B. S. ed., Scientific Publishers (Jodphur, India), ISBN 978-81-7233-619-6: 91-106.

Di Matteo, L., Dragoni, W. and C. Giontella. 2010b. Environmental and hydrological problems in times of climatic change and increasing water demand: the case of Bracciano Lake and its aquifer (Central Italy). In: *Book of abstract of 13th Biennial Conference ERB 2010 on Hydrological Responses of Small Basins to a changing environment.* Holzmann, H., Godina, R. and G. Müller (eds) (Austria), ISBN978-81-7233-619-6, 187-190.

Di Matteo, L., Valigi, D. and C. Cambi. 2013. Climatic characterization and response of water resources to climate change in limestone areas: some considerations on the importance of geological setting. *J Hydrol Eng* 18:7:773-779.

Dragoni, W. 1982. Idrogeologia del lago Trasimeno: sintesi, problemi, aggiornamenti. Geogr Fis Dinam Quat 5:192-206.

Dragoni, W. and D. Valigi. 1994. Contributo alla stima dell'evaporazione dalle superfici liquide nell'Italia Centrale. *Geologica Romana* 30:151-158.

Dragoni W. and D. Valigi. 1995. Un modello per la stima di deflussi mensili in bacini impostati su litotipi a bassa permeabilità. Quaderni di Geologia Applicata 2:279-286.

Dragoni, W. 1998. Some consideration on climate change, water resources and eater needs in the Italian region so the 43°N. In: *Water Environmental and Society in Times of Climatic* Change Issar, A. and M. Brown (eds). Kluwer, 241:271.

Dragoni, W., Speranza, G. and D. Valigi. 2003. Impact of climatic variations on the hydrogeological systems: the case of Pescara di Arquata spring (Umbria-Marche Apennines, Italy). *Geologia Tecnica ed Ambientale* 3:13-27.

Dragoni W, Piscopo V., Di Matteo L., Gnucci L., Leone A., Lotti F., Melillo M., Petitta M.: (2006): Risultati del Progetto di Ricerca PRIN "Laghi 2003-2005 - *Results of the PRIN Research Project "Lakes 2003-2005"*. Atti del 2° Congresso Nazionale AIGA 15-17 febbraio 2006, BARI, in *Giornale di Geologia Applicata*, Vol. 3, pp. 39 – 46.

Dragoni, W. and B. S. Sukhija. 2008. Climate change and groundwater: A short review. Climate change and groundwater, Dragoni W. and B. S. Sukhija (eds.). *Geol. Soc*. Spec. Publ., London, 288:1-12.

Dragoni, W., Melillo, M. and C. Giontella. 2012. Bilancio idrico del Lago Trasimeno. In: "*Tutela Ambientale del lago Trasimeno*", Martinelli, A. ed., Libri/Arpa Umbria, ISBN: 978-88-905920-03, 69-85.

Dragoni, W., Mottola, A. and C. Cambi. 2013. Modeling the effects of pumping wells in spring management: the case of Scirca spring (Central Apennines, Italy). *J Hydrol* 493:115-123.

Fiorillo, F. and F. M. Guadagno. 2012. Long karst spring discharge time series and draughts occurrence in Southern Italy. Environ Earth Sci 65:2273-2283.

Gao, X. and F. Giorgi. 2008. Increased aridity in the Mediterranean region under greenhouse gas forcing estimated from high resolution simulations with a regional climate model. *Global Planet Change* 62:195-209.

Gasperini, L., Barchi, M.R., Bellucci, L.G., Bortoluzzi, G., Ligi, M. and C. Pauselli. 2010. Tectonostratigraphy of Lake Trasimeno (Italy) and the geological evolution of the Northern Apennines. *Tectonophysics* 492:164-174.

Giorgi F. 2006. Climate change hot-spots. *Geophys Res Lett* 33, L08707.

Gualdi, S., Somot, S., Li, L., Artale, V., Adani, M., Bellucci, A., Braun, A., Calmanti, S., Carillo, A., Dell'Aquila, A., Déqué, M., Dubois, C., Elizalde, A., Harzallah, A., Jacob, D., L'Hévéder, B., May, W., Oddo, P., Ruti, P., Sanna, A., Sannino, G., Scoccimarro, E., Sevault, F. and A. Navarra. 2013. The Circe Simulations. Regional Climate Change Projections with Realistic Representation of the Mediterranean Sea. *American Meteorological Society*, January 2013:65-81.

IPCC. 1996. Technical summary. In: *Climate change 1995. Contribution of WG1 to the Second Assessment Report of the IPCC*. Houghton, J. T., Meira Filho, L. G., Callander, B. A., Harris, N., Kattenberg, A. and K. Maskell (eds.), Cambridge University Press 9-49.

IPCC. 2013. *WGI Fifth Assessment Report. Final Draft* (7 June 2013) Chapter 11.

Maugeri, M. and T. Nanni. 1998. Surface air temperature variations in Italy: recent trends and an update to 1993. *Theor Appl Climatol* 61:191-196.

McKee, T. B., Doesken, N. J. and J. Kleist. 1993. The relationship of drought frequency and duration to time scales. Proc. Eighth Conf. on Applied Climatology, 17-22 January 1993, Anaheim, CA, *American Meteorological Society*, Boston, MA, 179-184.

Moriasi, D. N., Arnold, J. G., Van Liew, M. W., Bingner, R. L., Harmel, R. D. and T. L. Veith. 2007. Model Evaluation Guidelines for Systematic Quantification of Accuracy in Watershed Simulations. Transactions of the *ASABE* 50:3:885-900.

Nanni, T., Brunetti, M. and M. Maugeri. 2008. Variazione nella frequenza e nell'intensità delle precipitazioni giornaliere in Italia negli ultimi 120 anni. In: *Clima e Cambiamenti Climatici - Le attività di Ricerca del CNR*, Carli, B., Cavarretta, G., Colacino, M. and S. Fuzzi (eds.). Consiglio Nazionale delle Ricerche: Roma.

Nash, J. E. and J. V. Sutcliffe. 1970. River flow forecasting through conceptual models part I - A discussion of principles. *J Hydrol* 10:3:282-290.

Regione Umbria. 2007. *Piano regolatore regionale degli acquedotti (P.R.R.A).* Supplemento straordinario al Bollettino ufficiale delle Regione dell'Umbria. Serie generale, n. 16 dell'11 aprile 2007.

Romano, E. and E. Preziosi. 2013. Precipitation pattern analysis in the Tiber River basin (central Italy) using standardized indices. *Int J Climatol* 33:1781-1792.

Romano, E., Camici, S., Brocca, L., Moramarco, T., Pica, F. and E. Preziosi. 2014. On the variables to be considered in assessing the impact of climate change to alluvial aquifers: a case study in Central Italy. *Procedia Engineering* 70:1430-1440.

Russell, S. C., Warrick R. D., Stephen P. C., Freddie S. M., Santosh A., Olga B. and G. K. Summerell. 2011. Differences in future recharge estimates due to GCMs, downscaling methods and hydrological models. *Geophysical Research Letters* 38: L11406.

Schneider, C., Laizé, C. L. R., Acreman, M. C. and M. Florke. 2013. How will climate change modify river flow regimes in Europe? *Hydrol Earth Syst Sci* 17:325:339.

Strzepek, K. M. and D. N. Yates. 1997. Climate change impacts on the hydrologic resources of Europe: A simplified continental scale analysis. *Climatic Change* 36:79-92.

Teutschbein, C., Wetterhall, F. and J. Seibert. 2011. Evaluation of different downscaling techniques for hydrological climate-change impact studies at the catchment scale. *Clim Dyn* 37:2087–2105.

Thornthwaite, C. W. and J. R. Mather. 1957. Instructions and Tables for computing potential evapotranspiration and the water balance. Publ *Climatol Lab Climatol*, Drexel Inst of Technol 10:3:185-311.

Todisco, F. and L. Vergni. 2008. Climatic changes in Central Italy and their potential effects on corn water consumption. *Agr Forest Meteorol* 148:1-12.

Vandewiele, G. L., Xu, C. Y. and N. L. Win. 1992. Methodology and comparative study of monthly water balance models in Belgium, China and Burma. *J Hydrol* 134:315-347.

CHAPTER 21

CLIMATE CHANGE RELATED RISKS PREVENTION AND MANAGEMENT IN UMBRIA REGION – POR-FESR 2007-2013 FUNDS

Angelo Viterbo[1], Nicola Berni[2], and Claudia Pandolfo[2]

[1] Umbria Region, Water Resources and Hydraulic Risk, Perugia, ITALY
aviterbo@regione.umbria.it

[2] Umbria Region, Civil Protection - Functional Center, Foligno (PG), ITALY
nberni@regione.umbria.it <> cpandolfo@regione.umbria.it

ABSTRACT

This Chapter provides an overview of the Umbria Region (Italy) policies and activities in the climate changes related risk prevention and management.

21.1 INTRODUCTION

It's almost certain that global climate change will affect the frequency and severity of extreme meteorological and hydrological events. In southern Europe, IPCC (2013) assesses that extended periods of drought and water scarcity are attended, while across all Europe extreme weather events such as intense rains giving rise to flash flooding are expected to become more common.

Also in Umbria region (8000 Km^2, central Italy), in the last fifteen years an intensification of extreme weather events with the alternation of droughts and floods has been observed.

At European level, the directives 2000/60 and 2007/60 deal with water availability, quality and flood risk issues: a new approach, shifting member countries policies focus to integrated and multidisciplinary management is set.

Especially in Italy, a common effort, of all institutions and stakeholders involved, to develop and apply new methodologies for the better understanding and prevention of weather induced extreme events effects is required.

In line with European Directives, Umbria Region policies in the two major water-related sectors (water resources and flood risk management) promoted several activities, mainly financed through POR-FESR structural funds.

21.2 CLIMATE CHANGE IN UMBRIA REGION AND RISK PREVENTION AND MANAGEMENT

In Umbria region, in the past fifteen years an intensification of extreme weather events occurred, with alternating dry periods (2001-2003, 2007, and 2012), which led to the declaration of national state of emergency, and flood events (2005, 2008, two in 2010, 2012 and 2013), with significant effects/losses. Moreover, this intensification of extreme events could undermine the system of regional drinking water supply with serious economic losses and inconvenience to the affected populations.

Therefore, the planning and management of water resources, with particular reference to the use of drinking water, can no longer rely, as in the past, only on the traditional time series analysis of supply sources (springs, aquifers, reservoirs, etc. ..).

Umbria Region policies were addressed to:

- water resources management, ensuring regional planning to be based on sustainability principles (priority for drinking water use);
- flood risk management, developing prediction and prevention actions in order to effectively mitigate impacts on the population (priority), mainly through non structural approaches.

The planning of water resources used for drinking water is regulated in the recent "Water Supply Master Plan" (Piano Regolatore Regionale degli Acquedotti - PRRA), which, concluding a process started in the '90s with preliminary studies, led to the development of management models of Apennine springs, aquifers and resources of reservoirs. According to criteria of rationality, quality and environmental sustainability sources of water supply have been identified in the limestone formations of the Apennines, in the volcanics, alluvial aquifers and large reservoirs. The Plan has provided, in particular, the rationalization of the aqueduct systems converting the 240 local aqueducts into eight large water systems interconnected and able to guarantee, in addition to the availability of the water resource, a higher quality of the water distributed and greater resilience in case of water scarcity. The availability of adequate funds allowed, in a short time, the construction of seven of the eight water systems. The Plan establishes an use of water resources based on seasonality: Apennines carbonate aquifers used mostly during winter season; reservoirs mainly used during summer seasons, in order to reduce effects of lack of water in the springs and a reduction of water taken from wells to protect groundwater. The Plan led the improvement of drinking water distribution systems efficiency, sustainability of water resources use and the solution of water quality issues. During the last water crisis of 2012, the criteria of resource management, as identified in the PRRA, and the realized structural interventions were tested with positive results: the systems have ensured the audience without too many problems.

For the flood risk management the Umbria Region realized both structural (main flood prone and high risk areas defense works design and realization, defense works maintenance, ...) and non structural measures (prevention and protection through flood management plans, territorial planning, early warning systems through QPFs, hydrological and hydraulic modelling, real time monitoring networks, meteo-radar, expected ground effects evaluation, dynamic risk scenarios for civil protection DSS, ..)

In the non-structural measures context, the Umbria Region Functional Centre (FC) plays a key role. The Italian national hydrometeorological early warning system is composed by a network of offices (called "Functional Centres") coordinated by the Central Functional Centre (CFC) at the National Civil Protection Department (DPC) in Rome. The Umbria FC provides technical tools able to support decisions when significant flood/landslide events occur, furnishing 24h support for the whole duration of the emergency period. The alert system is based on hydrometric and rainfall thresholds of three different increasing critical levels according to the expected ground effects, together with a detailed procedure for the management of critical events in which the different role of various involved authorities/institutions is defined. For the real time flood forecasting system, several hydrological and hydraulic models were developed.

21.3 Case Study: 2012

In 2012 a severe drought and a severe flood occurred in Umbria, and the national state of emergency was declared.

In the summer 2012 the Umbrian territory was effected by a period of water scarcity: during the hydrologic year (period October 2011 – April 2012) a deficit of precipitation approximately equal to 50% respect historical mean value was recorded and, by way of example, the hydrometric level of Trasimeno Lake had an anomalous negative trend with a minimum hydrometric level equal to -1.60 m. Most of the negative effects were mitigated by previous structural actions, however the 2012 Emergency Plan costs 289 Million Euros.

During the period on 11th to 14th November 2012, central Italy (Umbria, eastern Tuscany and northern Lazio) was interested by a significant rainfall event ("V-shape" mesoscale convective system) causing many floods and landslides (Servizio Protezione Civile – Regione Umbria, 2012). Intense and persistent rainfall invested Umbria: over 300 mm registered in almost 48 hours (corresponding to over 100 years return period). Almost every branch of Umbria hydrographic network was involved, especially the Nestore River and the Paglia River at Orvieto town. Fortunately no casualties were reported, but a lot of damages occurred (230 million euros direct damages).

21.4 POR-FESR ACTIVITIES

Since 2007, the Umbria Region carried out a lot of activities, financed by the structural funds POR-FESR 2007-2013, both in water resources and drought management and in flood risk management, and in particular for the climate change assessment and for defining a new drought/flood management regional approach.

In last years, POR-FESR 2007-2013 (Regional Operational Program funded by the European Regional Development Fund) is one of the main financial instruments for the promotion and support of economic regional development. Umbria Region planned on 2007 over 10 million Euros to develop in the Axis II "Environment and risk prevention" the following activities:

- "Plans and interventions for the prevention of natural risks" (Activities a1);
- "Plans and interventions for the prevention of technological risks" (Activities a2).

In the water sector, the planning had the purpose of defying if climate changes (CC) plays a role at regional scale and, if so, if CC can affect Regional policies, how and how much.

In particular, for the water resources management, the objective was studying what may be the impact on the availability of drinking water in Umbria, in the next future, taking into account non-stationary climate scenarios. There was the need of a study able to assess the validity of the assumptions and strategic choices of the Water Supply Master Plan in all its key points, and the need of a dynamic DSS in order to manage critical drought periods, also for other uses (not only drinking water) taking into account:

1. Climate Change assessment at regional scale;
2. Impact assessment of CC on regional planning for drinking water use;
3. Drought risk analysis;
4. Development of early warning models for risk scenario assessment;
5. Web-based “open source” DSS.

In this contest, the Umbria Region has sought through the project SECLI (“Siccità E Cambiamenti cLImatici” - Drought and Climate Change) to consider the effects of climatic change on the water resources management identified in the PRRA and to verify the availability of water resources planned until 2040.

The project, carried out in two tranches was coordinated by CNR-IRPI (National Research Council - Institute for Hydrogeological Protection) of Perugia and carried out in collaboration with CNR-IRSA (National Research Council - Water Research Institute) of Rome and two departments of the University of Perugia (DICA and DIST) as well as other European research institutes.

The study evaluated the possible climate change on a regional scale, starting from the climate scenarios presented by the IPCC commission (2013), which were similar to those obtained in the past by researches based on regional climatic series, that provide in the future a significant rainfall reduction and temperatures increase in the Mediterranean area (Gao and Giorgi, 2008).

Based on the analysis of meteo-climatic time series and hydrological data for the period 1917-2013 (rainfall, temperature, river flows, springs, wells, etc. ..) for the Umbrian Tiber River catchment (Camici et al. 2011) and implementing a database "open-source" and "web-based" (Pierleoni et al. 2013), procedures and numerical models have been developed aimed at assessing the hydrological response of water systems to future climate scenarios for a time horizon of 60 years, provided by Global Circulation Models downscaled at a local scale.

The analysis of the data showed generally a significant positive trend of temperatures and a significant negative trend for precipitation, although in some restricted areas this trend does not exist or is not significant.

The impacts of future climate scenarios on water courses, reservoirs, springs and alluvial aquifers identified in the PRRA were then analyzed. The criticalities related to the availability of the resource, in terms of frequency and intensity of the crisis, were identified together with possible criteria for management of crisis, also thank to Dynamic Decision Support Systems (DSS) implemented and calibrated for this purpose.

The analysis focused on some of the main sources of supply identified in the current planning: the alluvial aquifer formations of Petrignano di Assisi and Cannara; aquifers of the carbonate massifs of Bagnara, San Giovenale, Scirca, Valnerina and Lupa; the reservoirs of Montedoglio dam along Tiber River and Valfabbrica dam along Chiasco River (Casadei et al. 2014).

Hence, for the flood risk management, the major impact of CC is more likely the greater frequency and intensity of flash floods, so the objective of the planning was developing tools and approaches that can mitigate their effects, like:

1. Meteorological models (LAM) for QPF acquisition;

2. Improvement of the real time regional monitoring network operating in real-time;
3. Extension of flooding maps;
4. Flood damage assessment;
5. Development of early warning models for risk scenario assessment;
6. Web-based “open source” DSS.

Most of the results/products of these actions converged in the FC activity.

21.5 CONCLUDING REMARKS

During the last years, Umbria Region carried out a lot of activities, both in water resources and drought management and in flood risk management, and in particular for the climate change assessment and for defining a new drought/flood management regional approach. It was possible thanks to structural European/Italian funds POR-FESR 2007-2013.

21.6 ACKNOWLEDGMENTS

The listed activities were possible thanks to the passionate and active cooperation among qualified partners: National Research Council - Institute for Hydrogeological Protection (CNR-IRPI), University of Perugia, National Civil Protection Department, National Research Council - Water Research Institute (CNR- IRSA), Politecnico di Milano – University of Milano, Centre for Environmental Research (UFZ) of Leipzig, University of Palermo, Tech. University of Vienna, other Italian Regions.

21.7 REFERENCES

Camici, S., Romano E., Preziosi E., Tarpanelli A., Brocca L., Melone F. and Moramarco T. 2011, Analisi di trend e ciclicità di serie storiche pluviometriche per la modellazione stocastica di scenari di precipitazione in condizioni di cambiamenti climatici. *Le modificazioni climatiche e i rischi naturali*, Polemio M. Ed., 205-208, ISBN 9788890508806.

Casadei, S., Dragoni W., Moramarco T and Preziosi E. 2014. Rapporto Tecnico Finale, *Progetto POR-FESR Siccità e Cambiamenti Climatici* (SECLI).

Gao, X., and Giorgi F. 2008. Increased aridity in the Mediterranean region under greenhouse gas forcing estimated from high resolution simulations with a regional climate model, *Global and Planetary Change*, 62, 195–209.

IPCC - Intergovernmental Panel on Climate Change 2013. *WGI Fifth Assessment Report*. Final Draft, 11.

Pierleoni, A., Camici S., Brocca L., Moramarco T. and Casadei S. 2013. Climate change and decision support systems for water resource management. *CCWI2013*, Procedia Engeneering, DOI 10.1016/j.proeng.2014.02.146.

Servizio Protezione Civile – Regione Umbria. 2012. Evento alluvionale 11-14 Novembre 2012: Rapporto di evento. Available on line at: www.cfumbria.it

CHAPTER 22

FLOOD RISK TO THE POPULATION AND ITS TEMPORAL VARIATION IN ITALY

Paola Salvati[1], Cinzia Bianchi[1], Federica Fiorucci[1], Ivan Marchesini[1], Mauro Rossi[1], Fausto Guzzetti[1]

[1] CNR IRPI, via Madonna Alta 126, 06128 Perugia, Italy, tel.: +39 075 5014427, fax.: +39 0755014420, e-mail: paola.salvati@irpi.cnr.it

ABSTRACT

Floods pose a severe threat to the population of Italy. Quantitative flood risk levels to the population can be evaluated determining levels of individual risk. Using an updated version of an historical record of flood events with human consequences in Italy, we studied the temporal and the geographical distributions of events with casualties in the 1861 - 2013 period, and we determined individual flood risk for the whole Italy, and for five physiographical areas established classifying the Italian municipalities on their terrain elevation and the distance to the sea coast. To study individual flood risk, we determined flood mortality rates using Italian census data from 1861. To evaluate the temporal variation of flood mortality, we divided the historical record into three 50-year periods, and we calculated the average flood mortality for the periods, and their variations in the different physiographical zones. The results were compared to risk levels calculated for landslides, earthquakes, and volcanic activities, and with the perception that the population of Italy has of the risk posed by different natural hazards. The comparison revealed that the perception of the threat does not match the long-term risk posed to the population by floods, and by the other natural hazards.

Keywords: flood risk, individual risk, risk perception, Italy.

22.1 INTRODUCTION

Floods are recurrent events in Italy, where they cause damage and pose a severe threat to the population (Guzzetti et al. 1994; Guzzetti and Tonelli, 2004). Flood hazards and the associated risk were determined at various geographical scales in Italy, from the site specific (local) to the synoptic (national) scale. At the local scale, detailed investigations were used to produce zonations of flood risk ("*Piani di Assetto Idrogeologico*"), which are exploited to design defensive structures and to implement mitigation strategies. At the national scale, estimates of individual and societal risk levels were first determined by Salvati et al. (2003), and were revised by Guzzetti et al. (2005b). More recently, Salvati et al. (2010) have estimated societal flood (and landslide) risk levels for the 20 Italian Regions. The synoptic-scale assessments were conducted exploiting historical information on damage caused by flood events to the population of Italy, and used administrative subdivisions of Italy that do not correspond to morphological or physiographical subdivisions.

In this work, we re-examine flood risk to the population of Italy using an improved version of the historical record of flood events with human consequences prepared by Salvati et al. (2003), and updated by Guzzetti et al. (2005a) and Salvati et al. (2010; 2013). First, we update the existing estimates of individual flood risk in Italy, and in different physiographical zones of Italy. Next, we compare the national flood risk levels to the levels of risk posed by other natural hazards, including landslides, earthquakes, and volcanic activity, and we compare the results with levels of risk perception obtained through a nationwide survey (Salvati et al. 2014 in press).

In this work, we use the term "fatalities" to indicate the sum of the deaths and the missing persons caused by a harmful flood event. The term "casualties" indicates the sum of fatalities and injured people. "Evacuees" are people forced to abandon their homes temporarily, and the "homeless" are people that lost their homes. We use "human consequences" to encompass casualties, homeless people, and the evacuees. A "fatal" event is an event that resulted in fatalities. A "harmful" event is an event with human consequences. "Individual" risk is the risk imposed by a hazard to any unidentified individual.

22.2 RECORD OF HARMFUL FLOOD EVENTS IN ITALY

Using different sources of information, including archives, chronicles, newspapers, scientific journals, technical reports, and other bibliographical sources, Salvati et al. (2003), Guzzetti et al. (2005b), and Salvati et al. (2010, 2013) have compiled a comprehensive historical record of flood events with direct human consequences to the population of Italy. Details about the sources of information used, and the problems encountered in compiling the historical record are given in Guzzetti et al. (2005b). We make constant efforts to keep the catalogue updated. To update the catalogue, we: (i) search national newspapers available on-line and, where available, their digital databases, (ii) obtain daily information from Google Alert® (http://www.google.com/alerts) using pre-defined keywords, (iii) search blogs and other Internet resources for specific events, (iv) search digital libraries and catalogues of archive documents, (v) search regional and local catalogues of harmful natural events, and (vi) read chronicles and local history books. The sources of historical information include regional and local history books, transcriptions or translations of old chronicles, and reports prepared by regional and local technical offices that typically describe the effects of single events and their consequences. The old chronicles provide information on events in the early part of the record, chiefly for geographical areas of limited extent. The regional and the local catalogues provide abundant and accurate information on the date of occurrence of the events, the exact or approximate location of the events, and the type and the extent of the damage. The Internet-based resources provide abundant and accurate information on the location and the impact of the recent events.

Our updated record of historical harmful flood events lists 2,939 events that have resulted in deaths, missing persons, injured people, and homeless in Italy, from 589 to 2013, a period of 1425 years. For some events in the record it was not possible to determine the exact or the approximate number of the consequences, or their exact or approximate location. For the very old events the sources of information provided a qualitative description of the number of fatalities defining them as few, several, many or hundreds. For a number of events, the figures for the evacuees were obtained from the number of the families forced to abandon their homes, and when the exact number of evacuees is available most commonly it sums the figures from different sites. In the historical record, information on the location of flood events is generally accurate, but the geographical accuracy decreases going back in the record. Table 22.1 lists figures on the quantity and quality of the information listed in the record, classified based on the quality of the information measured by the percentage of the events for which the number of the consequences is known/unknown (C_K/C_U), and the geographical location of the consequences is known/unknown (L_K/L_U). In the historical record, for most of the events with casualties (92%) the number of the consequences is known, and for only 8% of the events the number of the consequences is undetermined. The percentage reduces (71%) for the events that have caused evacuees and homeless people. The geographical location of the consequences in known for most of the events with casualties (90%) in the historical record. The percentage is slightly higher (92%) for the events with evacuees and homeless.

Table 22.2 lists information on the human consequences caused by 2,939 harmful flood events in Italy in the period 589-2013, and for four sub-periods.

Table 22.1. Harmful flood events in Italy in the 1,425 year period 589-2013, classified based on the quality of the information.

	All	C_K	C_U	L_K	L_U
	#	%	%	%	%
Events with casualties	1352	92	8	90	10
Events with evacuees and homeless	1834	71	29	92	8

Table 22.2. Statistics of the human consequences of harmful flood events in Italy from 589 to 2013, and for four different sub-periods.

Period	589-2013	589-1860	1861-1910	1911-1960	1961-2013
Length of period	1,425	1,272	50	50	53
Death	40,964	37,782	665	1782	735
Missing person	92	6	-	18	68
Injured people	1,948	44	12	989	903
Evacuee and homeless	708,941	18,264	123,918	275,747	291,012

22.2.1 Temporal Analysis

Figure 22.1 portrays the temporal distribution of the harmful flood events in the historical record. In the figure, the upper chart shows the number of casualties per year for the entire record (1,425 years), and the lower chart shows the recent portion of the record, from 1861 to 2013 (153 years). The oldest flood event in the historical record occurred in late October 589. During this event, widespread inundation caused fatalities in the Veneto region, and probably in other regions of northern Italy, but the number and type of the human consequences remain unknown (open squares in Fig 22.1). The first event in the record for which the (approximate) number of fatalities is known (1000 deaths) occurred in 671 at Sarnico, Lombardia, northern Italy. Inspection of Fig. 22.1 allows for two general considerations. First, the number of reported events has increased significantly with time, and especially after about 1700. Second, the severity of the recorded events, measured by the number of the casualties, has changed with time. The largest number of flood casualties caused by single events was reported in the oldest portion of the record (before 1860).

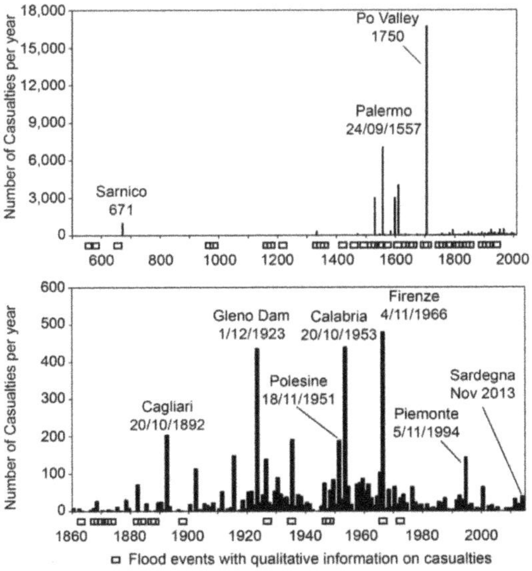

Fig. 22.1. **Historical record of harmful flood events with casualties in Italy from 589 to 2013 (upper chart), and from 1861 to 2013 (lower chart). Modified from Salvati et al. (2012a).**

Figure 22.2 shows the number of casualties per event, for the entire record and for the recent period 1861-2013. The two graphs measure the completeness of the historical record. The amount of information available increases moving forward in time, especially for the low intensity events, which are completely missing in the oldest portion of the record. In the recent period (after 1860) there is a decrease in the frequency of high intensity events, which is more evident after 1970. The difference in the number of casualties per event indicates that the oldest events reported in the record were mainly catastrophic, but also that the first part of the record is incomplete for the medium and the low severity events (Salvati et al. 2010, 2012a). The former is due to the lack of flood protection structures and appropriate warning systems. The latter is a known bias of non-instrumental records of natural events (Guzzetti, 2000). In the record, lack of occurrences in any given period may be due either to incompleteness or to variations in the conditions that led to harmful floods, including climate anomalies, rainfall events, land-use changes, and human actions (Glade et al. 2001; Guzzetti et al. 2005a). Using the recent portion of the record after 1860, we investigated possible trends in the distribution of harmful flood events. For the purpose, the number of events per year was normalized to the yearly average number of events in the 153-year period 1861-2013. Results are shown in Fig. 22.3.

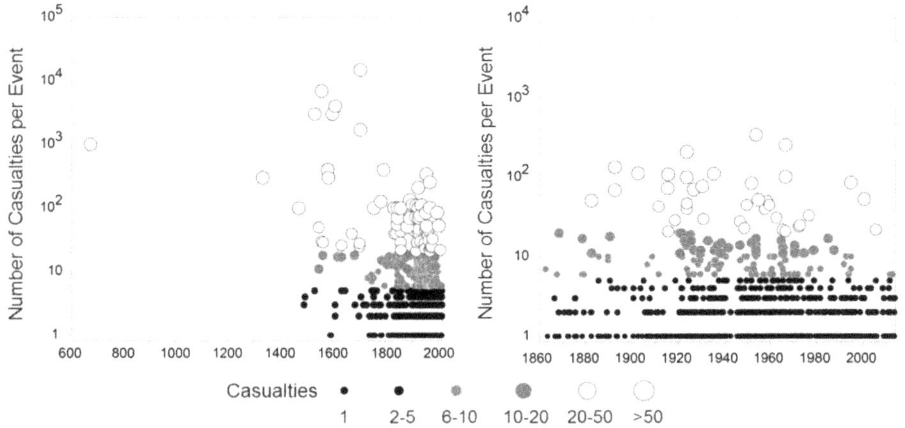

Fig. 22.2. **Number of casualties per harmful flood event in Italy from 600 to 2013 (left), and from 1860 to 2013 (right). Modified from Salvati et al. (2012a).**

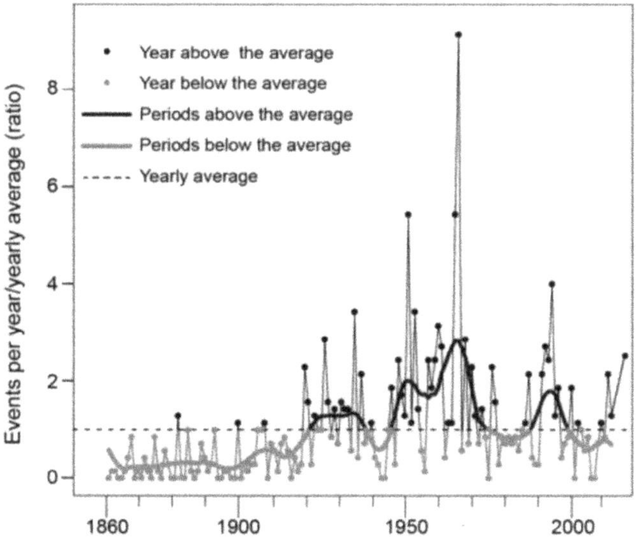

Fig. 22.3. Yearly average number of harmful flood events in Italy between 1861 and 2013.

In Fig. 22.3, the thin line shows the temporal pattern of a parameter that measures the magnitude of the harmful flood events. The parameter is given by the total number of flood events with human consequences per year, divided by the yearly average number of harmful flood events in the observation period (dashed horizontal line in Fig. 22.3). The thick curve was obtained using a low-pass Gaussian filter based on an 11-year kernel window and a three-year standard deviation. The grey parts of the line show periods below the long-term average, and the black parts show periods above the long-term average. Inspection of Fig. 22.3 reveals that the historical record is not uniform, with significant temporal variations in the frequency of the harmful flood events. Before 1900, the record is below the long-term average, even for the most destructive years (1868, 1882, 1885, 1893). After 1900, the trend rises until 1966, albeit with significant variations. Between 1920 and 1970, there were 34 years above and (only) 17 years below the long-term average. The trend dropped sharply between 1968 and 1990, peaked in 1994, and decreased again after 1996. In the 43-year period between 1971 and 2013, there were 20 years above and 23 years below the long-term average. Reasons for the variations in the temporal trend are manifold, and include incompleteness of the old part of the record, climatic and environmental variations, increase of the elements at risk including the population, and deployment of risk mitigation measures.

To identify seasonal trends, and the season with the largest concentration of harmful events, we investigated the monthly distribution of flood events with casualties in Italy in the period 1861-2011, and in the three 50-year periods 1861-1910, 1911-1960, and 1961-2011. Results are shown in Fig. 22.4, where the Table lists the monthly distribution of events in the last two years (2012, 2013) that were not included in the histograms.

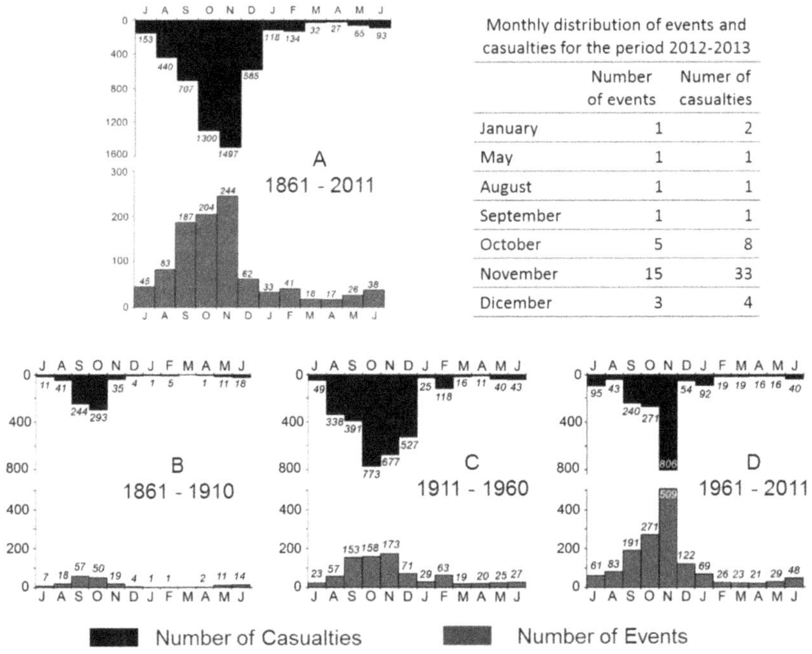

Monthly distribution of events and casualties for the period 2012-2013

	Number of events	Numer of casualties
January	1	2
May	1	1
August	1	1
September	1	1
October	5	8
November	15	33
Dicember	3	4

Fig. 22.4. Monthly distribution of flood events (grey) and flood casualties (black), for different periods.

Considering the entire 151-year period 1861-2011, the majority of the flood events (244) and of the flood casualties (1,497) occurred in November. Harmful floods were more common in autumn, from September to November, when 635 harmful events (63.7%) in the record have resulted in 3,504 flood casualties (68%). The same monthly distribution was registered in the 2-year period 2012-2013. During the last two years the majority of the flood events and of the flood casualties occurred in November (15, 33) and in October (5, 8). Considering the monthly distribution segmented in three 50-year periods (Fig. 22.4 B, C, D), the bar charts reveal a slight variation in the monthly distribution of flood casualties, with the peak value shifting from October to November. The same trend can be seen in the monthly distribution of flood events, with the peak value shifting from September to November. We explain the differences with possible variations in the distribution of the precipitation in the considered periods, possibly driven by climate changes.

22.2.2 Geographical Analysis

Information on the precise or approximate location of floods with human consequences in Italy is available for most of the events listed in the historical record (92% in Table 22.1).

Figure 22.5 portrays the location of 1,093 sites that have experienced one or more flood events with casualties (A), and 1,453 sites with homeless and evacuees caused by flood events (B), in the period 590-2013. The figures comprise sites for which the location is known but the date is unknown. Visual inspection of Fig.22.5 reveals that sites affected by harmful floods are not distributed equally in Italy. Harmful flood events occurred in all the Italian regions and provinces, and in 1,547 of the 8,102 Italian municipalities (19%) (Table 22.3).

We used order statistics (David and Nagaraja, 2003) to investigate the persistency of the events in the affected municipalities and sites using the procedure described in Salvati et al. (2010, 2012a). We performed the analysis for the harmful events, and for the events with casualties. We considered the single sites where flood consequences were reported, and we grouped the information in each municipality. The latter was performed because: (i) municipalities are an administrative subdivision important for land management in Italy, (ii) for some municipality more than one site with consequences is listed in the historical record, and (iii) for some event the exact site of occurrence within a municipality is not known. Results are shown in Fig. 22.6 where the four graphs portray the number of harmful events and of the events with casualties, in each municipality and at each site, against the rank calculated for the period

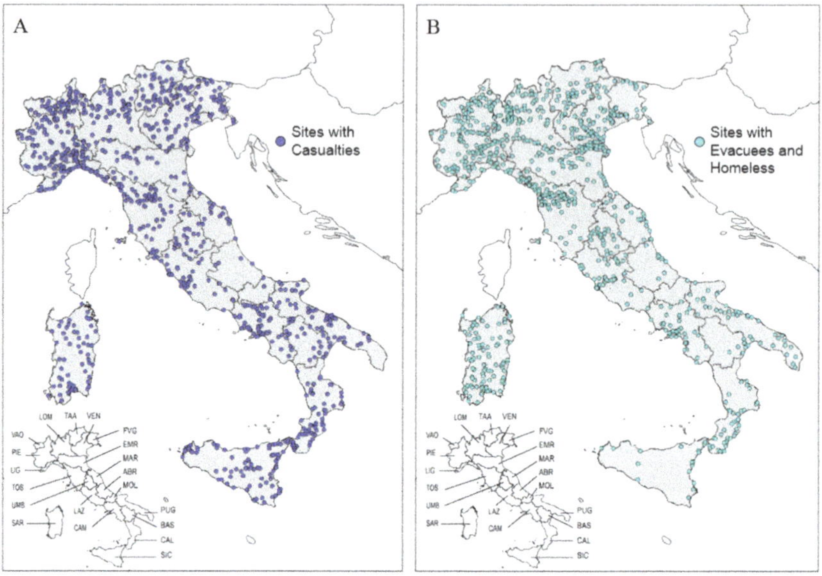

Fig. 22.5. Geographical distribution of sites with flood casualties (left) and flood evacuees and homeless people (right), in Italy.

Table 22.3. Number of regions, provinces, municipalities, and sites affected by harmful flood events in Italy, for different periods.

	589-2013	589-1860	1861-1910	1911-1960	1961-2013
Regions affected	20	18	17	20	20
Provinces affected	111	59	48	98	111
Municipalities affected	1,547	211	142	536	998
Sites affected	2,254	262	166	649	1,312

589-2013. The tables in Fig. 22.6 list the number of municipalities and the number of sites (and the corresponding percentages) affected one time, two or more times, five or more times, and ten or more times, by harmful flood events. Inspection of Fig. 22.6 reveals a difference between

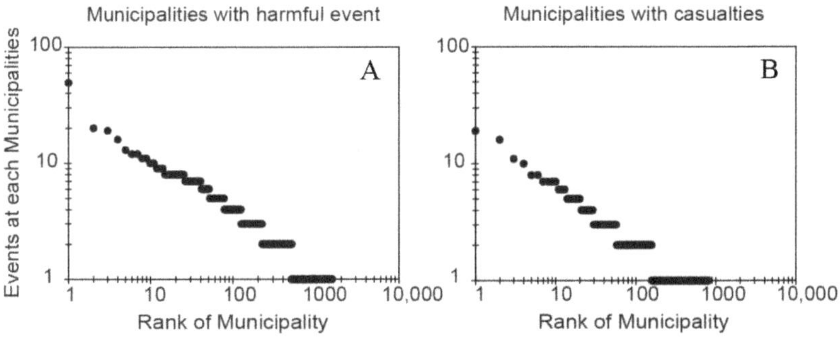

	Municipalities	%
One time	1040	67.18
Two or more times	507	32.75
Five or more times	77	4.97
Ten or more times	11	0.71

	Municipalities	%
One	652	80.59
Two or more times	157	19.41
Five or more times	20	2.47
Ten or more times	4	0.49

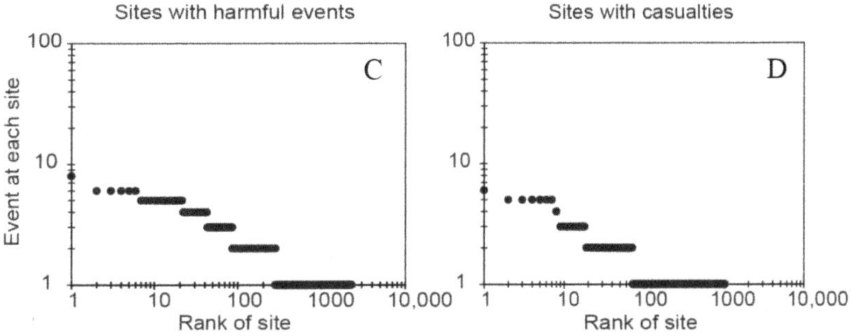

	Sites	%
One time	1975	87.72
Two or more times	279	12.38
Five or more times	21	0.93
Ten or more times	0	0

	Sites	%
One time	949	93.04
Two or more times	71	6.63
Five or more times	7	0.69
Ten or more times	0	0

Fig. 22.6. Geographical persistence of harmful flood events, and of events with flood casualties in Italy, in the period 589-2013.

the geographical persistence measured for the single sites and for the municipalities. Considering the harmful events, relatively few municipalities (77) have experienced frequent harmful flood events (i.e., five or more flood events in the considered period), and only eleven municipalities have experienced very frequent harmful flood events (i.e., ten or more events in the period (Fig. 22.6 A). Considering only the events that have caused casualties (Fig. 22.6 B), only twenty municipalities have experienced frequent harmful flood events (five or more events in the period), and only four municipalities have experienced very frequent harmful events (i.e., ten or more events in the period). These municipalities include major cities crossed by major rivers (e.g., the Tiber River in Rome) and by multiple torrents (e.g., Genova). Studying the persistency of the sites, we note that no single site was affected ten or more times by harmful flood events, or by events with casualties. We further note that the number of sites affected one time is higher than the number of municipalities affected one time. This confirms that flood risk to the population is widespread in Italy, that the sites where casualties occurred frequently are a limited number, and that there are municipalities where flood risk to the population is persistent because the (average) interval between harmful events is short (Salvati et al. 2010).

Table 22.4 lists the temporal variation in the total number of flood casualties in the 20 Italian regions. Considering the entire flood record, Sicilia experienced the largest number of flood casualties, due largely to the September 1557 event in the Palermo area that resulted in 7,000 casualties. The single most destructive flooding event occurred along the Po River in 1705, when up to 15,000 people were killed, went missing, or were injured at multiple sites by extensive flooding and the subsequent famine. For this event, the historical information is insufficient to locate the human consequences precisely, or even approximately (Salvati et al. 2010) and, consequently, the regions where casualties occurred remain unknown and the casualties were not counted in the Table 22.4.

Regions in southern Italy experienced the largest number of casualties until 1960 (Sicilia in the period before 1860, Sardegna in the period 1861-1910, Calabria in the period 1911-1960). In the most recent period (1961-2013) Toscana, central Italy, experienced the largest number of flood casualties, chiefly due to the number of deaths and injured people caused by the 4 November 1966 flood of the Arno River that inundated Florence. Figure 22.7 shows the cumulative number of casualties for the different periods, in northern, central, and southern Italy.

Table 22.4. Regional statistics of flood casualties in Italy, for different periods.

		Regions	589-2013	589-1860	1861-1910	1911-1960	1961-2013
North	1	Piemonte	4,992	4,493	78	187	234
	2	Valle d'Aosta	94	84	2	3	5
	3	Lombardia	3,458	2,920	18	466	54
	4	Trentino Alto-Adige	211	59	58	57	37
	5	Veneto	324	42	68	175	35
	6	Friuli-Venezia Giulia	154	13	-	81	60
	7	Liguria	231	72	29	22	106
	8	Emilia-Romagna	82	25	30	7	19
Centre	9	Toscana	865	378	1	33	453
	10	Umbria	63	38	-	11	14
	11	Marche	71	4	-	42	22
	12	Lazio	6,151	6,009	11	79	52
	13	Abruzzo	28	-	-	23	5
South	14	Molise	3	-	-	2	1
	15	Campania	1,039	682	17	222	118
	16	Puglia	449	5	5	334	105
	17	Basilicata	83	-	-	43	40
	18	Calabria	1,373	721	14	569	69
	19	Sicilia	7,847	7280	122	269	173
	20	Sardegna	487	7	220	156	104

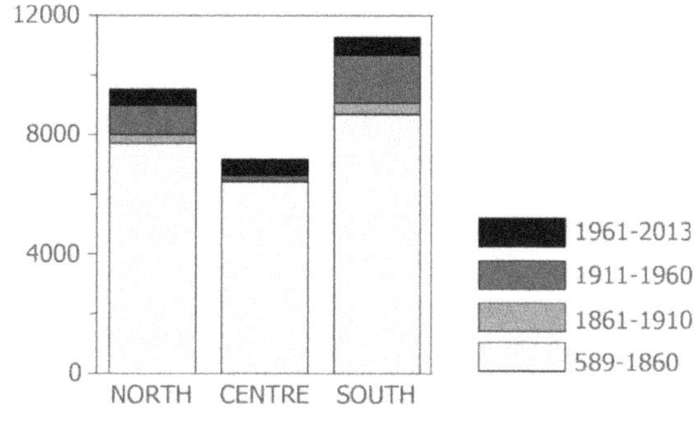

Fig. 22.7. Number of flood casualties in northern, central, and southern Italy, for different periods.

22.3 RISK EVALUATION

In this section, we study flood risk in Italy and its temporal and geographical variations. To determine individual risk criteria (Fell and Hartford, 1997; Guzzetti et al. 2005a), we investigate the number of fatalities in relation to the size of the population exploiting the updated record of flood events with casualties and the census data available in Italy for the 151-year period 1861-2011.

22.3.1 Individual Flood Risk

Individual risk levels are measured by mortality (or death) rates, which are given by the number of fatalities in a population, scaled to the size of the population, per unit time (Guzzetti et al. 2005b, Salvati et al. 2003). In this work, the number of flood fatalities per 100,000 people in a period of one year represents mortality. To calculate mortality, information on the size of the population and on the number of fatalities per year is necessary. We obtained the number of flood fatalities per year from the historical record of floods with human consequences in Italy. Beginning in 1861, the *Istituto Nazionale di Statistica* (ISTAT) has measured the size of the population of Italy every ten years through general censuses. The size of the Italian population has increased from 22.2 millions in 1861, to 47.5 millions in 1951, to 60.3 millions in 2011. The increase was largest in the plains, moderate in the hills, and lowest in the mountains. From the 1920s, and increasingly in the second half of the 20th century, there was migration from mountainous areas to urban areas, which are generally located in the plains and lowland hills. Consequently, the increase in the size of the population in urban areas was larger than in the rural areas, and some of the hills and the mountains suffered net population losses (Guzzetti et al. 2005b). Analysis of Fig. 22.8 reveals that in the ISTAT physiographical subdivisions the largest increase in the population occurred in the plains, with a maximum increase of about 13 millions in the Po and Veneto plains, in northern Italy (Fig. 22.8 A). Minor increases were observed in the hills, and chiefly in the inland hills of northern and central Italy (Fig. 22.8 B), and in the coastal hills of southern Italy (Fig. 22.8 C).

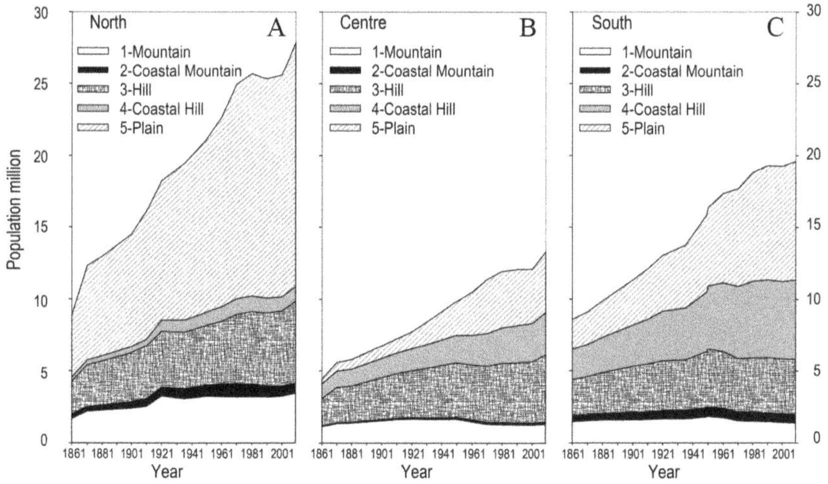

Fig. 22.8. **Variation of the Italian population in northern, central, and southern Italy, for different physiographical subdivisions. Source of population data: ISTAT (http://www.istat.it).**

Using yearly population data for each ISTAT physiographical subdivision, we calculated the annual flood mortality rates (FMR), and the corresponding average values, in the period 1861-2011 (Fig. 22.9A). The largest average FMR was recorded in the coastal hills (0.06), followed by the coastal mountains (0.05). The smallest FMR (0.03), was measured in the plains. The pie chart in Fig. 22.9 B shows the percentage of Italy in each ISTAT physiographical subdivision. To investigate the temporal variation of flood mortality, we calculated the average FMR for three subsequent periods of equal length, namely: 1861-1910, 1911-1960, and 1961-2011.

Results are shown in Fig. 22.9 C, D, E. In the north and in the centre of Italy mortality was highest in the coastal mountains, and reduced in the plains. In southern Italy, the highest mortality was in the coastal hills and in the plains, in the period 1911-1960. In the most recent period (1961-2011) mortality increased in the coastal mountains of northern and central Italy, and decreased in the hills of the north. In southern Italy mortality decreased slightly in the plains and significantly in the coastal hills. We conclude that in the last 150 years, in Italy, individual flood risk has increased significantly in the coastal mountains of northern and central Italy, chiefly in the Apuane Alps, and has decreased in the coastal hills of southern Italy. This is in contrast with the distribution of flood fatalities occurred in the last two years 2012 and 2013 that concentrated in the coastal hills of southern and central Italy.

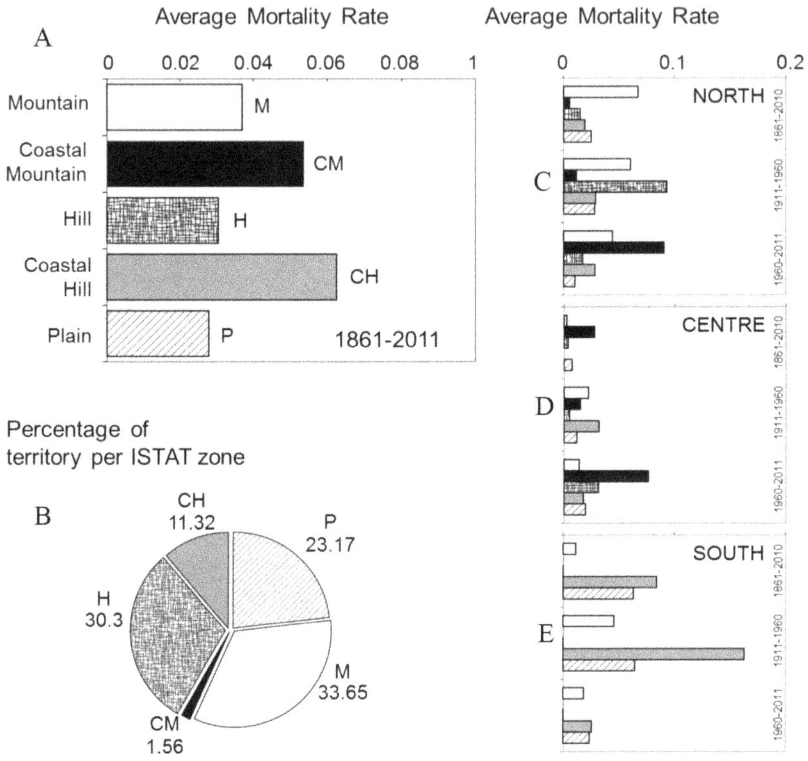

Fig. 22.9. Flood mortality rates in the ISTAT physiographical subdivisions, for different periods.

22.3.2 Comparison to Other Natureal Hazards

In addition to floods, landslides, earthquakes and volcanic activities are other types of natural hazards with human consequences that are frequent in Italy (Guzzetti, 2000; Salvati et al. 2003, 2012b; Guzzetti et al. 2005b). We used historical records of landslides, earthquakes, and volcanic events with fatalities in the 153-year period 1861-2013 to determine the individual landslide, earthquake and volcanic risk in Italy. Table 22.5 compares individual risk posed by different natural hazards (floods, landslides, earthquakes, volcanic activity) at the national scale. Inspection of Table 22.5 reveals that during the considered period earthquakes registered the highest average mortality. In the same period, the largest number of fatalities was caused by the 7.1 M_W Messina earthquake of 28 December 1908 that, together with the induced large tsunami, produced more than 100,000 fatalities, and the 7 M_W Avezzano earthquake of 13 January 1915, with 33,000 deaths (Boschi et al. 1995). The most disastrous earthquakes in the more recent period were the 6,4 M_W Friuli earthquake (6 May 1976) that caused 989 deaths, the 6.9 M_W Irpinia earthquake (23 November 1980) with 2,735 deaths (Boschi et al. 1995), and the L'Aquila 6.3 M_W earthquake (6 April 2009) that produced 308 deaths (http://istituto.ingv.it).

Table 22.5. Average mortality rates and number of fatalities for different natural hazards in Italy, in the period 1861-2013.

Hazard	Average Mortality Rate 1861-2013	Number of Fatalities 1861-2013
Flood	0.052	3,268
Landslide	0.084	5,844
Earthquake	2.295	126,129
Volcanic activity	0.005	275

To investigate possible variations of mortality with time, for each natural hazard we calculated the average mortality rates for three subsequent periods: 1861-1910, 1911-1960, and 1961-2013 (Fig. 22.10). We decided to extend the analysis to 2013 to include the 2012 Emilia-Romagna earthquake that produced 27 deaths. From 1861 to 1960 earthquakes were the natural events that caused the largest number of fatalities, and that produced the highest mortality. Since 1960 the trend has changed slightly; landslide mortality and earthquake mortality are very similar, and flood mortality has decreased slightly. After 1960, where earthquakes have occurred in Italy they have not caused very large number of fatalities. Beginning in the 1970s, seismic codes were introduced to reduce building vulnerability to earthquakes. As a result, the number of fatalities due to the total or partial collapse of buildings decreased (Salvati et al. 2012b).

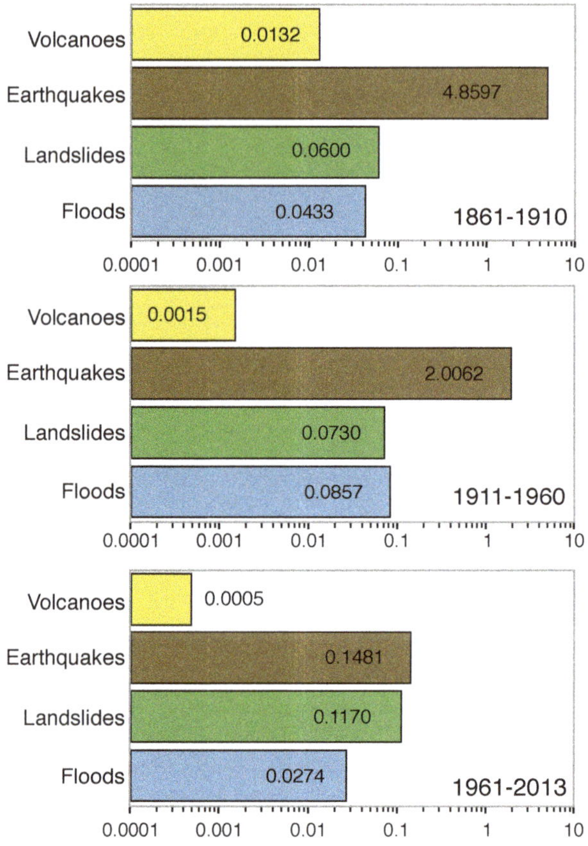

Fig. 22.10. Mortality rates for different natural hazards and for different periods in Italy.

Unlike for earthquakes, little information is available on the vulnerability of buildings and the infrastructure to geo-hydrological hazards, which are more frequent and abundant phenomena in Italy (Guzzetti et al. 1994; Guzzetti and Tonelli, 2004). Flood and landslide events produce damage to buildings, fatalities and injured people every year, thought since 1960 flood fatalities have decreased as a consequences of the widespread construction of flood protection structures and the deployment of warning systems.

22.4 RISK PERCEPTION

Despite the large number and widespread geographical distribution of historical flood events with human consequences, the population of Italy has relatively little information on the type, frequency, and severity of harmful flood events in the areas where they live, and has little awareness of the risk posed by floods. In Italy, efforts were made to determine flood risk at different scales, from the site-specific (local) scale to the synoptic (national) scale. Zonations of flood hazard and risk ("*Piani di Assetto Idrogeologico*") are available, and are used to design defensive structures and to implement mitigation strategies. Despite, little is known on the perception that the population of Italy has on the risk posed by floods, and in general by all natural hazards. This is confirmed by two national surveys conducted in 2012 and 2013 through computer-assisted telephone interviews to measure the perception of landslide and flood risk of the population of Italy (Salvati et al. 2014, in press). The two surveys compared the perception of the risk posed by different natural hazards, including landslides, floods, earthquakes, and volcanic eruptions, against the perception of the risk posed by road accidents and environmental pollution, two human-induced (technological) risks. Figure 22.11 shows the percentage of people that felt "considerably or somewhat" exposed to the different hazards. In 2012 and in 2013 people in Italy felt more exposed to technological than to natural risks. This was expected, considering that differences exist in the perception of natural vs. human induced risks (McDaniels et al. 1995; Rohrmann and Renn, 2000; Plapp and Werner, 2006). Among the different natural hazards, the interviewees felt more exposed to earthquakes, and then to flooding, landslides, and volcanic eruptions.

The two surveys allowed measuring the ability of individuals to evaluate the frequency (or the likelihood of occurrence) of natural hazards in the general area where they lived. Collectively, the interviewees felt that in the area where they lived earthquakes were the most likely risk, followed by floods, landslides, and volcanic activity (Table 22.6). This was in spite of the fact that in Italy geo-hydrological events (floods and landslides) are far more frequent than geophysical events (earthquakes and volcanic activity) (Salvati et al. 2012a).

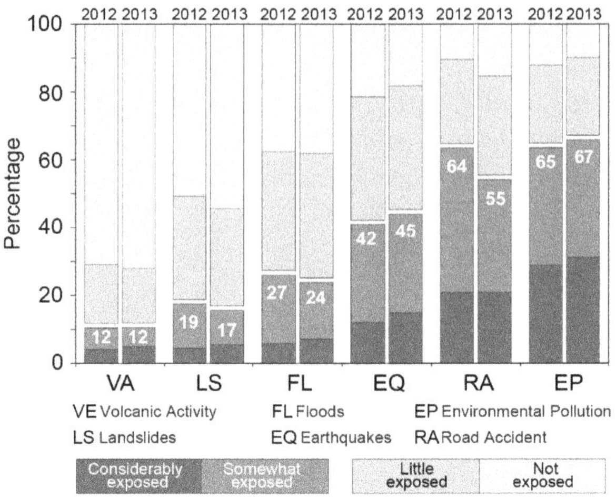

Fig. 22.11. Percentage of interviewees that felt exposed to technological and natural hazards in Italy in the 2012 and 2013 surveys.

Table 22.6. Percentage of interviewees that considered a specific hazard to be most frequent or likely in the general area where they lived.

Possible answers	Most frequent or most likely hazard	
	2013	2012
Flood	30%	31%
Landslide	10%	9%
Earthquake	41%	39%
Volcanic activity	2%	4%
None of these	15%	14%
Don't know	2%	3%

To test the understanding of geo-hydrological events, and to verify if the interviewees had a direct experience or an indirect knowledge of the occurrence of floods or landslides in the general area where they lived, a specific question was formulated. Analysis of the answers (Table 22.7) reveals that the majority of the interviewees (52% in 2013, 55% in 2012) had no direct experience or indirect knowledge of flood or landslide events occurring in the area where they lived. This is in contrast with the large abundance of flood and landslide events in Italy. All provinces in

Italy have experienced recurrent landslides or floods, and of the 8103 municipalities (in 1998), 6,475 (79.9%, 91% of the territory) have experienced at least one landslide or one flood (Guzzetti et al. 1994; Guzzetti and Tonelli, 2004).

Table 22.7. Percentage of people with direct experience or indirect knowledge of flood or landslide events in the area where they lived.

Possible answers	Direct experience or indirect information	
	2013	2012
Flood	27%	24%
Landslide	13%	14%
Flood and Landslide	8%	7%
No information	52%	55%

The two surveys revealed that the perception of the threat posed by floods does not match necessarily the distribution of flood events and of flood risk to the population of Italy. People fear most the earthquakes, and secondly floods and landslides. In the 52-year period 1961-2013 mortality rates due to earthquakes and landslides are similar. The underestimation of geo-hydrological risks by the population can have negative effects in the management of flood risk, and in the implementation of effective risk reduction and adaptation strategies.

22.5 CONCLUSIONS

We used historical information on flood events with human consequences to study flood risk to the population of Italy. Analysis of the geographical distribution of the sites where floods have caused deaths, missing persons, injuries, and homeless between 589 and 2013 confirmed that flood risk to the population is widespread in Italy, but that areas where harmful events are recurrent are rare. Analysis of the temporal distribution of flood events, and of flood casualties, indicates that the oldest events reported in the historical record were mainly catastrophic, but also that the first part of the record is incomplete for the medium and the low severity events. The historical record is not uniform, with significant temporal variations in the frequency of harmful flood events. Reasons for the variations in the temporal trend are manifold, and include incompleteness of the old part of the record, climatic and environmental variations, the increased number of the elements at risk (including the population), and the implementation of risk mitigation measures.

In Italy, flood mortality depends on the physiographical setting and the intensity of the events. In the period 1960-2011, flood mortality has increased in the coastal mountains of northern and central Italy, has decreased slightly in the mountains, and has decreased significantly in the coastal hills of southern Italy. A comparative analysis of the risk posed by different natural hazards, including floods, landslides, earthquakes and volcanic events, revealed that earthquakes is the natural hazard with the highest average mortality, but also that after 1960 earthquake mortality has decreased strongly whereas flood mortality has decreased slightly, and the landslide mortality has increased in the same period. To study the perception that the population of Italy has of the risk posed by natural hazards, including floods, we analysed the results of two telephone surveys conducted in 2012 and 2013. The survey revealed that earthquake is the natural hazard that the population of Italy considers most treating, followed by landslides and floods. We maintain that underestimation of geo-hydrological risks by individuals and communities has negative consequences for the management of flood risk, and for the implementation of risk reduction and adaptation strategies.

We expect the results of our study to improve the understanding of the risk posed by different natural hazards to the population of Italy, with emphasis on flood risk, and for designing of appropriate risk reduction / mitigation strategies. We maintain that disseminating information on the consequences of flood events, and of geo-hydrological events in general, is essential to increase the awareness of the population on the risk that floods and other harmful natural hazards pose to the population, and to help mitigate their consequences.

22.6 REFERENCES

Boschi, E., G. Ferrari, P. Gasperini, E. Guidoboni, G. Smiriglio, and G. Valensise (eds). 1995. Catalogo dei forti terremoti in Italia dal 41 a.C. al 1980, ING-SGA, Bologna.

David, H. A. and N. H. Nagaraja. 2003. *Order Statistics* (3rd edn). Wiley, New Jersey.

Fell, R. and D. Hartford. 1997. Landslide risk management. In: D. M. Cruden and R. Fell (eds). *Landslide Risk Assessment*, 51–109. Balkema, Rotterdam, The Netherlands.

Glade, T., P. Albini, and F., Frances (eds). 2001. *The Use of Historical Data in Natural Hazard Assessments.* Kluwer Academic Publisher, Dordrecht, 225. ISBN-10: 0-7923-7154-2.

Guzzetti, F. 2000. Landslide fatalities and evaluation of landslide risk in Italy. *Engineering Geology* 58:89–107.

Guzzetti F., M. Cardinali M. and P. Reichenbach. 1994. The AVI Project: A bibliographical and archive inventory of landslides and floods in Italy. *Environmental Management* 18:623-633.

Guzzetti, F., P. Salvati and C. P. Stark. 2005a. Evaluation of risk to the population posed by natural hazards in Italy. In: O. Hungr, R. Fell, R. Couture and E. Eberhardt (eds). *Landslide Risk Management.* 381–389. Taylor & Francis Group, London, UK.

Guzzetti, F., C. P. Stark, and P. Salvati. 2005b. Evaluation of flood and landslide risk to the population of Italy. *Environmental. Management.* 36:1:15–36.

Guzzetti F. and G. Tonelli. 2004. Information system on hydrological and geomorphological catastrophes in Italy (SICI): a tool for managing landslide and flood hazards. *Natural Hazards and Earth System Sciences.* 4:213-232, doi:10.5194/nhess-4-213-2004

Kolmogorov, A. 1933. *Grundbegriffe der Wahrscheinlichkeitsrechnung.* Julius Springer, Berlin.

Luino, F. 2005. Sequence of instability processes triggered by heavy rainfall in the northern Italy. *Geomorphology,* 66:13–39.

McDaniels, T., L. Axelrod and P. Slovic. 1995. Characterizing perception of ecological risk. *Risk Analysis*, 15:5, 575–590.

Newman, M. E. J. 2005. Power laws, Pareto distributions and Zipf's law. *Contemporary Physics* 46:5:323–351.

Plapp, T. and U. Werner. 2006. Understanding risk perception from natural hazards: Examples from Germany. In: Ammann W.J., Dannenmann S., Vulliet L. (eds), Risk 21 Coping with risks due to natural hazards in the 21th century. Taylor & Francis Group, London, 101–108.

Reed, W. J. 2001. The Pareto, Zipf and other power laws. *Economics Letters.* 74:1:15–19.

Rohrmann, B., and O. Renn. 2000. Risk perception research - An introduction. In: Renn and Rohrmann (eds) Cross-cultural risk perception research. Kluwer, Dordrecht.

Salvati, P., C. Bianchi, M. Rossi and F. Guzzetti. 2010. Societal landslide and flood risk in Italy. *Natural Hazards and Earth System Sciences* 10:465–483, doi: 10.5194/nhess-10-465-2010.

Salvati, P., C. Bianchi, M. Rossi and F. Guzzetti. 2012a. Flood risk in Italy. In Z. Kundzewicz (ed) Changes of flood risk in Europe. IAHS Special Publication 10.

Salvati, P., C. Bianchi, F. Fiorucci, I. Marchesini and F. Guzzetti. 2012b. Temporal and geographical variation of geo-hydrological risk to the population of Italy. *Proceedings of the International Disaster and Risk Conference*, Davos 26-30 August 2012, 625-628.

Salvati, P., C. Bianchi, F. Fiorucci, P. Giostrella, I. Marchesini and F. Guzzetti. 2014. (in press). Perception of flood and landslide risk in Italy: a Preliminary Analysis. *Natural Hazards and Earth System Sciences*, doi: 10.5194/nhessd-2-3465-2014.

Salvati, P., F. Guzzetti, P. Reichenbach, M. Cardinali and C. P. Stark. 2003. Map of landslides and floods with human consequences in Italy. CNR-Gruppo Nazionale per la Difesa dalle Catastrofi Idrogeologiche Publication n. 2822, scale 1:1,200,000.

Salvati, P., I. Marchesini, V. Balducci, C. Bianchi, and F. Guzzetti. 2013. A new digital catalogue of harmful landslides and floods in Italy. In C. Margottini et al. (eds.) *Landslide Science and Practice* 3:409-414, doi 10.1007/978-3-642-31310-3_56.

Smirnov, N. 1933. Estimate of deviation between empirical distribution functions in two independent samples. *Bulletin of Moscow University* 2:2:3–16.

White, E. P., B. J. Enquist, J. L. Green. 2008. On estimating the exponent of power-law frequency distributions. *Ecology.* 89:4:905–912, doi: 10.1890/07-1288.1.

Wilk, M. B. and R. Gnanadesikan. 1968. Probability plotting methods for the analysis of data. *Biometrika* 55:1:1–17.

CHAPTER 23

A METHODOLOGICAL FRAMEWORK FOR ASSESSING DROUGHT HAZARD USING OPERATIONAL INDICATORS

Maggie Kossida, Maria Mimikou

NTUA – National Technical University of Athens, School of Civil Engineering, 5 Iroon Polytechniou str., 15773 Athens, Greece, mkossida@chi.civil.ntua.gr

ABSTRACT

In view of elaborating an analytical framework for the characterization of drought and water scarcity risk, a methodology for holistically mapping drought hazard has been developed and tested in the Ali-Efenti catchment (Pinios River Basin) in Greece. The proposed methodology is based on the use of operational indicators, derived from precipitation data, that capture all the drought hazard components: recurrence, magnitude, severity, duration. The indicators' thresholds have been elaborated in the study area, while their relevant weights have been assigned using an Analytical Hierarchy Process (AHP). Using the 12-month Standard Precipitation Index (SPI-12) four sub-indicators have been estimated reflecting: the maximum intensity of the observed drought episodes, the maximum duration, the number of drought events, and the number of drought events with a duration larger than 24 months. These sub-indicators have been blended into deriving a Drought Hazard Index (DHI), mapped and classified in the study area. The results show that the area has experienced significant drought hazards. The DHI is an easily reproducible indicator, and easy to integrate with additional relevant indicators of environmental or socio-economic context, necessary for the analysis of relevant vulnerabilities and risk. It has an operational value and could be used to serve policy needs.

23.1 INTRODUCTION

Europe has experienced in the past 40 years drought episodes of various severity, duration and extend, with adverse impacts on both the environment and the society. A large scale comparison per region (North, Central, Eastern, South EU) between the periods 1971-80 and 2001-11 clearly shows that drought occurrence has significantly increased in the period 2001-11, not only in the South and Central EU, but also reaching now the North and Eastern EU (Kossida et al. 2012). Drought spells have been further recognized in the River Basin Management Plans, reported by the Member States under the Water Framework Directive (WFD), either as River Basin District wide phenomena or as local phenomena affecting parts of the entire basin in various cases (Schmidt et al. 2012).

Drought risk management is an essential aspect of water resources planning. The Water Framework Directive (WFD) (EC, 2000) has identified the need to develop Drought Management Plans (DMPs) as supplementary measures to the basic Programmes of Measures in areas prone to drought and water scarcity in order to improve risk management. At the EU policy level the need to tackle water scarcity and drought by developing adequate tools has been further highlighted in the 2007 EC Communication on Water Scarcity and Droughts (EC, 2007a) and the recently published 2012 EC Communication "A Blueprint to Safeguard Europe's Water Resources Blueprint" (EC, 2012). The main objective of Drought Management Plans is to minimize the adverse impacts on the environment, economy and society in the areas affected by drought, through adequate planning and adoption of a proactive approach. At the heart of DMPs lies the necessity to properly identify the onset of drought episodes, and further monitor the evolution of drought and water scarcity events. To this extent, indicators and thresholds, accurately capturing the onset-offset, duration and severity of such episodes are valuable tools. Yet, as drought characterisation differentiates based on the analysed effects (i.e. meteorological, agricultural, hydrological, socioeconomic) (CEDEX, 2004), it is inherently difficulty to use a single indicator in order to measure it, mainly because drought affects a wide variety of disciplines and operates on a diverse geographical and temporal multiscale distribution. One common thing to all types of drought though, is the deficiency of precipitation (MedWSD, 2007). The onset of a drought event is difficult to identify both in time and place, mainly because drought events do not begin with an extreme meteorological event, such as floods. In addition, drought slowly affects regions and usually with different severity and impact levels. Drought end is also difficult to identify due to the fact that the resulting effects may be evident even after the event's "end" for a long period of time.

Understanding and assessing drought and water scarcity risk is a complex multi-factor problem. The underlying exposure to stresses and threats may be similar even in quite different conditions, yet vulnerability is influenced by the priorities set, the economic and adaptive capacity of the affected area and population (sensitivity and margin of), the dynamic choices and response strategies adopted (Kossida et. al., 2012). At the European policy level, although vulnerability to floods has been defined and common risk assessment guidelines have been elaborated under the EU Floods Directive (EC, 2007b), no analytical framework has been suggested for drought hazard, vulnerability and risk. Recent efforts within the WFD Common Implementation Strategy (WFD CIS) Expert Group on Water Scarcity and Drought (EG WSD) focused on the development of general guidelines for drafting DMPs (EC, Water Scarcity and Droughts Expert Network, 2008) and on the development of commonly agreed awareness raising indicators for both water scarcity and drought (Faergemann, 2012). Policy actions have recently been intensified at the European and national levels in order to effectively implement drought management schemas that can support proactive risk management and increase the resilience of the affected regions. Yet, assessing the components of drought hazard, vulnerability and risk is a complex multi-factor problem still to be methodologically tackled.

In this study, a methodological framework for assessing and mapping drought hazard has been elaborated. Commonly available monthly precipitation data have been used to develop four operational indicators that capture all the drought hazard components: recurrence, severity, magnitude, duration. These have then been merged into a blended Drought Hazard Index (DHI) and mapped for different time periods (from 1980-2010) allowing also the identification of the hazard evolution in time and space. The results have been compared against observed impacts on streamflow in different gauges within the Ali-Efenti catchment of the Pinios River Basin (Central-eastern Greece).

23.2 METHODOLOGY

23.2.1 Overview of the Methodological Approach

The fact that water scarcity and drought (a) operate on many scales (spatial and temporal) and levels (moderate to severe), (b) are a complex result of both natural and anthropogenic factors, (c) have a wide variety of impacts affecting many economic sectors, and (d) mitigation is highly dependent on the prevailing socio-economic conditions and adaptive capacity of a system, makes it inherently difficult to frame a single pathway into assessing the nature and degree of vulnerability and subsequent risk (Kossida et. al., 2012). Nevertheless, as in all risks associated with climate change, key parameters. which hold a central role. do exist and need to be coherently and scientifically integrated (i.e. exposure, sensitivity, impacts etc.) into a framework which can support accurate communication and consistent analysis, eliminating ambiguous interpretation. Figure 23.1 presents a schematic of the key parameters that influence vulnerability and risk to water scarcity and drought and their interplay (adopted from Wood, 2011; Kossida et al. 2012) linking them to the DPSIR (drivers-pressure-state-impact-response) framework, which is commonly used to guide the development of environmental indicators (Smeets and Weterings, 1999; Niemeijer and De Groot, 2008). On the basis of this concept the current study investigates the development of drought hazard maps, focusing on the natural hazard characteristics, namely recurrence, severity, magnitude and duration in the Ali-Efenti catchment of the Pinios River Basin District in Greece.

The drought hazard maps reflect the spatial variability of the developed Drought Hazard Index (DHI) and have been compiled following six methodological steps:

1. Analysis of the meteorological drought episodes in the catchment using the Standard Precipitation Index (SPI) (McKee et al. 1993) as a basis. Calculation of the 12-month SPI (SPI-12) from monthly precipitation data, for each rain gauge of the catchment and for the 30-year period 1981-2010.
2. Post-processing of the SPI-12 results (meta-analysis) in each rain gauge in order to derive four new sub-indicators reflecting the frequency, intensity, magnitude and duration of the drought events.
3. Classification of the four sub-indicators and assignment of relevant scores for each rain gauge.
4. Blending of the four sub-indicators, using relevant weights, into a Drought Hazard Index (DHI) for each rain gauge.
5. Interpolation of the DHIs across the rain gauges to obtain coverage for the entire area and a corresponding drought hazard map of the Ali-Efenti catchment.
6. Repetition of steps 2-5 for two 15-year sub-periods (1981-1995, 1996-2010) in order to assess the evolution and change of the DHI in time. Comparison of the corresponding drought hazard maps.

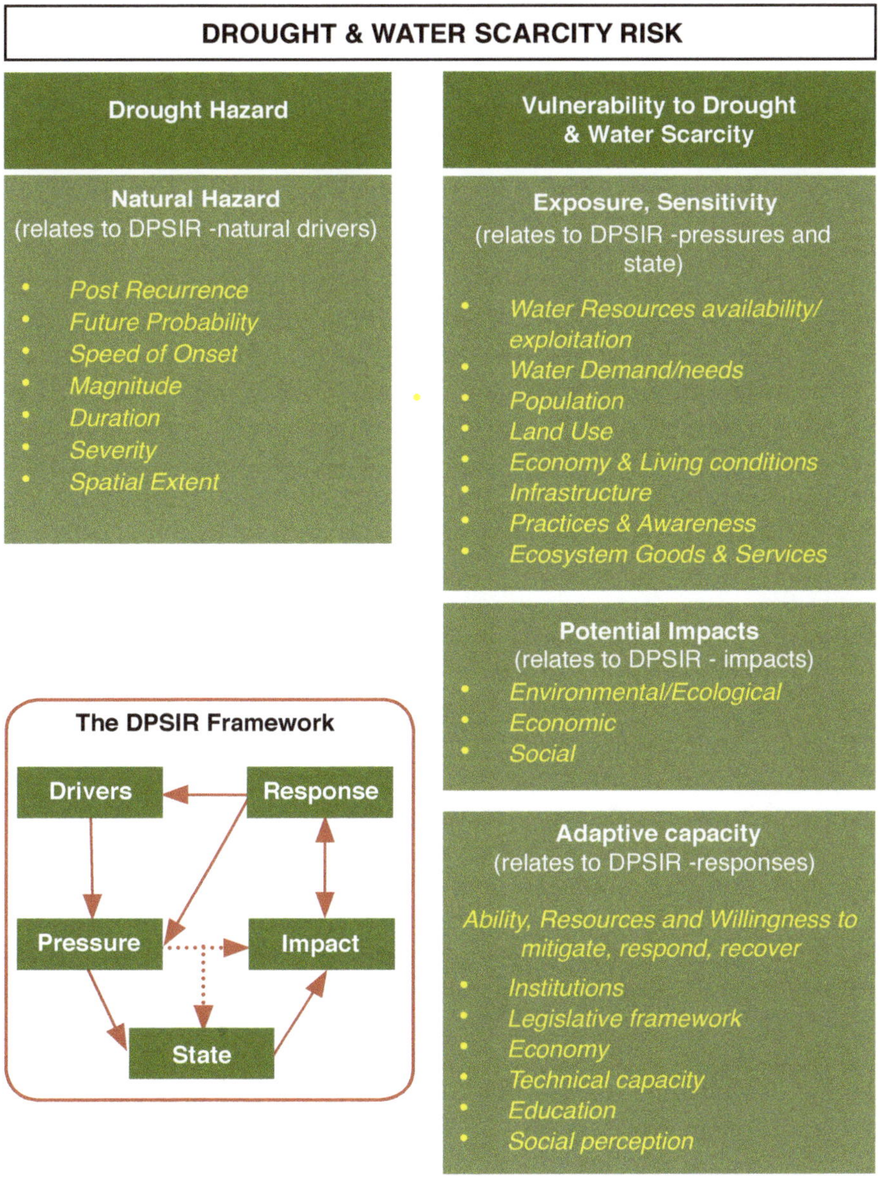

Fig. 23.1. Conceptual schema of the components of drought and water scarcity risk, and their relation to the DPSIR (adopted from Wood, 2011; Kossida et al. 2012).

23.2.2 Calculation and Post-processing of the SPI-12

The 12-month Standard Precipitation Index (SPI-12) has been used for the analysis of the meteorological drought episodes in the area as it can capture long-term precipitation patterns usually associated with streamflows, reservoir levels and in some cases groundwater levels. The SPI was proposed by McKee et al. (1993) to provide a functional and quantitative definition of drought at different timescales (1, 3, 6, 12, ... months). To calculate the SPI-12 a 2-parameter Gamma probability density function was fit to the 12-month cumulative precipitation for each rain gauge in the study area. The cumulative probability gamma function is then transformed into a standard normal variable with a mean of zero and a standard deviation of one, which represents the value of the SPI. SPI values lower than -1 indicate dry conditions, while values greater than 1 indicate wet conditions. The drought intensity gradually increases from moderate (-1.49<SPI<-1.0) to severe and to extreme, when SPI values fall below -1.5 and -2.0 respectively (McKee et al. 1993). The 12-month SPI used here allows for the comparison of the cumulative precipitation of 12 consecutive months every year within the 1981-2010 period. It presents the advantage of eliminating seasonality (applicable in smaller temporal scales) and capturing signals of distinctive wet or dry trends.

Based on the values of the SPI-12 the drought episodes from 1981-2010 have been identified in each rain gauge. These results have then been post-processed to derive four new sub-indicators that can reflect the magnitude, severity, duration, and recurrence of the drought hazard in each rain gauge. The focus of this meta-analysis is to derive operational indicators, each one reflecting common drought hazard characteristics (refer to Fig. 23.1), easy to reproduce and blend into a Drought Hazard Index. The following sub-indicators have been defined and computed at each rain gauge:

FRQ: number of drought episodes (events) observed within the reference period. This sub-indicator is used as metrics of "recurrence".

FRQ24: number of drought episodes with duration greater than 24 months, within the reference period. This sub-indicator is used as a sensible descriptor of prolonged drought and thus metrics of "severity".

DMmax: maximum drought magnitude observed within the reference period. This sub-indicator is used as metrics of "magnitude".

dmax: maximum duration (in months) among the drought episodes observed within the reference period. This sub-indicator is used as metrics of "duration".

A drought episode (necessary for the estimation of the FRQ, FRQ24 and dmax sub-indicators) is identified when the SPI-12 first falls below zero (onset of the episode) and continuous to increase reaching a value equal or less than -1. When SPI-12 reaches again its first positive value this event has ended. If an SPI-12 value equal or less than -1 has not been reached, then this event is not characterized as drought (i.e. it is just low precipitation event but cannot be characterized as a drought episode). To estimate the drought magnitude (DM) (necessary for the estimation of the DMmax sub-indicator), the relation suggested by McKee (1993) has been applied:

$$DM = -\left(\sum_{j=1}^{x} SPI_{12j}\right) \tag{23.1}$$

where j starts with the first month of a drought event and continues to increase until the end of the drought (x),

Following the calculation of the four sub-indicators for each rain gauge, a classification has been elaborated, assigning four classes and relevant 1-4 scores (less to more significant) across all gauges. The classification has been performed for the entire 30-year period 1981-2010, as well as for the 15-year sub-periods 1981-1995 and 1996-2010 to allow comparison. The thresholds for each sub-indicators' classes are elaborated in the study area by subdividing the values in equal intervals (Tables 23.1 and 23.2). In the case of the sub-indicators FRQ and FRQ24 the boundaries of the four classes defined for the 30-year period slightly differ from the ones defined for the 15-year sub-periods for consistency purposes (i.e. smaller number of episodes in each class definition hence the sub-periods have fewer years).

Table 23.1. Classification and scores of the sub-indicators for the 30-year period 1981-2010.

Classification thresholds for each sub-indicator				
FRQ *number of episodes*	**FRQ24** *number of episodes with d>24 months*	**DMmax** *maximum magnitude*	**dmax** *maximum duration*	**Assigned Score**
1 – 3	1	≤ 35.0	24 – 36	1
4 – 6	2	35.1 – 50.0	37 – 48	2
7 – 9	3	50.1 – 70.0	49 – 60	3
≥ 10	4	≥ 70.1	≥ 61	4

Table 23.2. Classification and scores of the sub-indicators for the 15-year sub-periods 1981-1995 and 1996-2010.

Classification thresholds for each sub-indicator				
FRQ *number of episodes*	**FRQ24** *number of episodes with d>24 months*	**DMmax** *maximum magnitude*	**dmax** *maximum duration*	**Assigned Score**
1 – 2	1	≤ 35.0	24 – 36	1
3 – 4	2	35.1 – 50.0	37 – 48	2
5 – 6	3	50.1 – 70.0	49 – 60	3
≥ 7	4	≥ 70.1	≥ 61	4

23.2.3 Calculation of the Drought Hazard Index (DHI)

An Analytical Hierarchy Process (AHP) was used to assign relevant weights to each sub-indicator and blend them into the Drought Hazard Index (DHI). The AHP, developed by Saaty (1977), is a well-known weight evaluation method, widely used in a variety of decision-making problems. For the estimation of criteria weights the AHP calculates the eigenvalues and the eigenvectors of a square preference matrix which contains the quantified preference information of all possible pairwise criteria constellations. According to the AHP the four sub-indicators are set in a priority order and are compared to each other in a pairwise comparison matrix, in which their relative importance is expressed by numerical values. Saaty and Vargas (1991) suggested a scale for comparison ranking from 1, which expresses equal importance, to 9, which expresses "extreme importance" of one factor over another.

Finally, a square preference matrix of fourth order is created, and the weights of the sub-indicators resulted from the eigenvectors of the matrix. The process resulted in equal weights for all four sub-indicators (i.e. they were all deemed of equal importance).

As follows, the sub-indicators are blended to derive a DHI value for each rain gauge for the entire period 1981-2010 (as well as for the sub-periods 1981-1995 and 1996-2010) based on the following equation:

$$DHI = (\theta_1 \times score_{FRQ}) + (\theta_2 \times score_{FRQ24}) + (\theta_3 \times score_{DM\max}) + (\theta_4 \times score_{d\max}) \quad (23.2)$$

where θ_i are the weights of the sub-indicators (all equal to 0.25 as resulted by the AHP).

The DHI values are expected to range from 1-4 (less to more prone to the drought hazard) since all the sub-indicators scores are 1-4 and their relevant weights are all equal to 0.25.

23.2.4 Compilation of the Drought Hazard Maps

The final step in the process is the compilation of the drought hazard map for the Ali-Efenti catchment based on the derived DHI. The DHI values of each of the 17 rain gauges and for each of three time periods (1981-2010, 1981-1995, 1996-2010) have been spatially interpolated using a two-dimensional minimum curvature tension spline technique. Four neighboring points (rain gauges) are used for local approximation. The resulting maps present the spatial variability of the drought hazard in the catchment, as well as its evolution from the 1981-1995 periods to the most recent 1996-2010 period allowing for cross-comparison and policy relevant assessment.

23.3 THE STUDY AREA

The study area is focused on the Northwestern part of the Pinios River in Thessaly plain, namely the Ali- Efenti catchment (Fig. 23.2). The drainage area of Ali-Efenti is about 2,920 km^2 (representing about one fourth of the Pinios River Basin), with a total permanent population of

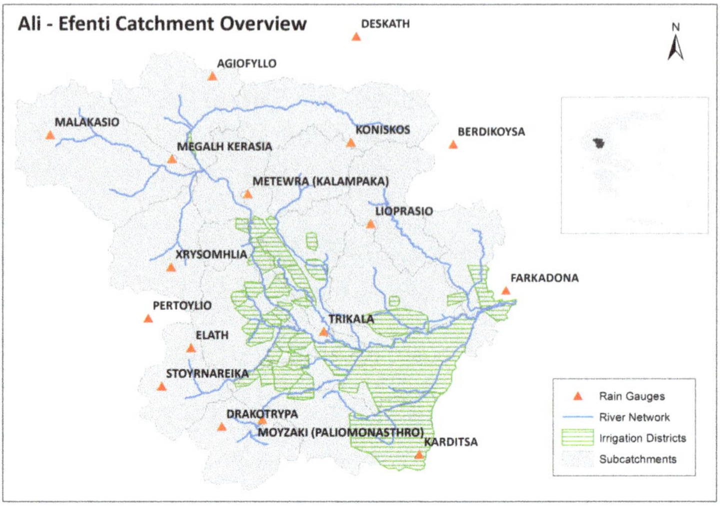

Fig. 23.2. Ali-Efenti catchment overview.

190,276 inhabitants. Two main urban centers, the city of Trikala and the city of Karditsa are within the catchment, while numerous significant peri-urban settlements are also present (e.g, Kalambaka, Mouzaki, Neoxori, etc.). In terms of land use the area is dominated by agricultural land and forests (about 33% each), followed by pasture (31%) and urban areas (2%), while wetlands and water bodies only account for 1% of the total area. With regard to the share of cultivated crops (within the agricultural land use) cotton is dominating 10% of the total basin area (corresponding to 44% share within the agricultural land use), followed by winter wheat, maize and alfalfa. Finally, the long-term annual average precipitation is highly variable from 1,000-1,600 mm (in the west part) to a lower precipitation of 460-800 (in the eastern part).

As previously mentioned, the Ali-Efenti catchment has extended irrigation areas (the main crop cultivated here is cotton), while irrigation efficiency is low. Imbalance between demand and availability (water stress) is frequent, and the unmet demand is highly pronounced during the summer period. As a result over-abstraction has led to environmental impacts, such as the degradation of the groundwater resources and declining groundwater levels. Drought episodes are frequent in the area, and desertification is becoming an issue. At the same time, as the area is among the most productive agricultural region of Greece and with significant competing water uses, the socio-economic impacts of drought are a major challenge. Yet, operational Drought Management Plans (DMPs) or other policy instruments are lacking, and drought management is currently based on "crisis management" rather than on a pro-active and preparedness approach.

For the analysis of the drought hazard in the Ali-Efenti, monthly precipitation data have been used from 17 rain gauges homogeneously spread in the catchment, covering the 30-year period 1981-2010 (Fig. 23.2. The analysis targeted the entire 30-year period, as well as the 15-year sub-periods 1981-1995 and 1996-2010 to allow for cross-comparison and conclusions on the evolution of the drought hazard.

23.4 RESULTS

In the following section, the results of the drought hazard analysis in the Ali-Efenti catchment are presented focusing on: i) the analysis of the drought occurrence and characteristics based on the SPI-12 and the four developed sub-indicators as discussed in the previous section 23.2.2; ii) the mapping of the drought hazard in the entire catchment area on the basis of the newly proposed indicator DHI, and the assessment of its spatio-temporal evolution and trends.

23.4.1 Analysis of Drought Characteristics in the Ali-Efenti Catchment

The SPI-12 has been used as a basis for the analysis of drought occurrence in the Ali-Efenti catchment with a view of eliminating seasonal variability and capturing signals of distinct drought events associated with hydrological drought. Based on the SPI-12 timeseries, calculated in each of the 17 rain gauges in the area, the drought episodes of the 30-year period 1981-2010 have been identified. Figure 23.3 presents an overview of the drought events per station, along with their durations and magnitudes. We can observe commonalities across the stations in terms of drought occurrence; with the main drought events noticed during the years 1988-90, 1992-94, 2000-02 and 2007-08. Some less significant events and of local character are additionally observed in some stations during 1983-85 and 2004-05. With regard to the main drought events, the respective magnitudes and durations range across the stations, from as low as 7 to as high as 58 for the magnitudes, and from 9 to 47 months duration. In some stations, we can observe that two or more episodes have been merged into a prolonged drought event of very high magnitude and duration (Agiofyllo: DM=92, duration 70 months within 1988-93; Deskati: DM=86, duration 75 months within 2004-10; Trikala: DM=66, duration 83 months within 2004-10). The overall statistics of the drought magnitude and durations across all stations are presented in Table 23.3. Overall, it can be observed that the event with the largest drought magnitudes has been experienced within the period 1981-1995, while the event with the largest durations has been experienced within the period 1996-2010. With regard to the prolonged events with durations of more than 60 months experienced in few locations, these are more apparent during the recent period 2004-2010.

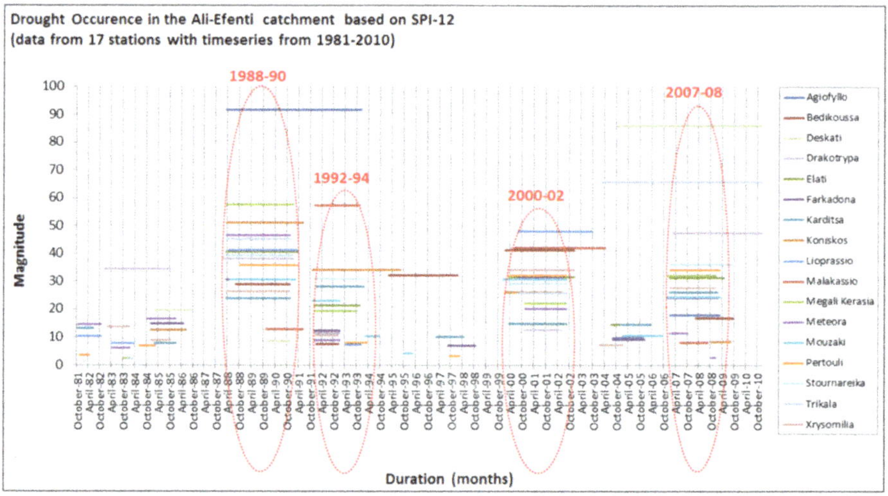

Fig. 23.3. **Drought occurrence in the Ali-Efenti catchment during the period 1981-2010 based on the SPI-12. The red circles denote the years with the main observed events.**

Table 23.3. Characteristics of the main drought events in the Ali-Efenti.

	Main Drought Events			
Statistics	**1988-90**	**1992-94**	**2000-02**	**2007-08**
Number (%) of stations	17 (100%)	15 (88%)	17 (100%)	16 (94%)
Average DM*	36.7	20.1	27.7	24.6
Max DM*	57.7	57.5	48.0	47.6
Average duration*	32	19	29	25
Max duration*	40	46	47	47

**The 3 prolonged drought events experienced in the 3 locations mentioned previously are excluded from these statistics.*

The above findings of the SPI-12 analysis are verified by the hydrological observations in the catchment. Although the available hydrological data do not consistently cover the whole study period, discharge data from the Kefalovrysso spring (located in center of the catchment) and the catchment outlet (Ali-Efenti streamflow station) also demonstrate low values during the periods of drought events as identified by the SPI-12 analysis (Fig. 23.4).

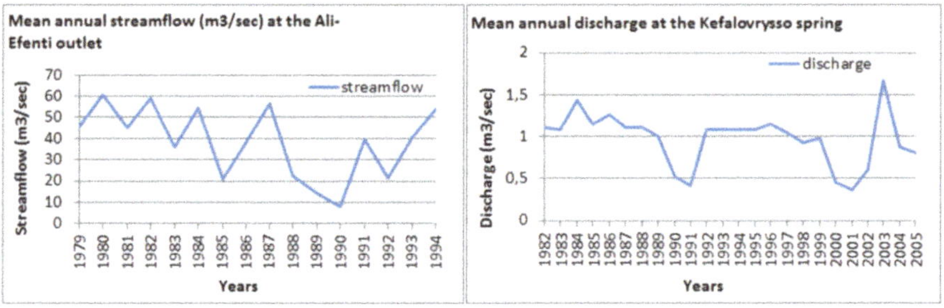

Fig. 23.4. Mean annual discharge at the Ali-Efenti outlet (left) and at the Kefalovrysso spring (right).

By post-processing the results of the SPI-12, the four sub-indicators (FRQ, FRQ24, DMmax, dmax) have been estimated for each rain gauge station, for the entire 30-year period 1981-2010, as well as for the sub-periods 1981-1995 and 1996-2010. The results of the classification for the period 1981-2010 are demonstrated in Fig. 23.5. With regard to the number of drought events (FRQ sub-indicator) in the majority of stations (about 60%) 4-6 events have occurred within the 30 years (thus roughly corresponding to one event every 6 years), while in 35% of them 7-9

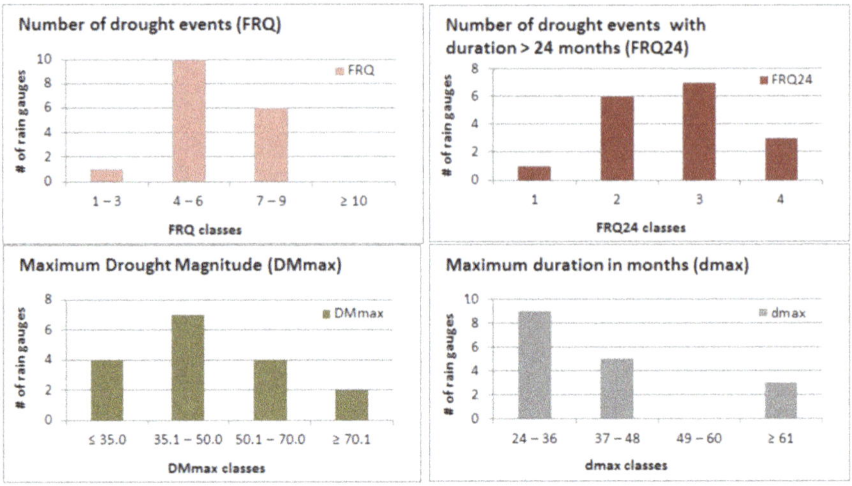

Fig. 23.5. Histogram of the four sub-indicators across the 17 rain gauges for the 30-year period 1981-2010. Each class relates to a relevant 1-4 score.

events have occurred (thus roughly corresponding to one event every 4 years) (Fig. 23.5). Almost half of those events had durations of more than 24 months. In some stations, all, or nearly all, of the events had durations of more than 24 months (100% of the events in Trikala station, 80% in Elati and Stournareika stations). With regard to the maximum drought magnitude (DMmax sub-indicator) in 41% of the locations maximum magnitudes ranged from 35-50, while higher ones occurred in 35% of the locations (Fig. 23.5). Finally, the drought events in the majority of the stations (53%) had an observed maximum duration 24-36 months, while 18% of the stations experienced events with more than 60 months duration (Fig. 23.5). A comparison of the sub-indicators classification between the sub-periods 1981-1995 and 1995-2010 is also provided in Fig. 23.6. Overall, we can observe that the number and the maximum magnitudes of the drought events decrease in the period 1996-2010 as compared to the previous period 1981-1995, yet their durations tend to increase and consequently the number of events with durations of more than 24 months. We can thus conclude that the occurrence of drought episodes may be a bit less frequent and/or intense in terms of magnitude, but they are now longer, and can evolve into prolonged drought of more than 60 months duration.

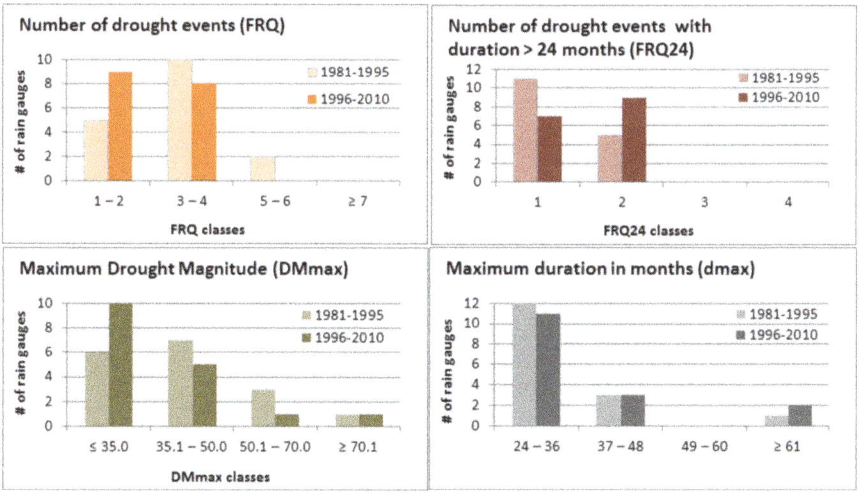

Fig. 23.6. **Histograms of the four sub-indicators across the 17 rain gauges for the 15-year periods 1981-1995 and 1996-2010. Each class relates to a relevant 1-4 score.**

23.4.2 Mapping and Evolution of Drought Hazard in the Ali-Efenti Catchment

The mapping of the drought hazard in the Ali-Efenti has been based on the interpolation of the DHI values calculated at each rain gauge for the entire 1981-2010 period, as well as for the individual sub-periods 1981-1995 and 1995-2010. The average DHI value across all 17 stations for the period 1980-2010 equals to 2.24, indicating that the Ali-Efenti catchment experiences overall a significant drought hazard (Table 23.4). The maximum DHI value (3.0) is observed in the Deskati station in the north, while the minimum (1.75) in Farkadona and Meteorta stations (Fig. 23.7). Zones of higher to lower exposure alternate from north to south in an east-west orientation. The zones with the highest DHI (2.00-3.16) prevail in the northern area and the south, while a lower risk zone lies in the center (Fig. 23.7).

When comparing the two sub-periods we can observe that the average DHI value across all stations has been slightly reduced by 5% during the 1996-2010 (as compared to the previous 1981-1995 period) (Table 23.4). The maximum observed DHI value (currently in Deskati station) is 10%

Table 23.4. DHI statistics across all rain gauges.

DHI statistics across all rain gauges	Time periods *1981-2010*	*1981-1995*	*1996-2010*	Δchange (%) (from 1996-10 to 1981-95)
Average	2.24	1.57	1.50	- 0.07 (-5%)
St. deviation	0.38	0.43	0.50	
Maximum	3.00	2.50	2.75	+ 0.25 (+10%)
Minimum	1.75	0.75	0.75	0.00 (0%)

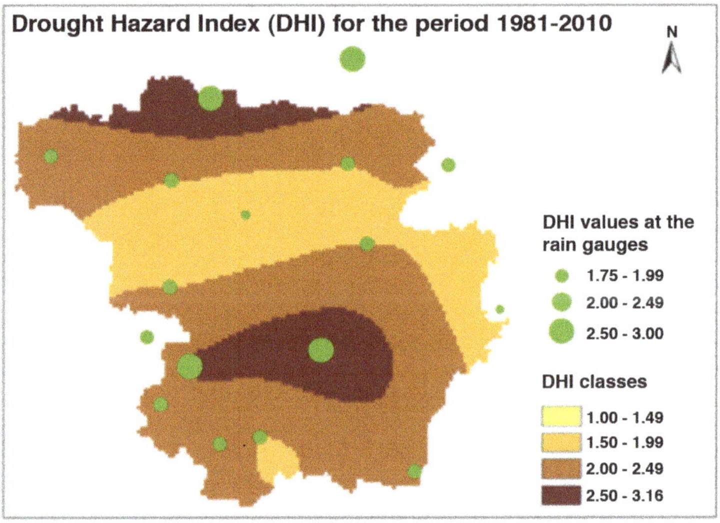

Fig. 23.7. Drought hazard map of the Ali-Efenti catchment based on the DHI calculated for the 30-year period 1981-2010. The green dots denote the location of the rain gauges and their size varies according to the respective DHI values.

higher than the previously observed maximum during 1981-1995 (in the nearby Agiofyllo station). Overall, in 9 stations the DHI has decreased during the 1996-2010 period, in 7 stations it has increased (pointing to more serious drought hazard conditions), while in 4 stations it remained constant. The observed decrease in the DHI values in those 9 stations ranged from -14% to -57%, while the increase observed in the 7 stations ranged from +20% to +267% in the Deskati station. What is most interesting to notice is the spatial evolution of the DHI in the area (Fig. 23.8). While in the period 1981-1995 the drought hazard was more significant in the northern part of the catchment (DHI > 2) it has now "migrated" to the southern part. Thus, the southeast of the catchment clearly experiences an increased exposure to the drought hazard, while in the north a decreasing trend can be observed. Unfortunately, the area where the DHI has increased coincides with the main irrigation districts of the catchment where a lot of the water is needed for agricultural purposes, thus making them prone to increased water stress conditions.

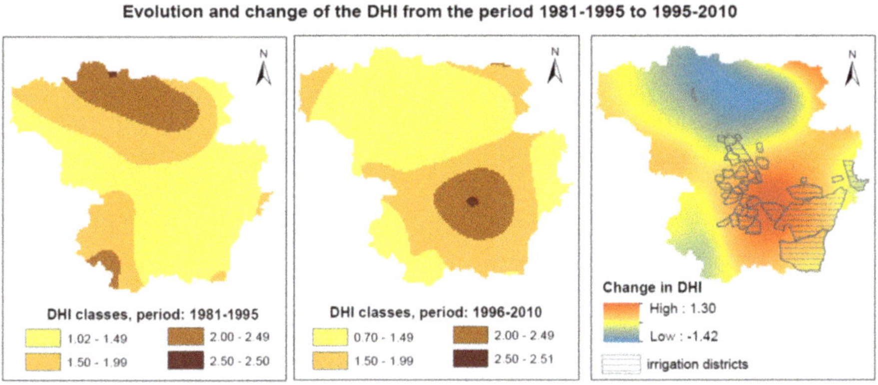

Fig. 23.8. Evolution of the drought hazard in the Ali-Efenti catchment based on the DHI (left: drought hazard map for the period 1981-1995, middle: drought hazard map for the period 1996-2010, right: change of the DHI between the two periods; red areas denote an increase in DHI in 1995-2010 as compared to the previous period 1981-1995).

23.5 CONCLUDING REMARKS

A methodology for holistically assessing and mapping drought hazard has been developed and tested in the Ali-Efenti catchment (Pinios River Basin) in Greece. The proposed methodology is based on the use of operational indicators, derived from precipitation data, that capture all the drought hazard components: recurrence, severity, magnitude, duration. Post-processing the results of the calculated 12-month Standard Precipitation Index (SPI-12) in 17 rain gauges with precipitation data from 1981-2010 four new sub-indicators have been estimated reflecting: the number of drought events within the reference period, the number of drought events with a duration larger than 24 months, the maximum intensity of the observed drought episodes, the maximum duration. These sub-indicators have been blended into deriving a Drought Hazard Index (DHI), mapped and classified in the study area.

The results show that the Ali-Efenti catchment has experienced significant drought hazards during the years 1981-2010, also verified from discharge observations. In most of the area high values of the DHI are observed (greater than 2.0) indicating substantial exposure to drought. Vulnerability to drought in 1996-2010 as compared to 1981-1995 has increased in the southern part of the catchment and decreased in the north. We thus observe a shift in the spatiotemporal evolution of the drought hazard across the two 15-year sub-periods. The observed increase of the DHI in the southern part can relate to increased water stress conditions since this area is dominated by agriculture; higher exposure to drought hazards signals elevated irrigation water needs which can consequently lead to unmet demand and water scarcity in the area.

The DHI is an easily reproducible indicator, based on commonly available precipitation timeseries, and flexible enough to be computed for multiple timescales relevant to the drought analysis. It is spatially consistent allowing for comparisons between different areas. It can be calculated for different time periods, permitting thus the assessment of the evolution of drought hazard over time and the detection of trends, while its historic context is suitable for decision-making. Finally, the DHI allows for easy integration with additional relevant indicators of environmental or socio-economic context, necessary for the analysis of relevant vulnerabilities and risk (e.g. population density, land use, water exploitation, unmet demand, etc.). It has thus an operational value and could be used to serve policy needs.

23.6 ACKNOWLEDGEMENTS

This research has been co-financed by the European Union (European Social Fund – ESF) and Greek national funds through the Operational Program "Education and Lifelong Learning" of the National Strategic Reference Framework (NSRF) - Research Funding Program: Heracleitus II. Investing in knowledge society through the European Social Fund.

23.7 REFERENCES

CEDEX. 2004. "Task Plan 2004. Task 11 (Activity 4.3.1) Development of background information sheets on potential issues/extreme events of public interest for inland waters. Event: Droughts", ETC/WTR Authors: Manuel Menéndez, Concepción Lallana (CEDEX), July 2004.

European Commission. 2012. 'A Blueprint to Safeguard Europe's water resources - Consultation Document'. European Commission - Directorate-General Environment ENV.D.1 - Protection of Water Resources. Available online: http://ec.europa.eu/environment/consultations/pdf/blueprint.pdf (accessed 10/10/2014)

European Commissin. 2011. Third Follow up Report to the Communication on water scarcity and droughts in the European Union. European Commission. Available online: http://eur-lex. europa.eu /Lex UriServ/ LexUriServ.do?uri=COM:2011:0133:FIN:EN:PDF (accessed 10/10/2014)

European Commission. 2007a. 'Communication from the Commission to the European Parliament and the Council addressing the challenge of water scarcity and droughts in the European Union'. Commission of the European Communities. Available online: http://eur-lex.europa.eu/LexUriServ/LexUriServ.do?uri=COM:2007:0414:FIN:EN:PDF (accessed 10/10/2014).

European Commission. 2007b. Directive 2007/60/EC of the European Parliament and of the Council of 23 October 2007 on the assessment and management of flood risks (Floods Directive) (OJ L 288 06.11.2007).

European Commission. 2000. Directive 2000/60/EC of the European Parliament and of the Council of 23 October 2000 establishing a framework for Community action in the field of water policy (Water Framework Directive - WFD) (OJ L 327, 22.12.2000). Available online: http://eur-lex.europa.eu/resource.html?uri=cellar:5c835afb-2ec6-4577-bdf8-756d3d694eeb.0004.02/DOC_1&format=PDF (accessed 10/10/2014).

European Commission, Water Scarcity and Droughts Expert Network. 2008. Drought Management Plan Report including Agricultural, Drought Indicators and Climate Change Aspects. Water Scarcity and Drought Expert Network, Technical Report – 2008-023. Available online: http://ec.europa.eu/environment/water/quantity/pdf/dmp_report.pdf (accessed on 10/10/2014)

Faergemann, H. 2012. Update on Water Scarcity and Droughts indicator development. Report to the EU Water Directors. Available online (accessed on 10/10/2014).

Kossida, M., Kakava, A., Tekidou, A., Mimikou, M. 2012. Vulnerability to Water Scarcity and Drought in Europe - Thematic assessment for EEA Water 2012 Report. ETC/ICM Technical Report 3/2012, ISBN 978-80-85087-13-0, ETC/ICM, Prague, 2012. Available online : http://icm.eionet.europa.eu/ETC_Reports/VulnerabilityToWaterScarcityAndDroughtInEurope (access 10/10/2014).

Kossida, M., Koutiva I., Makropoulos, C., Monokrousou K., Mimikou, M., Fons-Esteve, J., Iglesias, A. 2009. Water Scarcity and Drought: towards a European Water Scarcity and Drought Network (WSDN). European Topic Centre on Water (ETC/W) Internal Report, EEA.

Kossida, M., Mimikou, M. 2013a. A methodological framework for assessing drought hazard using operational indicators. Florisa Melone Memorial Conference, 10-11 October, 2013 in Assisi, Italy.

Kossida, M., Mimikou, M. 2013b. An indicators' based approach to Drought and Water Scarcity Risk Mapping in Pinios River Basin, Greece. Geophysical Research Abstracts Vol. 15, EGU2013-6349, 2013, EGU General Assembly 2013, 07-12 April 2013, Vienna, Austria.

Maccioni, P., Kossida, M., Brocca, L., Moramarco, T. 2014. An assessment of the drought hazard in the Tiber River Basin in Central Italy, and a comparison of new and commonly used meteorological indicators. *J. of Hydrologic Engineering* (in press).

Mckee, T.B., Doesken, N. J., Kleist, J. (1993). The relationship of drought frequency and duration to time scales, (January), 17–22.

MedWSD. 2007. "Mediterranean Water Scarcity and Drought Report", Technical Report - 009 – 2007, produced by the Mediterranean Water Scarcity & Drought Working Group (MED WS&D WG), April 2007.

Niemeijer, D., De Groot, R.S. 2008. A conceptual framework for selecting environmental indicator sets. *Ecological Indicators* 8(2008) 14– 25.

Saaty, T.L. 1977. Scaling method for priorities in hierarchical structure. Journal of Mathematical Psychology 15/3 (1977) 243-281.

Saaty, T.L. and Vargas, L.G. 1991. Prediction, Projection and Forecasting: Applications of the Analytic Hierarchy Process in Economics, Finance, Politics, Games, and Sports, Boston: *Kluwer Academic*.

Schmidt, G., Benítez, C., Seiz, R., Raso, J., Palacios, E., Hernández-Torres, J.M. 2012. Topic report on: Assessment of Water Scarcity and Drought aspects in a selection of European Union River Basin Management Plans, Version: 3.0, final delivery to DG ENV, 5 September 2012. INTECSA-INARSA, S.A., based on a previous draft by TYPSA.

Smeets, E., Weterings, R. 1999. Environmental Indicators: Typology and Overview. European Environment Agency, Copenhagen. Report No. 25, 19 pp.

Vanneuville, W., Werner, B., Uhel, R., Kjeldsen, T., Miller, J., Kossida, M., Tekidou, A., Kakava, A., Crouzet, P. 2012. Water resources in Europe in the context of vulnerability, EEA 2012 state of water assessment. EEA Report No 11/2012, ISBN 978-92-9213-344-3, EEA, Copenhagen, 2012. Available online: http://www.eea.europa.eu/publications/water-resources-and-vulnerability/at_download/file (accessed 10/10/2014)

Wood, N. 2011. Understanding risk and resilience to natural hazards: U.S. Geological Survey Fact Sheet 2011-3008, 2 p. Available online: http://pubs.usgs.gov/fs/2011/3008/ (accessed 10/10/2014)

www.ingramcontent.com/pod-product-compliance
Ingram Content Group UK Ltd.
Pitfield, Milton Keynes, MK11 3LW, UK
UKHW020427061026
15625UKWH00005B/310

9 781887 201858